Published By

Oxford Book Company

Regd. Off. "Bony Residency", Gate No. 2
Opp. Tilak Public School, Vishweswariya Nagar,
Gopalpura Road, Jaipur - 302018, India
Ph: 0141 - 2761280, Telefax: 2761381
email - oxfordbookjpr@gmail.com
website - oxfordbookcompany.com

Delhi Office: 102, 1st Floor, Satyam House,
4327/3, Ansari Road, Darya Ganj,
New Delhi - 110002, India
Ph: 011-45652440

Distributed by
KEYA INTERNATIONAL PRESS LLP
E-mail: keyainternationalpressllp@gmail.com

Published in India, 2025

ISBN : 9788189473426

Typeset by Shivangi Computers, Jaipur

Printed and bound in India by Replika Press Pvt. Ltd.

Principles of
Agriculture and Food Production

Principles of
Agriculture and Food Product

Archana Ruhela

Oxford Book Company
Jaipur ▪ New Delhi

Preface

Agriculture is responsible for providing virtually all the food and much of the clothing for world population which is increasing enormously. It appears that there will be a sizeable challenge facing not only those directly engaged in agriculture but also the politicians and scientists concerned with the many other activities that bear on the scale and productivity of the industry. Agriculture's share in national economies varies widely, normally as a function of their level of development. Historically, and without exception, agriculture has been the dynamo for subsequent industrial development.

The structure of agriculture and food industry is influenced not only by market forces but by cultural values embodied in inheritance relating to land ownership and tenure. An analysis of the structure of the industry reflects the use it makes of the basic factors of production—land, labour and capital. The development of industry is a story of growing interdependence as the contribution of each component has tended to become more specialised. Farms today rely on complex inputs to achieve high levels of output per unit area. The products they use pass through a variety of manufacturing and distributive activities before they reach the final consumer. Such linkages stretch across frontiers and are indicative of the extent to which economic behaviour in one region of the world may affect people who live in distant places. In many poor countries, a large part of the food supply is still consumed by the subsistence farm families who produce it. However, as urbanisation has spread, increasing numbers of the population of the spreading towns depend on food produced, harvested, preserved and transported from elsewhere. A major task of the development is to ensure that agricultural and food production must be sufficient to meet the human and animal needs.

Modern technology permits highly intensive forms of agricultural production provided capital is available. Such production can be genuinely low cost but there can be no guarantee that this is the case where an industry is protected from competition. However, intensive methods of production are also associated with undesired side effects. But looking to the future demand of agricultural produce increased intensification of agricultural practices involving heavy use of agrochemicals in crop and livestock enterprise, use of intensive machinery and equipments and chemical fertilisers has become and will be necessary in achieving the very high levels of productivity.

Archana Ruhela

Contens

1

The Food Chain in Developed and Developing Countries

The Human Food Chain

The development of the food and agricultural industry is a story of growing interdependence as the contribution of each component sector has tended to become more specialised. Farms today rely on complex inputs to achieve high levels of output per acre and per man. The products they use pass through a variety of manufacturing and distributive activities before they reach the final consumer. Such linkages stretch across frontiers and are indicative of the extent to which economic behaviour in one region of the world may affect people who live in distant places. In many poorer, 'developing', countries a major part of the food supply is still consumed by the subsistence farm families who produce it. However, as urbanisation takes place, increasing numbers of the population of the spreading towns depend on food produced, harvested, preserved and transported from elsewhere. A major task of development is to ensure that these linkages work well so that the growing numbers of those who have no land of their own can still eat. In rich developed countries the subsistence element in food provision has become very small, more a hobby than a necessity of survival. Here, not only do farms depend upon inputs made by, for example, chemical, machinery and feeding stuff factories, but their output is changed in form, shaped into a vast array of packaged and preserved food items and made available to consumers through a variety of wholesale and retail businesses. A significant part of total food supply is prepared by caterers. In the mid-1990s consumers' expenditure on eating out in France was 20 per cent of total food expenditure, in Sweden 15 per cent and in the UK 12 per cent. The figure for the USA was nearly 50 per cent around 2000. So the consumer buys not just the nutritional content of the food, but meals and the ambience in which they are offered. This extended set of inter-related businesses we call 'the human food chain'.

The efficiency of today's agricultural and food industry depends both upon the use of resources within the businesses which make up the chain and upon the satisfactory co-ordination of heir activities. In market economies, price, often influenced by government policies, plays a major role. In command economies, co-ordination depends upon the formulation and execution of a 'plan'. Prices may be used as one of the instruments through which the plan is given effect but the targets are determined not by the consumer but by the government. Recent experience suggests that while both systems have problems, the market economy has been more successful in making sure that there is, for most people, ready access to abundant supplies of food.

The human food chain provides, for those who can afford it, a wide choice of different food products. The shopper in developed countries is accustomed to being able to buy most foods regardless of season. The food processing sector not only 'preserves the harvest', a crucial need for all societies, it also offers a growing variety of foods which take much of the drudgery out of work in the domestic kitchen. Together with the food relating sector it supplies frozen, chilled, canned, dried as well as fresh food. The affluent housewife now has opportunities to prepare without difficulty meals which meet the differing preferences of the household. In the jargon of the industry, they add convenience and ensure that food is not just fuel but, for most people, fun.

At the beginning of the food chain suppliers of farm inputs such as manufacturers of fertilisers and suppliers of machinery and irrigation equipment have made it possible to bring previously unproductive land into food production. Products which control pests, weeds and diseases raise productivity and so lower costs of production. Farm machinery enables farmers to catch 'windows' in the weather to plant, cultivate or harvest and so escape some of the constraints which have traditionally limited agricultural production.

Within the agricultural and food industries the proportion of people working on farms, the traditional definition of agricultural employment, has fallen. In wealthy countries only a small proportion of the workforce is now engaged in farming (see Table 1.1). The smaller labour force of the richer countries now provides their citizens with a larger and more secure food supply than existed in the past or is now available in countries where a much greater proportion of the population continues to work in agriculture. The percentage of the population engaged in farming can even be used as a crude index of development. Such an index is, of course, too simple; we need to adjust it to allow for imports and exports and to take account of such special cases as desert-based oil producers. However, as the table demonstrates, the correlation between the share of population in agriculture and a familiar measure of relative income, Gross Domestic Product per capita, is remarkably close.

Table 1.1: Basic indicators of agriculture and development by major economic grouping, 2000

Region	Population (millions)	Labour force in agriculture (%)	GNP per capita ($)	Area ('000s km²)
Low income	2493.0	71	270	33608
Lower middle income economies	691.2	59	750	15029
Upper middle-income economies	577.2	29	1890	22238
Industrial market economies	741.6	7	12960	30935

Source: World Development Report.

The structures of the businesses which make up this food chain are very different. The production of individual farms is usually very small in relation to the output of agriculture as a whole. The average farm in the sample comprises only one-hundredth of a per cent of the UK total output. In contrast, in the food processing and distributive industries, although many tiny businesses still exist, a small number of very large international businesses may control significant shares of national or regional markets (in Great Britain the top four multiples held a 45% share of the trade sector in 2000 for packaged groceries: Burdus). This leads to unequal bargaining strength between the various links in the food chain. Large processors and retailers are actually aware of the need for reliable, good-quality supplies if they are to offer their customers precisely the product they want. Success in securing such supplies helps to make them competitive with their rivals. Food companies employ a wide range of techniques of sourcing supplies, ranging from the use of terminal markets to detailed production contracts in which production methods and the timing and quantity produced may all be specified. Since they represent the major market for many farm products they are in a strong position to persuade farmers to respond to their needs. In some countries farmers have set up their own co-operative marketing agencies to increase their bargaining power. In others, governments have intervened, for example, by creating marketing boards or state trading agencies.

The Structure of the Agricultural and Food Industry

The structure of farming is influenced not only be market forces but by cultural values embodied in inheritance customs and legal arrangements relating to land ownership and tenure. In some developing countries land is held in common rather than by individual farmers. In some South American countries large estates, often owned by absentee landlords and employing many landless labourers, are common. In centrally-planned economies the 'state' and some 'collective' farms form large agro-industrial complexes serving not only to produce farm products but to process them and provide the ancillary activities farming needs. The output of such enterprises is often supplemented by tiny 'workers' plots' which not only produce garden vegetables and intensive meat and livestock products for the household but also a significant part of total national supplies of these foodstuffs. In most developed market economies, although company farms exist, farming remains a family business. Its scale varies but there is a common sense of responsibility of the family and a shared concern with inter-generational transfer.

The structure of an industry reflects the use it makes of the basic factors of production, land, labour and capital. Data about farm area provide an inadequate view of the structure of the industry. In a market economy it is possible, at the margin, to substitute one factor for another. If, in real terms, labour becomes relatively more expensive than capital, it will be sensible to employ fewer people but to equip them with more machinery. If land is relatively abundant then an extensive system of

Principles of
Agriculture and Food Production

Principles of
Agriculture and Food Production

Archana Ruhela

Oxford Book Company
Jaipur ▪ New Delhi

production with low capital and labour inputs would be appropriate. To understand the structure of the industry we need to take account of all the factors it uses and their present disposition. (For broad regional groupings, some of the principal characteristics of the world's agriculture are shown in Tables 3.4-3.7 of Chapter 3).

Market pressures which modify the structure often originate outside the agriculture and food sector itself. Rising demand for labour in other growing sectors of modern economies has required agriculture to release labour for employment in other sectors. As economies grow, capital tends to become readily available compared with land or labour and its use in farming increases. Such changes alter the most profitable combinations of the factors of production and the best size of productive unit. Commonly they have led to greater specialisation among farmers, to larger-scale production units and to bigger farm businesses. Table 1.2 shows recent trends in this direction in the United Kingdom.

Table 1.2: The increasing size of farm enterprises in the United Kingdom

Enterprise	1990	1995	2000	2002
% total cereals in holdings with 50 or more ha of cereals	65.9	70.8	73.6	73.9
% total potatoes on holdings with 20 or more ha of potatoes	33.6	34.3	38	41
% dairy cows in herds of 60 or over	53.3	64.1	70.0	70.8[a]
% breeding sheep in flocks of 500 or over	37.6	40.8	43.8	47.6
% breeding pigs in herds of 500 or over	36.6	75.4	82.9	84.9[a]
% laying fowls in flocks of 20000 or over	47.1	60.0	65.7	69.8

[a] = 2001

Source: MAFF.

It is only external economic pressures which lead to change in the structure of the industry in market economies. New technology changes the relative productivity of the factors employed. Techniques which increase output by introducing new types of capital tend to diminish the relative value of labour and of land. This may change the international competitiveness of agriculture between the developed and developing world and the distribution of wealth within the rural communities of poorer countries. Cost-reducing methods which require new investment are most readily applied in capital-rich, developed countries. In poorer countries, where capital is less accessible, new technology may strengthen the richer, land-owning classes but undermine the incomes of small farmers and landless labourers. Because of barriers to international movements of both capital and agricultural goods, the need to invest, in order to employ techniques which make best use of natural resources, may reduce the comparative advantage in agriculture of developing countries. The advantage they have as suppliers of low-cost labour is diminished. Increased productivity in wealthy countries is likely to make it more difficult to sell to them. The lack of an external market also makes it less attractive to invest in the infrastructure needed to develop their agriculture and thus to raise the living standards of their rural populations.

Some common trends have tended to modify the structure of agriculture in the second half of the twentieth century. Improved technology has increased yields per man and per hectare. Fewer resources are now needed to supply a given level of output. Population and demand have grown but insufficiently to avoid long periods of weakness in world markets. The terms of trade for farm products have tended to deteriorate. Such market signals indicate that fewer resources are now needed in agriculture. However, governments in rich countries have protected their own farmers. As a result, many of the world's lowest cost producers have been forced to cut production. OECD notes that, had there been a general reductions in assistance in agriculture, then production in New Zealand would have been 0.5% greater in the period 1996-99.

Modern technology permits highly intensive forms of production provided capital is available. Such production can be genuinely low cost but there can be no guarantee that this is the case where an industry is protected from competition. Intensive systems of production are often associated with undesired side effects. Pollution, dissatisfaction with the animal welfare appearance implications of some farming methods and undesired changes in the appearance of the landscape have all led to demands for control of farming methods. Given the abundance of food in world markets, a more open world economy might make it less necessary to resort to intensive farming. However, the structure of agriculture in most protected economies would be forced to change. Although farming would be less intensive, it would also provide fewer rural jobs and result in farms which were part-time or required increased area. The wish to avoid such social consequences is one of the reasons for the continuation of agricultural protection which involves large economic losses to the world economy. A recent IMF study suggests that in 1994 the welfare effects of the Common Agricultural Policy to the Community were as high as 15.3 billion US$. For the developing countries, estimates of the costs imposed by the CAP range from 0.5 billion US$ to 10.5 billion US$.

The structure of the food processing and distributive industries has been less affected by protection than has farming. These industries have been forced to adopt both to attain technical economies of scale and because of operations in the international capital market. The result has been to transform the sector from a local small-scale industry into one which increasingly operates on an international scale and involves some of the world's largest companies. In UK grocery retailing, the market share of multiples (companies with ten or more outlets) has risen from an estimated 5 per cent in 1900 to over 50 per cent in 1990.

Technical economies in production and distribution have tended to strengthen the position of the largest players. The creation and maintenance of a 'brand' involves very large expenditure in design and advertisement. RHM recognised the enormous value of its brands by their inclusion in the 2002 balance sheet. A 'current cost' value of £678 million was placed on more than 50 brands. Innovations, needed to offer a fickle market a variety of novel products, require investment not only in the new products themselves but also in research and development into new processes and materials. Such overhead expenditures are difficult to justify unless they can be spread over a large volume of output. Processing and packaging machinery can achieve very high levels of productivity provided the 'runs' involved are sufficiently large. Improvements in information technology enable businesses to operate with smaller stocks, to adjust purchasing policy and input mixes rapidly in relation to changes in ingredient prices and to identify market opportunities at an early stage. They require investment in computerised data handling and training for staff. Such developments help to explain the advantages possessed by large firms in the food industries. Despite this the industry still has many more small than large businesses. In 2002, within the UK, 72 per cent of the food, drink and tobacco manufacturing industries employed between 1-19 persons (HMSO 2002). Such firms often occupy market niches where scale is less important and which large companies found unattractive, in the retail sector they provide 'out-of-hours' service, they may specialise in the food demands of ethnic or other minorities. In the food processing area they may offer variety, cater for small quantities or serve some 'traditional' markets which would be unrewarding for major national firms.

Companies quoted on the Stock Exchange may become the subject of acquisition or sale as a result of changes in the financial markets themselves. A wide variety of issues can affect the decision to launch a takeover bid. A company in the food sector and in any other, whose assets include sites of which the value is not fully reflected in its share price, may become a target for asset stripping. Major conglomerates may find food companies an attractive element in a balanced portfolio of shares. Although food companies may not be 'high performers' in terms of growth related to rises in personal incomes, they are likely to be relatively resilient when recession threatens. Thus food may figure as a diversified asset, major food companies can acquire easy access to a new technology or a new market by buying an existing operator in that field. The outcome, not unexpectedly, is that the food industry exhibits a wide variety of structural forms, including horizontally and vertically integrated businesses, specialised companies with a very narrow product base and others where food vies for capital with clothes, hardware and pharmaceuticals.

Food retailers and processors, who are often in the front line of modern consumerism, are increasingly anxious to assure themselves of the quality of their raw material and to influence the ways in which they are produced. This encourages them to source supplies directly from farmers. The costs of monitoring supplies tend to be negatively related to the quantity purchased, so buyers generally prefer to establish contractual relations with larger producers or producer groups. Farmers chosen as 'partners' may have preferential access to capital and to technical advice.

In product areas where this method of co-ordination is used, farmers excluded from such arrangements risk becoming residual suppliers to a lower-priced and more erratic 'arms length' food market.

The Contribution of the Food Chain to Human Welfare

The primary, although not the entire, job of the food chain is to ensure an adequate supply of food for the human population. The concept of 'adequacy' varies according to the circumstances of the people concerned. In conditions of famine the issue may be defined in nutritional terms; what is needed is a sufficient volume of food to feed those affected. Even when total food supplies more than meet the nutritional needs of a population as a whole, some individuals may be hungry because they are too poor to buy the quantities they need. A nutritional concept of adequacy is also relevant. In rich countries, 'adequacy' may involve much more than nutritional need. The task of the food chain is defined in terms of quality and variety of food rather than simply of volume, which it is assumed will be sufficient. Indeed, many people suffer from 'overnutrition'.

The most serious failures of the food chain occur where supplies fail and famines result in widespread malnutrition and deaths. In recent years such catastrophes have been confined to developing countries and often to remoter regions. However, improvements in information technology have increased awareness of such failures throughout the world. Television pictures of emaciated people and starving children arouse concern in countries where such sciences are unknown. Charitable agencies and governments have sought to respond both by the emergency shipment of food and by providing aids to increase the productivity of agriculture in the regions most affected. Relief in the form of food is important in emergencies even though the costs of its delivery may exceed the value of the food itself and some aid may not reach those who are hungry in time to alleviate their plight. For a more durable solution the underlying causes of hunger have to be removed.

The food chain contributes to human well-being not only by ensuring an adequate food supply but also by generating incomes for those it employs. This is a major source of rural incomes in all countries but especially in many poorer areas where most people still work in agriculture. The ability to earn incomes depends upon competitive strength and this, in turn, is greatly influenced by access to capital and markers. In poor countries domestic markets are restricted and local capital limited. Farmers in poor countries, given access to international markets, might profitably borrow capital from abroad and use local labour and land to earn substantial export revenues. The fact that, for many agricultural products, entry to the richer countries of the world is restricted thus contributes to rural poverty in a number of developing countries. The disadvantage facing these food industries is intensified by the need many governments feel to provide urban populations with cheap food. Such urban bias encourages the import of dumped foodstuffs from the world market and a neglect of investment in rural infrastructure. Farmers in such a situation face markets which provide a highly uncertain and often unattractive return. They are thus encouraged to concentrate on feeding their own families, selling for cash only those products which are surplus to their needs. In such a situation there is little incentive to develop the infrastructure of communications and education which is needed to enable a modern food industry to flourish. Rural poverty is unlikely to be alleviated and production methods which require injections of capital are unlikely to be applied anywhere in the food chain.

In many developed countries, notably in Europe and Japan, governments have attempted to support farm incomes. Faced by the tendency for the terms of trade to move against agriculture, by the volatile nature of world markets for most farm products and by the relatively weak bargaining positions of individual farmers, policies have protected domestic products from international competition. Measurement of the degree of support is complex but one estimate by the Organisation for Economic Co-operation and Development suggest that the extent of protection, in the United States, the European Community and Japan is very high (see Table 1.3).

The impact on international trade has been very serious. Protection has led to the displacement of imports by domestic production. As output has grown the EC has become a major net exporter of some commodities, relying on subsidies to make its products competitive in world markets. Through its Export Enhancement Programme the US has also subsidised agricultural exports in order to retain market share. This depresses world market prices, damages other exporters and, while benefits some poorer countries which import food, has discouraged investment in agriculture in all low-income countries. The cost of such policies to the consumers and taxpayers of the richer countries has been substantial. The OECD study quoted above shows that of the total cost of 219 billion ECU borne by the countries listed, half falls on consumers and half on taxpayers. Despite this expenditure, farm incomes have continued to lag behind other incomes in most rich countries, farm people have continued to leave the land and the contribution of agriculture as a share of national income has continued to fall.

Table 1.3: Estimates of the cost of agricultural policies (1998-2002 average)

Country	Total cost (billion ECU[a])	Agricultural population[b] ('000s)	Expenditure per member of agricultural population (million ECU)
United States	78.50	7531	10.42
Canada	3.20	1047	6.49
Australia	1.50	909	1.65
New Zealand	0.60	329	1.82
Japan	50.10	9535	5.25
Austria	1.90	453	4.20
EEC10	79.60	15387	5.17

[a]The European currency unit of the European Economic Community.

[b]Agricultural population is defined as all persons depending for their livelihood on agriculture. This comprises all persons actively engaged in agriculture and their non-working dependants.

Source: OECD.

The food chain provides many incomes other than those of farmers. Surrounding agriculture itself are a variety if support activities such as the provision of inputs, the services of bankers and insurance companies and the activities of advisers and educationists. Downstream jobs exist in processing, transport and retailing. Catering is a major industry which generates many jobs. Legislation engages the attention of the civil servants and politicians at the local and international level. It is impossible, on the basis of the available data, to provide an accurate estimate of total employment and incomes generated in non-farming firms in the food chain. One estimate for the United Kingdom suggests that total employment in these activities greatly exceeds that in farming itself. Wormell in explaining that a large proportion of the UK population is dependent upon UK agriculture for its employment explains that the 2.7 per cent of the population actually engaged in farming is supplemented . . . by 9 per cent who now rely upon the farmer's custom. *Food from Britain* calculates that over 3.1 million were employed by the food industry of whom only 0.7 million were directly employed in agriculture.

Traditional agriculture, in which the distances between production and consumption were short, attempted to ensure that food was safe by customary conventions and rules. These did not always work well and particular problem could arise as seasonal foods reached the end of what we would now call their 'shelf-life'. Recently concerns about food safety have grown as the food chain has become more extended and the food products themselves more complex. There are several reasons for this. The wide geographical distribution of food products means that a failure of food safety may lead rapidly to outbreaks of disease in a number of separate localities. Tracing the source of the trouble and preventing further incidence of disease amounts to a deceptive job which can involve several countries. Many foodstuffs have had ingredients added to them to prolong shelf-life or to improve texture, appearance or flavour. Some additives, provided they are carefully used, in fact diminish the risk of food poisoning. However, a minority of people are allergic to particular additives and others are unhappy about their possible long-run effects on health. Despite regulations which require labelling, there has been a sustained campaign to remove additives from food. Food processors have responded by offering a variety of 'additive-free' products. This in itself can cause problems if housewives assume that food without additives can be treated in the same way as the products to which they are accustomed. New forms of food retailing present their own risks. Frozen food which has been allowed to thaw during transport or storage may no longer be safe. Within the domestic refrigerator unsafe food can contaminate other items. 'Cool chain' foods, which may be attractive because of their taste, require careful handling at home as well as in the food industry if they are not to increase the risk of infection. As a result of such concerns, the food chain has become subject to increased regulation which may stretch across frontiers as governments seek to assure their populations that food on sale is safe.

Responsibility for the safety of food is a long established if increasingly complex function of government. The notion that governments have a duty to persuade people to eat a healthy diet is more recent. Rules about diet are included in many religions. Some of these, apart from their ritual significance, can be attributed to the need to avoid foods which might be unsafe. Current concerns are not with particular foods but with diet as a whole. As people live longer the consequences of inappropriate eating patterns become more evident. Recent evidence suggests that many diets in developed countries contain too high a proportion of saturated fats. This has led to official campaigns to persuade people to alter their eating habits. Obesity is regarded as a risk to health and unaesthetic. Again the remedy is a change in diet. Concerns of this type, which seem likely to grow as incomes rise, present both challenge and an opportunity to the food chain. For farmers this news is bad. The substitution of animal-based foods by vegetable foods means that fewer resources will be needed on the farm. It requires up to ten times the land area to provide an equivalent quantity of human food via livestock compared with that needed for the direct consumption of arable products (an explanation of the reasoning behind this can be found in Spedding *et al.* 1981). Food processors, in contrast, may perceive market advantage. They can seize the opportunity to offer low-fat, low-sugar foods which embody a high degree of added value. They may also benefit from the need to replace some traditional itmes by novel food products in order to relieve the boredom of restricted diets. Such problems are perceived as matter of concern in the world's richer countries. However, a shift from animal based foodstuffs may effectively increase the potential supply of cereals available to meet the needs of poorer people.

The Political Economy of the Food and Agricultural Industry

Agricultural policy has a long history. Governments have been actually aware of the dangers of food shortage and the power which has been associated with the ownership and control of land. In the modern world the sophistication of policies has increased as additional goals have been added to basic requirement to assure a sufficient supply of food. The result is world food industry which is significantly different from that which would emerge from the operation of market forces alone. In classical economic terms these distortions can be seen to impose avoidable costs on the world's population However, the reasons for incurring these costs are deep seated in attitudes to agriculture. Those countries whose policies create such costs have regarded them as a price worth paying in relation to the goals addressed.

A primary goal of most agricultural policies is to ensure a secure supply of food. This has been regarded as a justification for seeking a higher degree of domestic self-sufficiency than would result in a market exposed to unfettered international competition. Production at home is believed to be more secure because it cannot be cut in times of war or as a result of economic sanctions. Strategists might well question the validity of this argument in the contemporary world. Nuclear attack would not only destroy domestic food stocks but also the productive capacity of domestic agriculture. The events which followed the accident of Chernobyl illustrate how devastating even a relatively minor nuclear event can prove. Economic sanctions applied to products so generally available on world markets are extremely, difficult to make effective. Provided the importer has the means of payment, a route by which supplies can be secured will readily be found. Despite this the agricultural policies of most developed countries still seek to encourage home production even when it is more costly than imports.

Agricultural market can be extremely volatile. In the short-run there is little that can be done to alter the current level of supply. Consumers are reluctant to change the volume they buy of most food products even when prices rise. As a result, in periods of scarcity prices may rise well above the level needed to ensure adequate long-run supplies. When surpluses occur, the market will not clear until prices fall to levels far below those of even the most efficient producers. Governments regard this volatility as undesirable on social and economic grounds. High prices for food bear with greatest severity on the poorest people in the community. Price collapses can bankrupt farmers whose output is needed in normal years and lead to major dislocation in rural economies. Food prices weigh relatively heavily in most indices of retail prices;

a sudden surge in their prices can lead to demands for higher wages and thus to inflationary pressures in the economy. The alternation of boom and bust which may result from volatile prices, can lead to a wastage of other resources. When prices are high, excessive investment may be encouraged. When they collapse much of this may be rendered unprofitable long before its physical capacity is exhausted. Such considerations have encouraged governments to insulate their own farmers from fluctuations which occur as a result of crop failures or bumper crops. One result of this has been that the market which is unprotected, the world market, has to absorb an even large share of the effects of surpluses or shortfalls and is thereby made much more volatile itself.

An important argument for agricultural support in some countries has related to the need to strengthen the national balance of payments. This may take the form of import substitution or export promotion. In either case it has encouraged governments to create an economic climate for their own farmers in which production expands. Few economists would now subscribe to the view that protection for particular sectors is an efficient way of resolving balance of payment difficulties. The solution is seen in terms of macro-economic policy relating to the supply of money, the government's fiscal policy and the control of the exchange rate rather than in any attempt to stimulate artificially a number of industries by shielding them from competition. Such a policy, by ignoring the need to use domestic resources where, in international terms, they are most competitive, impoverishes the nation and is likely to intensify balance of payment problems until aggregate demand is cut to a level at which imports are reduced. Although balance of payment arguments for farm support are largely discredited, they are still used by some agricultural pressure groups. They are with greater justification an important part of the argument against protection by low-cost agricultural exporters who are prevented from using their own resources in a way which would maximise their incomes, and those of the world as a whole, by the obstacles to trade which form part of the protective mechanism of agriculture in most developed countries.

In developed countries, agriculture is a declining share of national economic activity. The consequences of this are particularly marked in those rural areas which have no other primary economic activity. As farm populations decline the costs of providing services to the remaining population tend to rise. Schools close, public transport may be withdrawn and local shops have to charge higher prices than urban supermarkets in order to cover their overheads on a smaller volume of business. Such changes may result in the further drift of people from the countryside and the decay of rural communities. In regions close to towns, village populations may be maintained by commuters or by an influx of retired people who possess their own transport and do their shopping in towns. The loss of the traditional population, which may be accelerated if house prices rise as a result of 'incomers', is widely regretted. Village communities are often highly valued and their plight may influence the voting behaviour not only of those who live in the country but of many urban dwellers. Governments have justified support for agriculture, particularly for traditional family farms, on the grounds of such anxieties. Since, in a competitive industry, higher prices lead to higher costs these are then claimed to be a reason for shielding domestic agriculture from international competition.

Agricultural protection has not stabilised rural communities and its disadvantages have tended to become more serious. Product price support has proved to be an inadequate device to support farm income. The additional revenue which results accrues mainly to the large farmers who are by most standards already among the richer members of society. Much of the higher price paid is capitalised into land prices and so raises farm costs rather than farm incomes. Benefits go not just to landowners but to the suppliers of farm inputs. Modern farming tends to substitute capital for labour. Price support encourages investment which hastens the departure of some farm workers, raises the cost of policies to taxpayers and consumers and changes the physical environment in ways which may be regarded as damaging. The resulting distortion of trade has led to protracted disputes in international trade negotiations with consequences which may have effects far beyond agriculture itself. These results of past agricultural policies have led some major agricultural producers to re-evaluate their operation. This process is still incomplete but some features of a new approach have become evident during the 1990s. Greater attention is now given to the environmental impact of agriculture. Policies which encouraged the exploitation of land resources in ways which were damaging to its long-run productivity are being changed and, to some extent replaced by policies which seek to sustain valued features of the environment. Supply control pollicies have been introduced to lessen the growth of surpluses. The current round of discussions in the General Agreement on Tariffs and Trade (GATT) involves a commitment to lessen agricultural protection, an approach which implicitly accepts the need for discipline on domestic farm policies. The problems of rural communities are now seen to involve much more than agriculture. Encouragement is given to farmers to diversify and to new small-scale industry to seek a rural location. The cumulative effects of such a shift in the approach of governments to agriculture will become more evident during the next decade. They are likely to be unevenly distributed among the countries of the world. However, where they are applied the shape of the agricultural and food industries will change.

The consequence of agricultural policy in developed market economies has been to raise the level of production above that which markets justify or for which they are prepared to pay. Elsewhere the political economy of agriculture has had the opposite effect. In centrally-planned economies policy has not provided a profit incentive to individuals to increase their production. Prices have been kept low in order to meet the wishes of urban dwellers. In many cases farms have been organised in very large units which control capital and impose a remote management which may not encourage personal responsibility. Farms have has to bear the burden of disguised unemployment since they have been unable to cut their labour forces in response to new farming methods. The outcome varies greatly among the centrally-planned economies but in many of them food remains in short supply and in some rationing has continued into the 1995s. The failures of farm policy have had serious political consequences. In these economies, too, a re-orientation of agricultural policy is taking place. This seems likely to lead to a much greater reliance on market signals and incentives to individual producers. Since these economies possess very large agricultural resources, a successful solution to their production problem would have a major impact on world food supplies.

Among the developing countries of the world, the political economy of agriculture is closely linked to the overall problem of development and poverty. Conflicting pressures have influenced the stance of government. The need to provide low-priced food for the politically powerful urban communities has favoured the use of food imports, especially if these are available on concessional terms. A once popular belief that development required the expansion of manufacturing industry discouraged investment in agriculture, indeed agriculture was regarded as a source of both labour and capital for industrial expansion. Barriers to entry to the world's richest markets have discouraged investment in agriculture targeted on exports. Despite these discouragements agriculture has made great progress in most developing countries. Production has risen and food supplies per capita improved (see Table 1.4). Improvements in food production owed much to the use of new seed varieties linked to modern farming methods, including the application of fertilisers and the use of chemicals to control pests and diseases. Development strategies have also changed. Greater emphasis is now placed on the need to improve the incomes of the rural poor and agricultural development has assumed new importance in this context. Far from neglecting agriculture, development strategies now see it as central to wealth creation. Much work is done to discover new appropriate technologies which relate to the resource base of the developing country rather than that of the developed world from which most recent innovations have stemmed. Such changes in the perceived economic role of agriculture in development influence policy making and the incentives given to agricultural producers. Their success will depend upon the ability of the food chain to meet the need of a growing population in many relatively poor countries.

Table 1.4: Changes in food supply per capita by major economic regions

Region	Average index of food production (1979-81 = 100) Index for 2003-04
Developing economies	110
Low-income countries	114
Middle-income countries	103
Industrial market economies	103

Source: World Development Report.

❑❑❑

2

Meeting Human Needs

Introduction

When man was a hunter gatherer, his main concern during the two hours or so he spent each day looking for food must have been whether he would get enough to satisfy his hunger. Such must still be the case for much of the developing world, as indeed it was until the last century for the so-called developed world. Knowledge about the particular dietary needs for protein, energy or vitamins was certainly not commonplace until after the First World War and many would argue that nutritional knowledge is sadly lacking even today. It is, however, not more than two or three generations ago that most people had some contact with the land whether directly or indirectly and everybody has a view about food and agriculture.

Agriculture may be the world's largest business, but its role is changing dramatically and producing food is only a part of the story. Farming was once considered to be a way of life, and so it may remain for a decreasing few, but the complexity of modern agriculture cannot be so simplistically dismissed. Agriculture today encompasses, on the one hand, enormous multinational endeavours reliant on institutional capital and, on the other, enterprises held together by national policies designed to keep people on the land. All, however, benefit to some extent from subsidy which on average, for example for OECD countries, is of the order of 40 per cent of the value of production and in Norway rises to about 70 per cent. It is not surprising therefore that the European Community's Common Agricultural Policy, which swallows so much for the Community's budget, is the major focus for Community affairs.

The state's involvement in agriculture worldwide may provide the insurance to ensure a stable agriculture in case of emergency, or a social policy to maintain rural communities. In the developed world it is difficult to justify a role for agriculture simply in terms of the provisions of food, fuel and apparel. Agricultural surpluses are *de rigueur*, which fosters political sensitivity but also establishes food as a commodity which becomes subject to the usual forces of the market-place. Nowadays consumers require their food to be of consistent quality, wholesome, health-giving and free from potentially injurious substances (see Table 2.1).

Table 2.1: Customer concerns about food buying

	Percentage of survey responses			
	Extremely concerned	Very concerned	Somewhat concerned	Not concerned
Price	45	36	18	1
Safety	41	32	21	3
Nutritional value	33	43	18	2
Wholesomeness	36	37	21	4
Non-fattening	37	31	25	6

Source: American Association of Cereal Chemists.

As the market for food and food products moves into fewer hands (in the UK, five or six major retail chains supply the majority of the population with groceries, meat and vegetables), the power of consumers to dictate their needs and the profit incentive for supermarkets to meet them, ensures that product quality in the widest sense is met whether it is soundly based or simply fashion. There are, for instance, good medical reasons to reduce the contribution which saturated fats make to dietary energy, yet traditional perceptions, and the thinnest scientific evidence, of the role of intramuscular fat in deterring the eating quality of meat, may well invite butchers and meat retailers to offer meat with greater marbling fat and thus undermine people's health.

The demand for a growing variety of new products which are convenient to use and easily prepared, has stimulated food processors and manufacturers to use novel technologies and approaches to provide an infinite variety of new foods, some of which may require careful storage and preparation if the risk of their caryying food-borne disease is to be minimised.

It is against this complex background that meeting human needs for food must be considered. There are implications in this for the spectrum of endeavour between the conception and consumption of plant as well as animal crops and this chapter will review them for the most part in relation to their importance in the food and agriculture sectors

Food Production and Requirements—a Global View

Achieving the balance between food production and need depends not only on the food production and nutrient requirement per head of population, but also on the food purchasing power of individuals which varies enormously both within and between communities. The developing world is characterised by low output, and a population that is large and growing, with minimal purchasing capacity. Wealth is the protection against starvation. In the developed world it is still poverty which is the main cause of hunger, for population growth may be static or even declining and food surpluses are commonplace. Population pressure in developing countries is a major influence which is likely to become an even greater challenge. By 2010, the proportion of global population living in the Third World will have increased from its present 73 to 78 per cent.

The enormous advances made in agricultural science and practice have allowed food production per head to be increased on average in both developing and developed countries (0.4%/year and 1.4%/year, respectively) but this masks the massive declines in Africa, for instance, and the smaller deficiencies of eastern Europe, Russia and parts of Oceania.

Much of the improved output is attributable to greater cereal production which has allowed livestock production to expand. But the availability of cereal grains for livestock feed in developing countries is still far behind their use in the western world (13% versus 72%). If livestock production is to be stimulated it will need to be based on cereal production, for the potential global meat output from grazing land is only about 25 per cent of the total.

Increases in food consumption have outstripped the rise in domestic production of the poorer countries of the world and for some the shortfall may have to be made good by the purchase of food imports. This may be a possibility for eastern Europe and the Russia, for example, but the lack of foreign exchange and purchasing capacity of the poorest countries means that, at the least, there will be less food on offer and the risk of hunger and famine will increase.

The increased food production of recent times in most countries of the world has relied on the increase in yield of plant and animal crops brought about by intensification. This process demands greater energy inputs via fertilisers, mechanisation, irrigation and so forth and the use of new and higher-yielding crop varieties or livestock which are superior in milk or meat production. The western world has used all the advances and its high investment capacity to stimulate agricultural production into considerable surpluses. The Third World, on the other hand, has been much slower to incorporate such modern developments because of financial constraints but also because of the lower pace of technology transfer in these countries.

It is not enough, therefore, to focus on improving agricultural production in the poorer countries for it may simply obscure the need to reduce debt and poverty and strengthen institutions and economies. What seems to be needed is a more comprehensive and equitable investment not only in agriculture and food production, but also in industry, commerce and infrastructure. It may well be the only way of protecting the environment, controlling population and growth and bringing to all the quality of life which presently seems to be the prerogative of the First World.

Man's Needs and Demands for Food

Danton's memorial in the Place de l'Odeon in Paris reads: "Education is the greatest need for the people, but first they must be fed". Few of us would find fault with such a sentiment though individually we might profoundly disagree about how 'they must be fed'. In the developed world today, the concern is no longer with getting enough to eat so much as controlling intake and ensuring that what we eat is safe, healthy and wholesome. Increasingly the agricultural and food industries are concerned with the latter rather than with meeting human needs defined in terms of recommended dietary allowances. Most of the problems and opportunities for meeting western man's requirements for food are related to animal foods and the following sections will highlight them.

The implications of diet for health, survival or mental development provide endless copy for scientific journal and popular magazine alike. The consensus among nutritionists is probably that we should seek to eat a well-balanced diet, but what the constituents should be and in what proportion, are hotly debated.

The adequacy of a diet is commonly judged by how we grow. Though there are many reports of growth retardation caused by lack of specific nutrients, including protein, it is undoubtedly the intake of energy which exercises the main control over growth for it is difficult to consume a varied diet which meets energy needs without consuming an adequate amount of protein. Given this, the conclusion must be that it is poverty and ignorance which are the prime causes of malnutrition.

The secular trends in the growth of height and weight among many nations of the world over the last 70-80 years have been well documented and most authors suggest that this is attributable to improvements is nutrition and environmental circumstances. But the trends are not confined to the lower classes although the size of the change tends to be less in the wealthier groups of society. An interesting fundamental aspect of it all is the interaction between rate of growth and period of growth. It is matter of historical record that slow-growing individuals may reach their final bights six or seven years after their faster-growing contemporaries but the mature size of the two groups may not be different. There does seem to be a chronoglical age beyond which further growth is impossible and if the usual adult size has not been reached by that time, it never will be. Today's children, however, become taller because they grow faster to a greater height even though their growth period may be truncated.

There is a secular trend for age at menarche in girls and it is usually explained in similar terms to that for growth. The association between mutation and reproductive function has long been known in animal production. Animals in poor condition, for example, may not breed whereas animals kept in better condition or even in poorer condition but 'flushed' with a short period of plentiful feed before mating, could show better fertility and fecundity. Frisch believes that it is particular threshold relationship between fat and lean body mass which determines the onset and maintenance of reproductive ability and she links the size of the fat associated store and its energy value with the energy needed to maintain a pregnancy and lactation. The fine setting of this balance is easily disrupted by strenuous physical activity and controlled food intake and some athletes and ballet dancers to whom this applies may suffer delayed menarche and amenorrhoea.

All of this goes to show that the relationship between nutrition and growth is articulately resilient in the longer term, and it is only towards the extremes of experience that problems occur and even those may be corrected by appropriate nutritional adjustments.

Today's Diet

It should not be surprising to discover that over the years there have been many revisions of the nutrient requirements for man to accommodate moves towards more sedentary, less-strenuous employment, or central heating of homes and workplaces, or the greater use of cars. Despite the exquisite control of food intake by neuro-humoural mechanisms (Forbes 1988), fat and thin people can still be found. Overeating, leading to obesity, is a serious hazard of contemporary life which contributes to the high incidence of cardiovascular disease and diabetes in the western world. *Anorexia nervosa* is seen as a psychiatric problem which causes serious loss of weight, especially in women, and, in some cases, death. Neither of these conditions is strictly a function of particular diets but rather a consequence of eating too much or too little. Of course, the composition of the diet, whether it is energy-dense or energy-deficient, may exacerbate the problems, but it is only in more recent years that clinical and scientific investigations have implicated specific components of western diets in the aetiology of cardiovascular and coronary heart disease and initiated the diet/health debate.

Diet and Health

The usual arguments which are propounded on this issue are that high blood cholesterol signifies an individual to be at risk from heart attacks. The consumption of saturated fats leads to a high blood cholesterol. In those countries, for instance China, where there are few dairy and land animal products for the populace to consume, there is only a low incidence of cardiovascular disease (Peto 1990). The conclusion we are asked to draw is that moves towards diets which are low in animal fats and rich in plant carbohydrate and fibre will be beneficial to health. There would seem to be benefits in further analysis of this, admittedly simplified, thesis, for in it there are profound implications for agriculture and animal production in particular.

Implications for Animal Production

Meat and dairy production are enormous endeavours on which the economies of regions and countries may be almost totally dependent. The developed countries contribute about two-thirds of the world's output of meat (see Table 2.2). In the UK it is valued at about £ 3 billion or approximately one-third of total agricultural output. There is considerable interlocking of meat production with other farming activities; beef production is closely associated with the dairy herd and the success of the pig and poultry sectors can reflect the supply of home-produced cereals. The predominant meat varies from country to country. Beef ranks highly as the chosen meat of many countries whereas lamb is popular in only a few. Pigmeat is growing in importance and poultry, as a consequence largely of price competitiveness, most of all.

Table 2.2: Production of meat (× 10^6 tons). Contribution (%) to total production

			Percentage contribution	
	World	UK[a]	Developed countries	Developing countries
Beef	47	1	24	11
Sheep and goat	7	0.25	2	14
Pig	51	0.9	25	14
Poultry	28	0.75	15	6
Total	133	2.9	66	34

[a] Meat and Livestock Commission.

Source: Food and Agriculture Organisation.

The contribution which meat makes to a nation's diet is predominantly a reflection of a country's economy. In the western world meat may account, on average, for one-third of dietary energy whereas 6 per cent would be the figure for the developing world. For protein intake, the respective values are about 59 per cent and 15 per cent. Growing affluence in a country is marked by an increase in meat consumption and it is this, and the associated change of lifestyle, which some observers conclude to be the predisposing factors in the rise in the incidence of heart disease and cancer. It is, however, now commonly agreed that the most important cause of death is smoking, for which there is a simple solution.

But there is a perceived need to modify diets to improve health and reduce mortality from degenerative disease. Most exports have concluded that reductions in the saturated fatty acid intake could be beneficial but great improvements in national health could be achieved by a simple reduction in total fat intake. All meat is charged as suspect in the diet/health debate, and a wealth of information from surveys and experimental investigations exists to show that animal fat not only provides a rich source of calories and an easy route to becoming overweight but that specific components may be deleterious to the cardiovascular system and a predisposing factor in coronary heart disease and strokes. The consequence has been that many national medical authorities have begun to recommend and promote diets in which the energy available from animal fat contributes less than 35 per cent to the dietary energy. Whether or not the scientific basis for such recommendations is secure, there is now strong governmental pressure to reduce the intake of animal fat: the animal and meat industries ignore this at their peril and it is here where the great debate starts. In the UK, the Committee on Medical Aspdcts of Food Policy and the National Advisory Committess on Nutrition Education proposed certain guidelines on diet. The 'dietary goals' proposals made in the USA were rather more specific but have been significantly modified in recent years to recommend the consumption of a variety of foods of nutritional quality and in quantity to maintain reasonable body weight. More emphasis is placed on reducing the intake of all fats, without specific concern for their composition or source, and on the energy density of fat and its potential for effecting weight gain. This, in any case, has been the tendency of recent years (see Table 2.3), but it must be pointed out that the 'NACNE' recommendations come close to what people were eating 50-60 years ago. Today's consumers prefer their meat to be much leaner though this does not prevent them from indulging in paté, ice-cream and cream cakes which contain large amounts of animal fat.

Table 2.3: Food energy available and source

Years	Percentage contribution from			
	Energy available (k) (kcal)/ (person per d)	Protein	Fat	Carbohydrate
USA				
1909-13	14600 (3490)	11.7	32.1	56.2
1935-39	13680 (3270)	10.9	36.3	52.8
1965-70	13220 (3160)	12.1	41.0	47.0
UK 1980	12680 (3031)	11	38	45
NACNE[a] 1983 preliminary recommendation	12680 (3031)	11	34	50

[a] NACNE (National Advisory Committee on Nutrition Education).

Source: Olson.

All this does not bode well for any national policy on diet. Given the high correlation between serum cholesterol and the incidence of heart attacks there might be advantage in reducing dietary cholesterol. The evidence available, however, shows no specific surgical advantage to this approach for though fewer die from heart disease on low-cholesterol diets, more individuals die from other causes. In Australia, where the mortality from heart disease has fallen, the average serum cholesterol has not. In the USA there has been a uniform fall in the incidence of coronary heart disease across race and class which is unlikely to be explained solely by changes in diet.

An outstanding and common finding is that social class is highly associated with how responsive people are to advice on changing lifestyles to lower morbidity and mortality from heart disease. Such recommendations will certainly include reduction in smoking, curtailing the consumption of saturated fat and replacement by carbohydrate as fruit and vegetables: overall to eat less. There is accumulating evidence to support the greater consumption of polyunsaturated fats, especially those containing linoleic acid, and of vitamins E and C which may provide antioxidant protection of depot fats and prevent the release of potentially pathogenic 'free redicals'.

The suggestion that social class is closely identified with health education success, the professional classes responding quite dramatically, identifies the overwhelming need for public education on all aspects of improving lifestyle whether it be towards improving diets, stopping smoking or taking more exercise.

While there has been some reaction of the public to the diet/health debate in terms of their food purchasing, other consumer requirements have additional impact on food production, marketing and sales. The UK National Food Survey examines how people spend their money on food. Recent surveys have reflected the changing lifestyles and food habits associated with greater amounts of disposable income and different work patterns and employment. More wives are now in paid employment which supplements family income but also limits the time they spend in food purchasing and preparation. They will certainly now more often share these activities with other members of the family. Food processors and supermarkets recognise this and the range and types of foods and food products on offer have snowballed. They must also be processed, packaged and presented to appeal to potential purchasers.

The success of poultry meat in recent years (Table 2.4) exemplifies how that industry has met the perceptions of consumers. Poultry products also contribute to the growing market for convenience foods (Table 2.5) which is reflected in the sales of freezers and microwave cookers.

Table 2.4: Daily consumption of meat (g)

	UK			USA[a]
	1965	1975	1985	1979
Beef and veal	30	31	26	47
Mutton and lamb	24	18	12	<1
Pork	12	12	14	21
Poultry	15	24	28	75
Total	81	85	80	153

[a] Olson 1981.

Source: National Food Survey 1986.

Table 2.5: Expenditure on meat convenience foods in the UK (not adjusted for inflation)

	Pence/person per week		
	1975	1980	1985
Canned meat and meat products	6.2	9.4	11.8
Frozen meat and meat products	2.5	8.2	14.3
Other 'convenience' meat and meat products	16.6	33.5	49.9
Total	25.3	51.1	76.0

Source: National Food Survey 1986.

Poultry meat is successful because it offers many advantages, notably price, over its competitors. The poultry industry has, of course, been highly successful in improving efficiency of production through improved strains and types of birds, better nutrition, substabtional investment in systems and buildings and new facilities for slaughter, processing and innovation for the development and production of new lines. The whole is a highly integrated operation in the hands of a small number of large companies who have substantial financial backing and highly organised marketing arrangements with retail chains. Poultry, among the meat species, is successful because of its potential for industrialised production and cheapness of production which is associated with the leanness of the birds, rate of growth and consequent energetic efficiency. Beef and lamb compete poorly on all these counts. Pigmeat fares relatively well compared with beef and lamb and its production provides a useful monitor of the ways in which the meat sector or the animal production industry has needed to face market, economic and political challenges and to react to consumer pressure by having great concern for product quality, humanness of production and environmental matters. The pig industry of western Europe provides a valuable perspective on these matters.

The European Pig Industry

Self-sufficiency in pigmeat varies among the EC Member States from about 70 per cent in the UK, Greece and Italy to almost 300 per cent for the Netherlands and 350 per cent for Denmark. The shortfall (wholly of bacon) in the UK is met predominantly by imports from Denmark and an increasing, and currently, almost matching, contribution from the Netherlands. Denmark historically exported a large proportion of its pigmeat (presently about 48%) to the UK whereas the Dutch export trade to the UK is a relatively new venture, at least in the volumes presently traded. Both Denmark and the Netherlands operate trade in live pigs, particularly sows to West Germany from Denmark and weaners to Italy from the Netherlands.

Within the EC, West Germany produces most pigs and Ireland and Greece fewest. (In passing it is worth noting that even West Germany's pig population lags considerably behind those to Brazil, USA and the Russia: China's pig herd is fifteen times larger.) The striking differences in numbers of pigs produced do not, however, correlate with the size of herds in which they were raised. For example, 78 per cent of Irish pig production comes from large units holding more than 1,000 pigs. In West Germany only 6 per cent pigs come from holdings of that size. Industrialised pig-keeping is more characteristic of the UK, Netherlands, Denmark and Ireland and small-scale operations are typical of Greece and Italy. Even so, in these countries the small amount of intensive, large-scale production accounts for >40 per cent and >50 per cent respectively of all the pigs produced in those countries.

The rationalisation of pig production on to fewer, large farms is typical of developments worldwide and is at the heart of the untoward consumer perceptions about pigmeat. In the USA, for example, the number of pig farms has declined by 8 per cent per annum during the 1980s. There are now 350000 farms producing about 70 million pigs for slaughter, 90 per cent of which are raised on 27 per cent of the farms. It is expected that 500-sow units will become increasingly common.

The impetus for these changes stems from the increasing need, as profit margins narrow, to improve efficiency of production, marketing and the financial management of businesses. The economies of scale provide for better investment potential, the bulk purchase of feed ingredients rather than compounds, and the greater use of technology including computer control and data handling systems. The use of forward pricing, futures markets, the better management of financial and contractual arrangements with packing plants which increasing size of business allows, all add to the opportunities for business success.

Responses to Consumer Needs

Much of the foregoing also occurred in the development of the poultry industries, though the speed of change was much faster than is likely to be possible for the pig industry. The are also new problems which will have to be faced which are only just being met by the poultry sector. The social, environmental and consumer considerations which are becoming increasingly important may become the most crucial for the future pig and poultry industries and will undoubtedly have major implications for agriculture generally in the developed world. The importance of each of the different challenges varies from country to country but the situation observed in the UK is reasonably representative.

The growing importance of supermarkets rests not only on their value in signalling consumer requirements but also, because of their considerable purchasing capacity, in determining what is considered to be acceptable in how animals are raised, slaughtered and processed. It has been pointed out already that the wealthier people become, the greater their consumption of meat. But this is not the whole story for such people become increasingly discriminating in their tastes and have the money to demand and pay for their particular requirements.

Consistency of quality is one such demand and is sought by consumers and processors alike who need, for example, consistency of carcass composition for processing, packaging and product development. Consumers nowadays also look for value for money in their purchases and for foods to be 'wholesome' and 'health providing'. Convenience and speed of preparation become prime characteristics for modern foods, and the opportunity to create added-value products is avidly sought by food processors and manufacturers. Meat is rapidly becoming another supermarket line like cornflakes and toothpaste and like these is required to possess the applauding features which attract purchasers and invite repeat purchase. Poultry meat is perceived to have most of the desired characteristics; good value, predictable quality, acceptability to young consumers. It is also offered in a great variety of forms and products from whole carcasses to joints and pieces, precooked delicatessen itmes and dishes which use lean meat as the base for more exotic treatment. Features such as these are clearly also of the highest importance in the securing and developing of markets for pigmeat and pigmeat products.

The pig industry has worked to induce greater leanness in its stock through breeding and improved husbandry and nutrition but most of the carcass fat of pigs is located in one subcutaneous depot which may readily be removed. New processing technologies, the use of male animals, hormonal and non-hormonal growth promoters and immunological manipulations are all effective to some extent in improving carcass leanness or reducing the fat content of products. All, however, are subject to some constraint. Male carcasses risk being tainted with lipophilic pheromones; public opinion and legislation prevent the widespread use of exogenous hormonal agents and problems with adjuvants and delivery systems hamper the successful practical introduction of inmmunological approaches. The gratuitous use of drugs for blanket therapy, the illegal use of growth promoters and antibiotics and the movement of diseased animals across national frontiers are all detrimental to an image of safety and wholesomeness for pig and other meat.

The containment of large populations of pigs on small areas of land leads to problems of pollution and environmental contamination and curbs on pig production have been established, especially in the Netherlands, but also in the East Yorkshire and Humberside areas of England. In the Netherlands, 30 per cent of the phosphate and 15 per cent of the ammonia from animal production comes from pigs. There are now quota restrictions on adding to the volume of these pollutants and major practical difficulties in disposing of surpluses of manure, including their transport and spreading at certain times of the year. There are legal requirements to meet and new techniques, for example, the direct injection of slurry into the soil, are being developed as more 'benign' ways of dealing with it than spreading on the land.

There is no doubt that these are serious impositions on the pig industries of Europe which can put their future in jeopardy. It is not so long ago that the recommended approaches for successfully meeting market and consumer requirements for pigmeat rested on the provision of carcasses and meat of uniform and consistent quality in terms of leanness, absence of PSE (pale, soft and exudative), DFD (dark, firm and dry) and soft fat. It goes without saying hat these are still prime requirements for the wholesale and retail trading of meat but the pig industries of the developed world have infinitely greater challenges to face in terms of animal welfare, food safety and environmental pollution. To fail in these areas will not simply disturb the cyclicity of appeal of pigmeat, but will decide whether or not there is a future for pigs. This rather telling picture of the pig industry is not without parallel and though there are local and regional variations in the extent to which they are adopted and adhered to, the principle underlying philosophies are common to all.

Food Safety and Wholesomeness

The epidemiology of food-borne disease in the UK as elsewhere has changed dramatically in the twentieth century. Bovine tuberculosis, brucellosis and infestation by food-borne helminthes have been virtually eliminated in the UK. Some diseases including cholera, typhoid and paratyphoid are nowadays spread wholly or predominately via persons infected abroad and dome illnesses may have imported material such as herbs and species as their source. Shellfish, harvested from polluted waters, are often responsible for outbreaks of hepatitis A infections, but the most dramatic epidemiological occurrences of the 1980s were the rises in the incidences of Salmonellosis and Camploybacter infections (Figs 2.1 and 2.2).

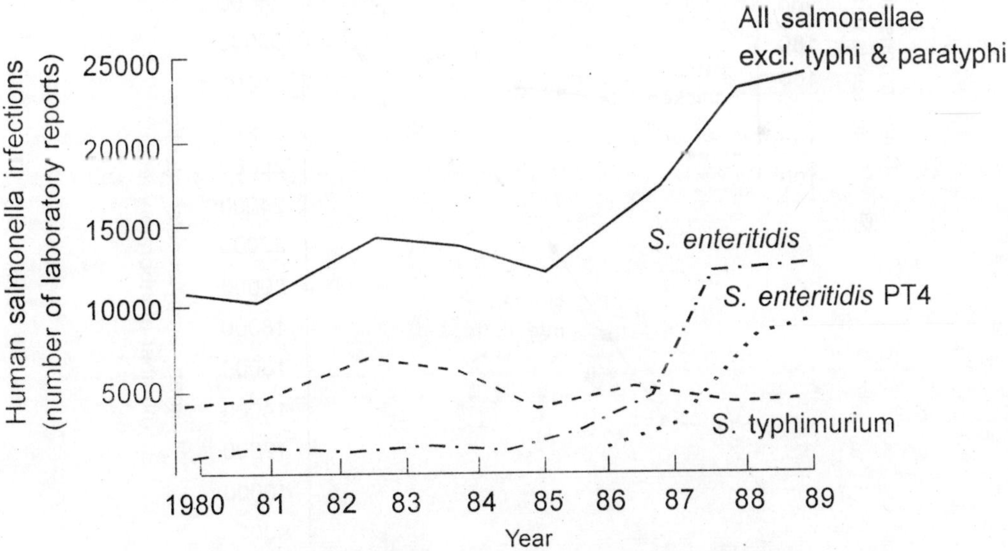

Incidents in animals (number of Zoonoses Order reports)

S. typhimurium in cattle

708 845 877 1215 1174 1208 958 679 663 431

S. enteritidis in fowls

7 8 8 1 0 15 36 111 401 581

Source: PHLS, MAFF.

Fig. 2.1: Trends in *Salmonella* infection for England and Wales 1980-89. (Source: Galbraith 1990).

S. typhimurium and *S. enteritidis* PT4 give rise to disease which may be symptomless or lead to systemic complications which can result in a patient's death. More commonly the disease is characterised by fever, abdominal pain and diarrhoea. *S. typhimurium* was the cause of an epidemic in the 1980s and thought to be attributable to contaminated meat and meat products which were widely distributed, giving rise to sporadic outbreaks. The incidence of the disease matched the changing size of the cattle populations over that period. *S. enteritidis* PT4 epidemics emerged in the mid 1980s and have continued to be widespread. There is a close association between the incidence of the disease in man and contamination and consumption of broiler chickens. Shell eggs have also been implicated. There is little doubt that Salmonellosis in man stem mainly from infection in poultry which are transmitted to man, directly or indirectly, by contaminated poultry meat and the contents of intact shell eggs.

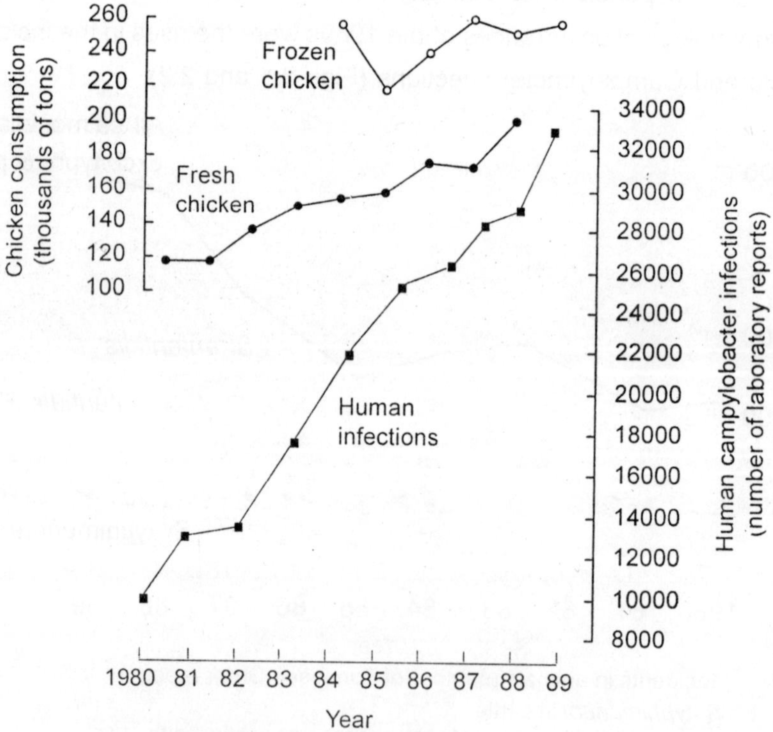

Source: PHLS.

British Food Information Service.

Fig. 2.2: The relationship between chicken consumption and human Campylobacter infections. (*Source*: Galbraith 1990).

Campylobacter infections in man are also characterised by diarrhoea. The causative agents are almost certainly *C. jejuni* which has been recognised as important in this regard since the late 1970s. During the 1980s the incidence of *Campylobacter* illnesses increased frantically (Fig. 2.2) in line with the consumption of fresh chicken. Poultry meat has been shown, through surveys in several countries, to be contaminated frequently with *C. jeuni* and *C. coli* and case-control studies have linked *Campylobacter enteritis* and eating chicken.

Listeriosis caused by *Listeria monocytogenes* is one of the so-called 'emerging' food-borne infections. It may appear as cutaneous infection in people who handle animals. It can have more damaging sequelae such as meningitis and septicaemia, especially in the very young and elderly, and there can be 30 per cent mortality. There can be an even higher mortality among foetuses infected *in utero. L. monocytognes* contaminates a wide variety of foods from salads to patés and cheeses. It grows only slowly at temperatures below 5°C and good refrigeration and process in practice are most effective in containing the bacterium.

A new and somewhat distressing twist to food-borne disease issues has arisen with the current epidemic of bovine spongiform encephalopathy (BSE) in the UK. BSE is probably caused by a scrapie-like agent which was thought to have been introduced into the cattle population through compounded feeds which contained protein from scrapie-infected sheep. Scrapie has been in the UK sheep population for more than 250 years without seeming to have crossed the species barrier. Apparently it has now done so which suggests that there has been a major change in the organism or that the processing methods for carcasses and offal do not destroy or inactivate it. There has been a move from batch to continuous systems of rendering animal products which can involve a change in the time/temperature requirements of the process, though it is uncertain what consequence this has had for the pathogenicity of the purported agent. A BSE-like disease has been diagnosed in confined deer, in mink and domestic cats and it has been transmitted to a single pig in laboratory investigations (though not by feeding). Concern has been expressed that a variant of BSE might be transmitted to man to cause Creutzfeldt-Jakob disease (CJD), a rare but widely-spread neurological disease with some resemblance to scrapie. It is, however, of note that

CJD occurs in countries where scrapie is not found, as well as in those where it is prevalent.

The Ministry of Agriculture, Fisheries and Food's (Southwood) Committee, established to investigate the disease and the risks to other animals and man, concluded that there was no risk to human health from consuming beef. Measures have now been taken, however, to ensure that BSE cannot enter the food chain. These start with the slaughter and destruction of infected cattle and have now spread to the banning of the incorporation of brain, spinal cord, spleen, thymus, tonsils and intestines of even healthy cattle into human food. Legislation, first to ban the feeding of ruminant protein to other ruminants, has now been widened to block its use in diets for monogastrics. There is no evidence of transmission of BSE between dam and progeny either via placental mechanisms or milk but, in view of the serious implication for the maintenance of this disease from generation to generation, these possibilities are being examined in a major research programme currently being conducted in the UK.

The complexity of issues surrounding food-borne disease requires that they be examined in a multidisciplinary manner involving agriculturists, veterinarians, food scientists and technologists. It is vital to be able to anticipate and, by so doing, avoid new risks of food-borne disease and to establish surveillance methods for animals, food and man to identify potential problems at an early stage so that steps can be taken to control them.

Animal Husbandry and Hygiene

Animal welfare is a clarion call for many, but the potential risk from zoonoses and food-borne diseases is perhaps more disquieting to people in general. There is no doubt that intensive production system risk the rapid spread of any disease to large numbers of animals and a focus for widespread infection of man. Outbreaks of food-borne diseases attributable to *Salmonella, Campylobacter* and *Yersinia* organisms have made consumers cautious in their choice and consumption of meat and animal products.

The move towards fewer production units with higher capacity is not confined to the pig and poultry sectors; average dairy herd size has increased by approximately five-fold in the last thirty years. Only sheep production has attracted more participants. Increasing unit size almost always involves an increase in the intensity of operation and with it the need for greater husbandry skills and vigilance. The problems associated with housing, management and health are multiplied and the risks of infectious diseases devastating large populations of stock are serious. But turning to more extensive, 'natural' or organic' systems of production does not remove all risk and hazard. Abandoning prophylactic measures, say, vaccination or anthelmintics, without adopting the necessary husbandry or management practices, can prove disastrous. Zoonoses, i.e., diseases transmissible from animals to man (though the reverse can apply also) become increasingly important in regard to food safety. *Salmonella* and *Listeria* are bacterial diseases which are food-borne to man, but contaminated sludge of effluent from sewage works or chicken waste spread on farmland can establish infections in animals using that land.

Diarrhoea and enteric disease may be of bacterial or viral origin and there are many instances where both may be found together. For instance, calf diarrhoea frequently (≈ 50%) involves rotavirus and enterotoxogenic *E. coli*. The parasite *Cryptosporidium* is found in the faeces of all species of animals, often without symptoms. It is, however, a significant cause of diarrhoea in young animals when it may be found together with *Campyloybacter* and *Salmonella*.

These organisms, either singly or in combination, account for many of the gastrointestinal infections of young animals. The prevention of disease rests to a large extent on the consumption of colostrum by neonates. Vaccination of sows with *E. coli* carrying the K99 pili provides passive antibody protection for piglets until such time as active immunity can be established. There are no good vaccines against *Salmonella* and *Campylobacter* and control is largely to be achieved by good management practices and the provisions of clean and disinfected housing.

Prophylaxis based on improved husbandry and management is seen increasingly as the preferred means of controlling disease in populations of farm stock. Not only is it seen as the avoidance of expensive veterinary charges but it also gives the products of the enterprise a better image among consumers.

While every effort can be made to provide environments and management regimes which reflect the physiological needs of animals, it is frequently difficult to translate those needs into practical facilities and routines. For instance, keeping animals in small groups confers distinct advantages, such as disease control, or maintaining comparability of size or age. It is not usually practical, however, to do this, and problems found in larger colonies such as adequate control of temperature, humidity or air movement may need to be overcome by improved ventilation.

The provision of appropriate flooring and bedding materials will enable animals to remain free of dung with obvious benefits in the prevention of the spread of disease, food contamination, or contamination of fleece, hide or feather prior to slaughter.

Building design and the materials employed are important in this context. Smooth wall surfaces are easily cleaned and also prevent damage to animals. Partitions between pens prevent cross-contamination and the spread of disease.

Young animals are always at risk from enteric disease and the maintenance of hygienic environments may be life-saving. But zoonotic diseases may be controlled most effectively by the same approaches for older stock, as well as contributing to a better image for farming.

Exposing stock to contaminated grazing or holding accommodation provides for the continuing life-cycle of parasites such as *Cysticercus bovis* or *Taenia saginata*. Modifying cattle grazing practices or vigilance at meat inspection and removal of affected caresses, or parts, provides effective control. *Salmonella* and *Listeria* contamination from sewage is not so easily monitored or combated but techniques such as direct injection of slurry into soil have been used with success. *Listeria*-contaminated soil may, however, be incorporated into big-bale silage during grass harvesting and, if the fermentation conditions are not appropriate for good silage making, the *Listeria* bacteria may survive to infect the ruminant which subsequently eats it.

Compound feeds are not totally free from pathogenic contaminants. *Salmonella*-

infected feed has traditionally been blamed for outbreaks of Salmonellosis and legislation was introduced in the UK to counter the rising incidence of such infections. Feed processing and regorous monitoring of the product has had considerable impact on this source of infection, but it may not be possible to use effective measures for the control of some pathogens without damaging the feed itself. The use of formic and propionic acids has been considered as an approach to avoid the use of damaging high temperatures for destroying or inactivating feed contaminants. New processing techniques must, however, be closely evaluated to ensure that they not only serve the processing needs but also effectively sterilise the material. There is the risk that this loophole was responsible for allowing the agent responsible for BSE to enter cattle production.

Animal Welfare

Man's attitude to his companion animals has rarely been passive. Cats were revered as gods in ancient Egypt, and dogs are thought of as an Englishman's 'best friend'. Cattle are of religious significance to Hindus and are singled out for special consideration. Until recently though, the farm animals of the western world have been looked upon primarily to provide food for man or draught power for his machines. They were housed, managed and husbanded to promote the highest efficiency of production or service and to improve the profitability of an enterprise. This is not to say that all farmers considered their stock simply as animal machines but certainly there was, and still is, a range of attitudes among stockmen which includes that one. Humanitarian concern for animals is part of the growing awareness of the impact of farming on the country-side and on consumers. Animal welfare has always been important to the good stockman but there is a strong feeling in society that it should not simply be left to the altruism of individuals and that it should become a responsibility for government and legislation.

It has to be recognised, however, that establishing and adhering to principles of animal welfare will always be a matter based on the priorities which individuals or societies in general are obliged or choose to set for themselves. In countries where a low value is attached even to human life, consideration for the welfare of animals is hardly likely to command a high priority. Where these issues receive a sympathetic hearing, arguments will still be heard among producers, for example, that meeting welfare codes of practice or installing the necessary equipment will prove prohibitively expensive or will lead to higher prices for food which consumers will not wish to pay.

The straight economic argument may be that the benefits of animal welfare ought to be considered as being no different from other benefits, that is, something to be enjoyed whenever people get something they prefer or want. It follows that if there is a growing demand for livestock products from systems perceived as kinder to animals, any attempts to meet these demands will be rewarded, probably through higher prices. In practice, however, it will be necessary to know from society, the technical and cost implications of providing them and, perhaps most importantly, how far political constraints will be able to cope with the choice between the interests of animals and people in the lower-income groups of society.

At the present time, countries differ substantially even in, say, western Europe and North America in the steps they are willing to take to improve the lot of farm animals. Sweden chooses to set very high standards for animal welfare by law and has set its stance against intensive animal production systems. The UK, through its Farm Animal Welfare Council, has made recommendations about the husbandry and management of animals from conception to slaughter on which codes of practice and even law may be based. The European Community is reviewing animal welfare in relation to systems of production, transport and slaughter, for example, and regulations and directives for community-wide application have been drawn up. The political implications, often based on national attitudes to animals, make the job of lawmakers particularly difficult and what may appear as inertia and foot dragging on the part of politicians becomes the focus for extremist and partisan attention.

European animal production is thus under considerable pressures in all these areas and the legal framework now being established by governments is likely to permit animal production only under conditions which offend neither animal nor stockman, which provide food which is safe and when the systems are not environmentally polluting. The pig industry is under particular scrutiny for its sow husbandry and management practices involving the use of stalls and tethers but concern is equally raised about the care, housing, transport and slaughter of growing and finishing pigs. Legislation may be the action which ultimately controls many of the more questionable practices, but there is a growing public feeling in many countries, especially among the young, that if animals have to exist in what are considered to be inhumane conditions then many people will simply stop eating meat.

What ultimately matters to the success of the animal production industry is that meat and animal products continue to be sold. There are signs, however, that as meat is no longer seen as a necessity in the diets of the developed world, the meat industry will need to concern itself more with the perceptions of consumers. Quality for the consumer no longer simply means leanness, colour, hue, taste, texture or wetness of meat. Questions are now regularly asked about the ways in which food animals are reared, whether they are humane, whether the risk carrying and spreading organisms responsible for the food-borne disease and whether the system of production is environmentally friendly. These criteria will be more important in future than the traditional quality attributes. The weight given to them via public protest, and now increasingly legislation, makes the consequences of not responding to them of cataclysmic consequence to the food animal sector.

Animals and Meat

There is a large section of the public that prefers to believe that meat is a commodity that we buy in a butcher's shop and has nothing to so with animals that are kept on farms. This convenient perception allows some meat eaters to close their eyes to animal production systems which offend animal welfare and leads some animal welfarists to demand husbandry practices which, for a variety of practical and economic reasons, are incompatible with producing meat. But there are participants even in the production, slaughter and processing sectors who fail to recognise that the delivery of wholesome, tender, juicy and flavoursome meat to a consumer's plate is the end point of a continuum of activities which began with the conception of the animal from which it came. So, bringing a lamb, pig or beef animal to the point at which it may be shipped from the farm for slaughter is only part of the way towards delivering a safe, wholesome and tasty cut of meat.

The quality of meat in its broadest sense is determined by the interaction of an animal's physiology with a variety of environmental conditions which animals encounter before, and their carcasses after, slaughter. The stress imposed on animals during grouping on the farm, loading and transport to an abattoir, and their experience in lairage, all exert influence. The sex of an animal is also important: bulls, for example, can be particularly nervous and disruptive during transport and lairing and their own carcasses and those of accompanying animals may suffer. The usual consequence of this is the so-called DFD or 'dark cutting' condition of beef which is explained by the inadequate acidification of meat *post mortem* (to pH >6.0.). This follows from nervous- or exercise-induced reduction in the living animal of muscle glycogen, the substrate for anaerobic glycolysis, and the curtailment of muscle acidification *post mortem*. More rapid but complete muscle acidification (to pH ≈5.5) may induce the PSE condition, primarily in pigmeat. PSE meat results from acute muscle stimulation in the immediate pre-slaughter period or as a physiological predisposition in certain breeds and strains, for example, Pietrain and Landrace pigs.

The degree of acidity in meat not only affects the appearance of the meat but also modifies the pattern and effects of microbial spoilage. In dark cutting beef, *Pseudomonas spp*. of bacteria degrade amino acids from muscle instead of glucose and create sulphurous off-odours. Vacuum-packed dark cutting beef spoils at a rapid rate because in the absence of glucose substrate the multiplication of *Enterobacteriaceae*, which include pathogens such as *Salmonella*, is promoted.

The spoilage process begins with the contamination of meat by spoilage microflora at some point in the meat chain. But there is no spontaneous development of spoilage organism on meat! On the farm, contaminated pasture or housing allows the recycling of excreted pathogens and symptomless carriers present sizeable complications. They may, for instance, be present in groups of animals being transported and under those conditions stress may lead to massive excretion of pathogens to contaminate the transporter or the lairage and any stock passing through to slaughter and the plant itself. Withholding feed, which is a common practice in the perislaughter period, may induce changes in the gut flora population and the excretion of *Salmonella*.

If contamination is the prime cause of spoilage, then practices in the slaughter chain may limit or promote bacterial contamination and meat spoilage. Washing animals, scalding and spraying carcasses, disinfecting equipment and preventing gut spillage all minimise the effects of contamination and must be the standard control practices. The training of operatives to maintain good working practices is a first priority.

The spoilage of meat has been observed for centuries and aspects of it have been controlled and used to preserve meat. Yet its scientific basis is a relatively modern discovery. Because meat is nowadays usually kept at temperatures of 5°C or below, the common surface spoilage organisms are those which thrive at such temperatures and are usually present as contaminants of the carcass from the slaughter and dressing procedures.

The storage atmosphere largely determines the microflora population. Chilled meat in air will have *Pseudomonas* as the major spoilage organism and their growth rate at the prevailing temperatures will dominate other bacteria such as the *Enterobacteriaceae*. But the predominant populations depend for their success on the specific conditions that are encountered on the meat surface-aerobic/anaerobic, high/low water activity, high/low ambient carbon dioxide. This is complicated further by the influence of 'imposed' environments which result when meat is packed in, say, oxygen-permeable/impermeable materials, in gas-flushed atmospheres (CO_2 or N_2 or combinations of the two) or simply from the competition between species of microflora under the ambient conditions.

Packaging meat gives it appeal and creates marketing advantage but it also improves the keeping quality of meat. The simplest process is to package cuts of meat under vacuum in films of varying permeability to oxygen. The lower the oxygen permeability, the longer the shelf-life of the product, but the less cherry-red the colour on account of the induced but reversible conversion of oxymyoglobin in muscle to the darker metmyoglobin.

There is usually some oxygen retained in the headspace of an evacuated package of meat and this is eventually metabolised by the muscle and carbon dioxide is produced. The CO_2 so produced delays the growth of *Pseudomonads* and allows *Lactobacilli* to multiply. *Enterobacteriaceae* are prevented from growing by growth inhibitors produced by the *Lactobacilli*.

The retardation of spoilage brought about by CO_2 provides the rationale for packing meat in a CO_2-rich atmosphere. The gas-flushing technique using up to 20 per cent CO_2 is now widely used and is considered to be as effective as vacuum packaging in inhibiting the growth of *Pseudomonads*, and an additional benefit is the retention of better colour in the packaged meat.

The development of spoilage micro-organisms on meat is clearly not a haphazard process. Specific conditions may protect or foster the growth of specific organisms but spoilage also provides a degree of production for the potential consumer. The loss of shelf-life, objectionable smells, off-flavours, and so on, may deter individuals from eating spoiled meat, but spoilage microflora compete effectively with pathogens, so that the microbial population on the meat may be relatively benign, which is why 'hung' game may be eaten in safety.

The consumer historically has in any case been protected against harm (except from microbial toxins) by the cooking process. There is now, however, a further, albeit controversial, alternative. The irradiation of food, either before or after packaging, with pasturising dose of ionising radiation has been an option for more than 30 years. The process has not found favour, however, although authorities can find little evidence to support banning its use. The usual arguments against the approach are the concerns over off-favours, probably resulting from the formation of free radicals, the expense, and logistical problems of using high-security irradiation facilities.

So the protection of consumers against food-borne disease is largely a question of avoiding contamination of the live animal or carcasses and cuts of meat, packaging to protect and control the growth of pathogenic organisms, or the use of preservative processes which restrict the growth of potentially hazardous gram-negative bacteria and, finally, some form of thermal processing.

Eating Quality of Meat

Having successfully brought a wholesome and safe product to market, the final provision is that the meat will bring satisfaction to the consumer. The culinary arts are not the concern of this chapter but the raw material with which to cook is. Once again, we discover that the inherent quality attributes of meat are determined by a combination of animal physiology and process technology.

The growth of an animal is marked by changes in its physiological chemical and anatomical characteristics, as well as its size. A very young animal produces a carcass which has little fat, meat which is pale, tender and succulent to eat, but lacking in flavour. Veal and sucking pig are typical example. An older animal will have a fatter carcass, darker meat and, because of changes in the amount and chemical constitution of the connective tissue it contains, it is likely to prove tougher to eat.

The traditional view is that fat endows meat with desirable eating quality. This is true to some extent but it is now evident that the threshold amount of fat required to confer an acceptable degree of tenderness and juiciness is very much lower than was previously considered necessary. The critical amount of intramuscular fat or 'marbling' is of the order of 1 per cent – 2 per cent of the fresh weight of muscle.

Fat is perhaps of greater concern in its contribution to flavour and it is recognised that the species from which a piece of meat came is difficult to discern if it is not cooked with its own fat. It is the flavour volatiles derived from fat during cooking which interact with the protein and minor constituents of muscle and confer on meat its flavour fingerprint. Flavour becomes more intense in the meat for older animals.

Tenderness is a characteristic of meat, the basis of which is still undeciphered. While it is true that the connective tissue of old cows is more heat stable, insoluble and chemically cross-linked, it is not sufficient to make the meat from discarded dairy cows, a major source of British beef, unassailable. It is quite evident from research carried out over recent decades that differences in the age or composition of meat animals at slaughter, or breed or sex, cannot account satisfactorily for the wide range in the eating quality, and especially tenderness, of meat which consumers normally experience. Increasingly the view is taken that it is the peri-slaughter handling of animals and their carcasses which determines how acceptable is the eating quality of a sample of meat.

Stress imposed on animals during the period preceding slaughter affects the appearance of fresh meat and microbial spread. The production of PSE meat is a consequence of falling pH in muscle after slaughter, while its temperature is still close to that of the living animal. The resultant paleness and wetness of the meat is attributable to the denaturation and precipitation of the soluble and structural proteins in muscle which reduces their ability to bind or retain water and allows the meat to drip: the optical properties of the surface layer of the meat are affected causing it to become pale.

Ambient temperature is also responsible for controlling the rate of growth of spoilage and pathogenic bacteria. There is potential advantage, therefore, both to reduce 'drip' and bacterial spoilage by rapidly cooling carcasses after slaughter. The over-zealous employment of this practice can, however, quickly cause other problems.

International trade in New Zealand lamb has traditionally involved frozen carcasses. When refrigeration technology allowed the very rapid chilling of carcasses to about +2°C, there was a dramatic and significant increase in lamb toughness. The condition, now known as 'cold shortening', was quickly identified to occur when carcasses were allowed to develop *rigor mortis* at temperatures below 10°C. The fundamental explanation was that the contractile units of muscle, the sarcomeres, could shorten by 30 per cent or more and lead to toughening of the meat.

The early solution to this problem was simply to allow muscle acidification to progress *post mortem* to pH 6.2 or less before the carcass was refrigerated. As more information was gathered about the process it became clear that it might be possible to modify *post mortem* changes in muscle and avoid cold shortening. Nowadays the use of electrical stimulation of carcasses shortly after slaughter enables the rate of acidification of muscle to be controlled and predicted and an optimally effective refrigeration programme to be adopted.

The small size of carcasses and the very effective chilling and freezing capacity of the New Zealand meat works were responsible for disclosing this major cause of toughness. But it is now evident that parts of beef and pig carcasses can also be cold shortened and toughened by the chilling practices common in modern slaughter plants.

New Specifications, Products and Innovations to Meet Consumer Needs

When food is no longer scarce, selling it takes on a quite different perspective. The lonely tins of sausage or meat products so characteristic of the window displays in eastern European food stores are a far cry from the 'show-biz' hype surrounding food marketing in the capitalist world. The temptation to buy product A rather than product B carried through to a successful sale is the essence of a market outlook and the approach which has been recognised as they key to success in the rapidly changing food industry of recent decades. The need for food itmes to be fresh, safe, wholesome and attractively presented at the right price has implications for all involved in the food chain. Nowhere is this better recognised than in the meat and dairy industries.

Animal scientists and technologists have achieved considerable success in improving the efficiency of production, yields and qualities of meat and dairy stock. Changing the ratio of lean to fat in carcass leads not only to greater consumer acceptance but also to improvements in the efficiency of production, and this continues to be a focus of attention for scientists and producers alike. Because body composition at a particular weight is largely a function of the mature lean body size which an animal achieves, there has been a significant move towards breeds and strains of livestock which develop a large size and delayed fat deposition. Limousin, Charolais and Simmental are typical of the breeds of cattle which show extremes of leanness at the usual slaughter weights. The leanness of male animals is also attributable principally to their larger final lean size and later maturity. The same biological principles apply to the other animal species and provide the basis for the genetic selection and breeding approaches for leaner types of animal, most notably pigs.

Pigs also provide a useful example of how slaughter weight is instrumental in determining not only carcass composition but also the comparative efficiencies of production. The energetic costs of changing various metabolic, reproductive and weight characteristics of 90 kg pigs have been evaluated by Flower (1976). He calculated that if the cost of protein synthesis or heat production could be reduced by 10 per cent the overall saving of energy costs is of the order of 2 per cent. Reducing the slaughter weight by 10 per cent, or better still, slaughtering pigs at 60 kg, could save as much as 15 per cent of energy. The explanation for this is simply that it is not until pigs reach weights greater than 60 kg that fat deposition (and its extra energetic cost) begins to accelerate.

Providing suitable diets for livestock and controlled feeding have been the other major contributors to decreasing the fatness of carcasses. The quality of diets and methods of feeding become increasingly important as the specifications for performance and carcass quality are more tightly defined. There are signs, for example, that the eating quality of pigmeat, especially its succulence, may be improved by *ad libitum* feeding. But this may lead to excessive fatness in some strains of pig and such a feeding regime may only be advisable for boars, even through there is a considerable labour saving from *ad libitum* feeding for all.

These are the kinds of strategies which need careful evaluation to meet the market requirements for lean meat. There is, however, one further approach which is to trim fat from carcasses or cuts or portions of meat on the plate. This, of course, has always been an option but it has been regularised of late by marketing practices such as 'Lean Choice' which has been promoted by the Meat and Livestock Commission to offer a variety of levels of leanness to consumers.

The need and appeal of techniques to limit the intake of animal fat have been felt just as keenly in the dairy sector. The rise in milk production in the EC in the late 1970s and early 1980s was not matched by increased consumer demand and stocks of butter and skim-milk powder became worrying surpluses which led to the imposition of quotas on production. Quotas caused the output of milk to fall but sales of milk and milk products have also declined despite the healthy and wholesome image which milk has traditionally enjoyed. Changes in family eating habits, the recognition that milk is not an essential part of the diet and consumer concerns about possible links between dairy products and heart disease have all taken their toll. The dairy industry's response has been to offer a greater range of low-fat milks and modified dairy products. Flavoured milks and milk shakes are growing in popularity and the introduction of low-fat spreads which mix vegetable and butterfat has been successful though to the determent of sales of butter and margarine. There is seen to be an opportunity to expand the market for cheese and many new varieties have found their way to super-market shelves.

Despite the voiced concerns about dietary fat and health, sales of cream continue to grow! There is a diverse and growing market for it which includes fresh and UHT (ultra-high temperature treated) creams, as a constituent of cream cakes and for domestic and external catering generally. Yoghurt is increasingly popular in both high- and low-fat recipes and there is a growing demand for a number of dairy desserts which sell on taste and quality rather than on price.

Ice-cream in the UK is typically based on vegetable fats. There is, however, a move towards the more widespread sale of premium dairy ice-creams comparable with the mainland European varieties.

All of these approaches attempt, with varying success, to balance the supply and demand for milk, to retain toeholds in traditional markets and to identify and satisfy opportunities for new products in new and different markets. The meat industry has also had to develop approaches like these in the hope of increasing its market share or at least retaining its present share. The poultry industry has led the way in this as it has done historically in employing nutrition, environmental and engineering science in production and slaughter technology and in the integration of many parts of the food chain.

The main initiatives have been taken to counter the cyclicity and seasonality of production and the market. Chicken and turkey sales peaked traditionally at Easter and Christmas. Massive marketing investment backed by formidable scientific, technological and practical approaches to improve production efficiency, product uniformity, quality and price, together with the benefits of economy of scale, have made poultry meat an increasingly competitive product worldwide. A move from sale of whole chicken to portioned cuts and joints sold fresh or frozen or as added-value itmes in the form of delicatessen and ethnic products has opened up new markets. Price is still an important determinant of the commercial success of poultry meat but innovation in production, processing and marketing has been of crucial importance and has had further repercussions in the red meat sector.

The wealth of opportunity created by offering poultry meat in guises other than the traditional whole chicken stimulated the search for the more widespread use of deboned meat and cuts. Food technologists identified ways of fabricating deboned turkey or chicken into meat portions of consistent composition and quality to which could be added sauces, garnishes and dressings for a great variety of convenience products for sale at considerably enhanced prices. Continuous fabrication of 'cylinders' of the product could also be undertaken. Further technological breakthroughs allowed an even greater range of convenience, added-value products to emerge which were based on lamb, beef or pork or even mixtures of two or all three species.

New product development is perhaps the major research activity in the large food companies and hundreds of new products and recipes may be tested each year though the success rate is very small indeed (<1%). Such interest is, however, a measure of the way in which the major food companies see their competitive position in food markets being sustained and it provides the lead to which the primary industry should be ready to respond.

Biotechnology in Food Production

There is nothing new about the application of technology to biological systems for food production (see Table 2.6) but the pace of advance and the potential for further developments grow daily. Thus while the selection and breeding of plants or animals to produce superior stock have been recognised for generations, only in the last ten years or so has it been possible to identify specific genetic control mechanisms and ways in which they, or the hormonal messenger system involved, could be manipulated (see Table 2.7).

Table 2.6: Family tree of food biotechnology – some roots and buds

	Plants	Animals	Micro-organisms
	Gathering	Hunting	
4000 BC	Sowing	Domestication	Brewing
	Selection	Selection	Cheese
	Crossing	Crossing	Tempeh
AD 1920	Hybridisation	Artificial insemination	Antibiotics
	Colchicum treatment	Vaccination	Enzymes
AD 1970	Tissue culture	Embryo transfer	Single cell proteins
	Biological pest control	Cloning	Bioconversion
	Gene transfer	Gene transfer	Hormones
	– disease resistance	Immunomodulation	Amino acids
	– nutritional composition		

Source: Wal, van der 1990.

Table 2.7: Control of growth and product quality via the somatotropin (ST) axis

Genome modification	Selection for higher ST levels Insertion of ST gene
Affecting messengers	Administration of ST or growth hormone releasing factor Potentiation of ST with monoclonal antibodies or vaccine

Source: Wal, van der 1990.

Molecular biology has opened up a variety of approaches which, with or without conventional breeding techniques, may lead to improvements in the performance of all stock whether animal or plant. The insertion of specific genes into the genome of recipient animals offers a selective means of improving milk yield or growth performance. The genes of interest are those coding for hypothalamic, pituitary and liver polypeptides which foster protein anabolism. The desired effects have been created in rodents but the techniques have not proved to be wholly without complication in farm animals.

The immunisation of animals against endogenously produced growth-hormone-release-inhibiting-hormone (somatosatin) to promote unfettered the effects of somatotropin and stimulate growth was examined in the late 1970s and early 1980s. It met with little success and the increased availability of somatotropin derived by recombinant DNA technology offered a demonstrably successful means of promoting lean body growth and milk yield in farm animals and prospective worldwide use. But, like anabolic steroids and β-agonists, which have been successfully used to manipulate growth and carcass composition, their use is banned or uncertain to be approved because of fears about the safety of consumers of meat from treated animals, or on grounds of animal welfare. In all these cases, efficacy of the agent or treatment is not sufficient to permit their employment universally let alone naturally. That some countries may permit their use while others do not is a basis for trade barriers which will prove difficult to accommodate or dislodge.

The use of biotechnology in the arable sector offers more prospect of both success and employment, though even here all is not ideal. The approaches envisaged cover the improvement of nutrient supply to plants by nitrogen fixation and fertilisation by algae, creating genetic resistance to disease and environmental hazards such as cold, drought or salinity, and tissue culture to provide for the rapid multiplication of improved varieties. It has become clear, however, that food crops are less responsive to biotechnological manipulation than was hoped for. Cereals, for instance, appear to be less amenable to genetic engineering and even where success is achieved the classical evaluation and selection techniques, which may take 5–15 years, still must be employed prior to the release of superior strains.

The use of micro-organisms to produce feedstuffs and nutrients for animals or man, either directly of indirectly, is a development for which there can be some optimism. Essential amino acids such as lysine, methionine, tryptophan and threconine can be produced by fermentation technonlgly and the commercial production of single-cell protein is a reality. There is interest also in the use of micro-organisms to recover nutrients from agricultural wastes, forestry products and organic residues. As yet, however, developments are constrained by the enormous financial provisions which are required and by the complexity of the research needed.

The Wider Issues

The perceptions of consumers about the quality and wholesomeness of their food have rapidly been translated into the purchasing, technical and marketing policies of the major food retailers. As the selling of food is concentrated into fewer hands, the competition for trade among producers and suppliers ensures that consumer demands quickly filter through to them and any intermediate agencies along the way. Tight contractual specification for the supply of meat, fruit or vegetables is part of the standard trading repertoire of supermarkets. But market pressure provides a basis for change in agriculture well beyond the simpler notions of consumer satisfaction and the perceived equality of products. Consumerism may now be looked upon as the power of customers to buy a better world and that power may extend from unwillingness to purchase particular products, for example, furniture made from Brazilian rain forest timber, or readiness to buy pork produced from 'outdoor sows'. Investment in companies which might contribute to environmental pollution, allow experimentation on animals or promote political extremism may be prejudiced. The 'green movement' as exemplified by Friends of the Earth and Greenpeace have been at the forefront in stimulating the momentum of the underlying message and promoting it internationally.

There appears to be no doubt that the world has the capacity, both physical and economic, to produce enough food to meet substantial increases in demand, but there are provisos. In the first place, the food sector must grow at near record rates if only to keep pace with current levels of consumption. Secondly, because of the increased energy costs of more intensive system of production, there will be a rise in the real cost of producing food which may be increased further by political uncertainty in some oil-producing countries.

In Europe and North America agricultural surpluses have caused widespread criticism of the traditional support framework. There is great pressure for adjustment to the European Community's Common Agricultural Policy from members of the Uruguay Round of talks on liberalising world trade. But it is quite unrealistic to think about the world's agriculture devoid of subsidies and strife barriers with crops being produced only where it is most efficient to do so. Certainly there would be a dramatic fall in commodity prices following such a move, but it is doubtful whether governments, let alone the farming communities, could weather the longer-term social, political and economic consequences. This is not to say that there is no room for revision of the support mechanisms for, or the organisation of, agricultural activity. Set-aside schemes are a case in point.

This situation of food plenty contrasts starkly with that pertaining in the developing world where the prospect is for food self-sufficiency to decline and the importation of food to become a necessity if widespread under-nutrition is to be avoided. This is a solution which Second World countries may, with difficulty, contemplate but it is an option which may not be open to the poorest countries.

The problems of over-production by the agricultural sector of the developed world constitute only part of the challenge facing the agriculture of the present and beyond. The basis of such success, considered by many as the means of salvation for the Third World, is now being questioned by rich and poor countries alike. There is no doubting the efficacy of investment in fertilisers, pesticides, herbicides, irrigation and machinery in increasing the productively of land. But the huge reliance on finite energy sources and the environmental consequences of intensive agriculture have caused many observers to assess the energetic efficiency and sustainability of these approaches.

While the consumption of commercial energy in the developing world may be one-quarter to one-third that in the wealthier countries, it still allows the production of about half of the world's cereal, roots and tubers and identifies the overwhelming contribution of manual labour to food production. But the energy equations are not easily balanced for while the energy consumption per unit of arable land might be very much greater in the developed countries, the difference is reduced for some developing countries when the consumption of energy by agriculture is expressed relative to the countries' gross domestic products (GDP). This reflects a shift in investment of the GDP towards agricultural machinery and fertilisers (Leach *et al.* 1986).

The efficiency and cost of food production is also related to the particular food being produced. The yield of energy from grain crops per unit area of land is roughly 3 to 4 times that expected from milk and at least 12 times that for beef. Here is therefore merit in using land to grow grains and legumes rather than animals where land is scarce, though to provide a nutritionally balanced diet may require a greater use of energy in order to increase yields of both plant and animal crops.

The marginal benefits of increasing the energy input into the agricultural sector is greater in the developing countries and the general belief is that the energy input in poor countries is very low. This may not be the case when all the available forms of energy, for example, human, animal, wood, dung and fossil fuels, are included. The real energy efficiency and the marginal benefits from additional energy may not, therefore, be as large as might, at first, be imagined. Moreover, the simple proposition that pooper countries need only to put more commercial energy into the agriculture sector to improve their food supplies and standard of living to that found in the USA or Europe is flawed. It has been calculated that the total *per capita* energy provision for food production, processing and distribution in the USA is equivalent to 1900 littres of gasoline per year and an energy inefficiency of about 9 : 1, i.e., 9 units of petroleum energy for every unit of food energy produced. The required area of land for this food provision is 0.52 hectare. On this basis, a world population of 5 billion people would require annually approximately of 5 : 1 (comparable value for the UK) 4 per cent of oil stocks would be required annually. A further limitation is the area of land available globally for production purposes, which is about 0.3 hectare or nearer half that deemed to be necessary. It is clear, therefore, that the energy strategies which are currently used in the USA, Europe and Japan are inappropriate for global use even if solution could be found to the environmental problems which are becoming all too evident in intensive farming systems.

The soil erosion and salinisation which are now a serious problem in the southwestern regions of the USA are linked to the long continued employment of large-scale mechanical cultivation and irrigation. Nitrogen fertiliser use not only leads to the increased yield of crop plants, but in excess contributes to the pool of nitrogen present in soils. Warm temperatures, moisture and aeration during cultivation activate the soil microbes to synthesise nitrates from soil nitrogen, not necessarily for use by the crop but available for leaching by rainwater thence to appear in groundwater and drinking water supplies. This becomes a severe problem in the rich and often wet arable areas of eastern England and the wetter areas of western Europe. The run-off from these areas into the North Sea contributes substantial amounts of nitrogen to foster the growth of algal blooms and plants. This has important consequences in, for instance, the Baltic Sea where algal populations may be so great as to restrict light passage into the sea and inhibit the activities of plants and animals on the seabed. The availability of oxygen may fall dramatically. The contamination of drinking water supplies with nitrites can cause nitroso-haemoglobinaemia (Blue Baby Syndrome) in neonates and there are purported links with gastrointestinal disorders.

The pollution of waterways and seas with phosphates and nitrogen also occurs in association with intensive animal production. Pig production in the Netherlands is now being severely restricted to the extent that increases in pig production can only be approved in certain areas of the country. Such curtailment of animal production is considered to be one of the most challenging problems facing Dutch agriculture.

Pesticide use has been criticised since the earliest introductions of DDT, and herbicides and fungicides are subjected to the most careful scrutiny and licensing procedure before they can be approved for practical application. Nevertheless there is a constant search for more acceptable alternatives and the concept of integrated pest management (IPM) has been seriously considered of late in developed countries. IPM involves the employment of the selective application of pesticides, the development of pest-resistant crop varieties, biological control through natural predators and sexual sterilisation of pests, improved forecasting of pest infestations and changes in agricultural practices (FAO). All of the expertise and technology for IPM will be available in the developed countries but the serious counter to its widespread take-up is the cheapness of insecticides.

Over the last 20 years or so many prosperous and vocal urban dwellers have made it plain that they are no longer willing to eat food that is produced under conditions which they consider unwholesome, cruel, unnecessary or polluting, and have persuaded some farmers to consider more 'natural' ways of rearing stock and growing crops. Some systems may be low technology and highly labour intensive. Others may demand that plant nutrients are supplied from organic waste (from livestock or green crops) and weed and pest control are achieved by husbandry methods which rely on cultivations, crop rotations, biological control and IPM approaches. It is generally considered that yields via 'organic farming' are lower than with conventional intensive methods but higher prices may be possible than which, together with the low cost of energy input, may lead to similar net returns. While they are not seen as replacements for contemporary farming practice, many of the energy-efficient procedures used in organic farming may make useful contribution to modern agricultural systems.

There was view that the protection of the environment was a rich man's prerogative and that poor countries should be allowed to develop their economies even at the expense of the environment. This view is no longer tenable, but how to balance the interests of conflicting groups remains a serious problem. It is not enough simply to produce an economic balance sheet, for although some measures can be 'market based', others, such as the depletion of natural resources, whether they be oil and mineral wealth or the destruction of biological species and habitats, cannot. In other words, to protect the interests of future generations there must be adherence to policies which have become described as 'sustainable'. Indeed it is considered by some to be the only solution to the dilemma of meeting human needs in the face of a rapidly increasing global population. Sustainable food systems require that dramatic changes be engineered, not just to the practice and technology of agricultural and food production, but to the social, economic, institutional and behavioural fabric of societies.

All of this identifies a new level of public involvement in agriculture, which is no longer seen simply as the provider of man's food, fuel and fibre but increasingly as a, if not the, major potential contributor to the destruction of the planet.

❑❑❑

3

World Agriculture

In terms of usage of land and employment of population, agriculture is the world's largest industry. As the predominant provider of food and related products, as well as fibre and other raw materials, for transformation into forms suitable for consumption, an efficient and sustainable agricultural industry is essential for the well-being of mankind now and in the foreseeable future.

This chapter seeks to present an overview of world agriculture to provide the perspective needed for an understanding of the role of the industry in meeting human needs. It examines the size and structure of agriculture worldwide, its economic, social and ecological importance, international trade in agricultural products, food aid and the motives that govern its provision. The world's major crops and animals are briefly reviewed prior to a concluding section on land use and production systems.

Size and Structure

The size of the world's agricultural industry may be measured in many different ways but the simplest and most satisfactory overall is in terms of the interdependent factors that constitute the essential elements for production: land, labour, climate and capital. Data series are available by country and region for the first three. Figures for capital investment in agriculture are more difficult to come by and to interpret except on a micro scale because of the fungibility of labour and capital. Thus, given an adequate labour supply, farmers may change from the use of herbicides to hand weeding or *vice-versa* in the production systems of many middle-income developing countries depending on the relative prices of labour and capital.

Land

Table 3.1, which shows worldwide movements in major categories of land use in the 1970s and 1980s in millions of hectares, indicates the increase in arable land and permanent crops at the expense of permanent pasture and forests. These numbers indicate the pressure already placed on the land resource implied by increases global population (Table 3.2). The steady increase in land use for 'other' purposes demonstrates by inference the impact of urbanisation on land availability. Questions of land quality are considered in relation to climate below.

Table 3.1: Global land use

Category	1972	1977	1982	1987	% change
			(millions ha)		
Land area	13074	13074	13077	13077	–
Arable land	1223	1342	1365	1373	+3.8
Permanent crops	91	97	100	100	+9.9
Permanent pasture	3226	3214	3215	3214	–0.4
Forests and woodland	4196	4417	4101	4069	–3.0
Other	4238	4275	4296	4320	+1.9

Totals may not add due to rounding.

Table 3.2: Global population

World population	1975	1980	1985	1988	% change
			(million)		
Total	4090	4450	4855	5115	+25.4
Economically active	1764	1957	2163	2285	+29.5
Economically active in agriculture	932	994	1053	1083	+16.2

Labour

The figures in Table 3.2, while demonstrating the increase in population and the relative decline in numbers of the economically active engaged in agriculture worldwide, mask the enormous difference between countries in different geographic regions in population growth and numbers engaged in agriculture. Table 3.3 illustrates this point.

Table 3.3: Population in selected countries

	UK		USA		China		Ecuador		Malaysia	
	1975	1988	1975	1988	1975	1988	1975	1988	1975	1988 (million)
Total	56.4	57.2	216.0	246.1	9.8	14.2	7.0	10.2	12.3	16.6
% change		+1.4		+13.9		+44.9		+45.7		+35.0
Economically active	26.2	28.2	97.9	121.1	3.8	5.2	2.7	3.1	4.5	6.6
% change		+7.6		+23.7		+33.3		+47.6		+46.7
Economically active in agriculture	0.7	0.6	3.8	3.1	2.7	2.7	0.9	1.0	2.1	2.3+
% change		–14.3		–18.4		+22.7		+11.1		+9.5

The numbers in Table 3.3 demonstrate that whereas only about 1 per cent of total population worked directly in agriculture in predominantly industrial countries such as the UK and the USA in 1988 compared to a world figure of 2.1 per cent, much large proportions worked in the agriculture of the developing countries. The fact that the USA is a significant agricultural exporter, notwithstanding its low agricultural population, illustrates its productivity and the fungibility between labour capital as factors of population in the agricultural industry.

Climate

Climate is a major factor affecting the size and structure of agriculture at all levels. Further, climate is vitally important in terms of such day-to-day phenomena as precipitation, temperature, humidity and winds, all of which can vary widely over comparatively small distances. The effect of past climate impact on soils and land forms also plays a major part in determining the physical limitations of present-day farming systems across the world. That climatic aspects are not constant is well illustrated by the Sahelian drought of the 1970s in West Africa and mankind's concern with the possible changes to agricultural systems necessitated by global warming resulting from the so-called 'greenhouse effects'.

Thus climate was and continues to be significant element in the vast wheatlands of North America, the sheep and dairy industries of New Zealand, the oil palm and rubber plantations of South-East Asia, and the mixed crops and livestock units of European countries, to name only a few examples of agricultural enterprises by which mankind sustains itself.

Capital

The interdependence between land, labour and climate in determining size of and structure of agriculture worldwide is further impacted by financial capital. On the one hand, a Nigerian farmer typically cultivates two hectares of yams, cassava and maize for family subsistence on land to which title is based on customary usage and not a monetary-based leasehold or freehold ownership. Such a farmer will have made no capital investment beyond the hoe and cutlass used for clearing, cultivating and weeding his or her crops. Shortage of labour will limit family cultivation capacity to about two hectares. Quality of the soil and climate will restrict the choice of crops to be grown. A combination of these factors and especially labour availability for key operations such as cultivation, timely planting and weeding will limit yields. The likely failure to produce a marketable surplus above subsistence needs will prevent purchase of such basic productivity enhancers as improved planting material or fertiliser. This farmer is thus in the classical, if caricatured, poverty trap as a result of the interdependence of land, labour, climate and capital.

Other things being equal, access to more land, labour or capital by a farmer in the same village would provide the opportunity, given adequate managerial skills, to escape the poverty trap. Another half hectare of land, more vigorous and skilled labour, or a small injection of capital might provide the opportunity to increase production above subsistence needs, thus yielding a cash surplus which can be used to gain access to yet more of the principle production factors which in turn can lead to more disposable income, thereby springing the jaws of the poverty trap.

The 'green revolution', based on short-strawed, high-yielding rice and wheat varieties, occurred because governments in such Asian countries as India, Pakistan, Indonesia and the Philippines improved the terms of trade for agriculture by changing their pricing and marketing policies while concurrently investing heavily, with external assistance, in irrigation infrastructure. These actions facilities the uptake of a production technology based on newly-developed varieties of the cereals that responded dramatically to fertiliser and irrigation. Among the consequences has been achievement, within 20 years, of self-sufficiency and periodic surpluses of rice and wheat in countries that had been in varying degrees of deficit for these commodities. This is a striking achievement under any circumstance. It is made more striking in the context of population growth rates in excess of 2.5 per cent annum. Equally striking was the response of farmers to the opportunities presented by the new policy and technology package. Few had access to more than one or two hectares of land. Most, by virtue of their poverty, had long followed traditional low input/low output production systems in order to minimise risk, and many had little experience of irrigated crop farming. Yet, within the space of a very few years, millions of such farmers had embraced the changes required to bring about the revolution in question.

In some areas the 'green revolution' resulted in changes to land ownership structure, with the richer and more efficient operators acquiring the land of their less fortunate or less efficient fellows, thereby promoting the development of a landless labouring class and, frequently, an acceleration of urbanisation as the landless moved to the towns in search of livelihoods other than farming. Efforts to kindle a 'green revolution' in other areas have met with no more than mixed success partly because of the shortage of technologies seen by farmers to be relevant to their needs, and more often by the inability or unwillingness of governments to create the terms of trade that farmers felt they needed to change their traditional practices.

West Africa provides a useful example of the developments of potentially highly effective technology for smallholder cotton production, apparently well-suited to many countries. In its essential elements, high-yielding varieties of uniform staple length and thickness, fertiliser responsive, with a degree of pest and disease resistance, and a requirement of early planting for maximum yield, the technology differed little between countries in the region. Soil and climatic conditions were broadly similar; size and structure of holdings and traditional farming practices also showed few differences. But adoption of the technology from the 1960s through to the late 1980s varied enormously largely between the francophone and anglophone countries that make up the West African cotton-producing zone. The technology had a production potential of approximately 2500-3000 kg seed cotton per hectare under carefully controlled experiment station conditions that has been demonstrated consistently across the zone. Leading farmers obtained yields at or above 50 per cent of these levels, an acceptable differential under developing country condition, again with little significant difference between countries. However, the similarity between anglophone and francophone countries ceased totally at the level of the typical cotton grower. In anglophone countries, such as Nigeria, natural yields rarely reached 250 kg/ha, whereas in francophone countries such as Mali, national yields regularly topped 750 kg/ha. The principal reasons for this consistent threefold yield difference were superior services to cotton growers in Mali compared with Nigeria in respect of access to high-quality seed, appropriate fertilisers and pesticides, extension advice and marketing arrangements. Farm gate price were somewhat more favourable in Mali but nowhere near so critical as the other factors in determing technology adoption.

These cases illustrate a further element affecting the size and structure of the agricultural industry. That is the fact that the industry is composed of millions of enterprises, each controlled by individuals making their own decision on their form and operation within the limits imposed by external factors. Agriculture is a unique industry in this respect. The sheer numbers and heterogeneity of the world's farmers underline the role of social and political influence of agriculture locally and worldwide. *Prima facie*, change could be expected to be slow, but as these other examples demonstrate, this is certainly not always the case.

In summary, size and structure of agricultural operations are governed by a range of interdependent factors of which land, labour, climate and capital are of major importance. But within each climatic zone or major soil type, social, political and/or economic factors have an overriding influence on technology development and adoption, and hence on production system, agricultural productivity and the economic status of the country concerned.

Importance and Role

Agriculture is responsible for providing virtually all the food and much of the clothing for a world population estimated at over five billion in 1988. World population is projected, by the United Nations and other authorities, to double by the early part of the next century. It thus appears that there will be a sizeable challenge facing not only those directly engaged in agriculture but also the politicians and scientists concerned with the many other activities that bear on the scale and productivity of the industry.

Agriculture in the Economy

Agriculture's share in national economies varies widely, normally as a function of their level of development. Historically, and without exception, agriculture has been the dynamo for subsequent industrial development. This fact is clearly illustrated by the weakness of the economies of many Eastern Bloc countries on the one hand and the wealth of the western industrialised countries on the other. The latter built their industrial wealth from a strong agricultural base: the former following Marxist ideology, sought and failed, to do the reverse.

The following tables (Table 3.4 and 3.5) illustrate the structure of production between agriculture in industry in countries grouped by levels of income, and geographic regions as a percentage of Gross Domestic Product (GDP) in a time series over two decades between 1965 and 1986.

Until the end of the period, agriculture dominated the share of GDP in low-income economies with middle- and high-income economies illustrating a consistently obverse picture. Table 3.4 also shows the increased share of GDP taken by industry at the expense of agriculture over the period, except in the high-income economies.

The general lesson presented by Tables 3.4 and 3.5 is that agriculture has been and continues to be a dominant part of the economies of low-Income countries, but that it has declined in importance relative to industry in all countries. However, whereas agricultural growth rates have increased significantly in high-income economies they have declined in all developing country regions, except East Asia. This outcome has serious implications for the well-being of a very large proportion of the world's population.

The Human Factor

As indicated earlier, agriculture is characterised by a substantial heterogeneity arising from differences in major factor endowments. Their interdependence and changing nature require constant adjustments to farming practice in all locations and in all production systems if investments in agriculture are to be optimised. Thus since the human factor has the potential to be the most dynamic in changing agricultural practice, it is appropriate that this section focuses clearly on the importance of people as producers and consumers. It is also important that due consideration is given to the question of sustainability of present ad future agricultural practices. Such consideration is mandated by increasing awareness of the lack of sustainability of

Table 3.4: Structure of production between agriculture and industry – percentage of GDP

Country	1965		1973		1980		1986	
	Agric.	Indust.	Agric.	Indust.	Agric.	Indust.	Agric.	Indust.
Low-income economies	41	26	37	31	32	36	31	33
Middle-income economies	19	32	15	35	12	38	13	36
High-income economies	5	40	5	38	3	37	3	34
Sub-Saharan Africa	40	18	33	24	28	32	33	23
East Asia	37	34	31	40	26	44	23	43
South Asia	42	19	45	18	35	22	29	25
Europe, M. East and N. Africa[a]	22	32	16	38	14	41	14	–
Latin America and Caribbean	15	32	12	33	9	36	11	35

[a] Excluding the high-income economies of the region.

Table 3.5: Average annual percentage changes in sector growth rates from 1965-1987

Country group	Agriculture			Industry			Services		
	1965-73	1973-80	1986-87	1965-73	1973-80	1986-87	1965-73	1973-80	1986-87
Low-income economies	3.0	2.1	4.0	10.6	6.9	8.6	5.9	4.9	5.2
Middle-income economies	3.3	3.3	2.5	8.0	4.0	2.9	7.5	6.9	2.8
High-income economies	1.4	0.5	2.5	3.9	2.2	1.9	4.5	3.4	3.0
Sub-Saharan Africa	2.4	0.3	1.2	13.5	4.7	–1.2	4.1	3.6	1.5
East Asia	3.2	3.0	5.9	12.7	9.3	10.1	9.2	6.4	6.4
South Asia	3.4	2.4	1.4	3.7	5.4	7.2	3.9	5.7	6.1
Europe, M. East and N. Africa	3.5	3.1	2.4	8.7	1.6	1.9	8.3	8.5	3.2
Latin America[a] and Caribbean	2.9	3.7	2.2	6.9	4.8	0.8	7.1	6.3	1.3

[a] Excluding the high-income economies of the region.

many farming systems and the fact that population increases will perforce, place heavy and, arguably, insupportable burdens on the natural resource endowment of many countries.

Consideration of the population issue requires close scrutiny at subglobal levels for a variety of reasons. These include questions of productivity of land and labour, incomes and differing population growth rates, present and projected, between countries and groups of countries. It would be inappropriate and unnecessary here to attempt to cover the issue on a country by country basis. Purely geographic groupings on a, say, continental basis could be misleading. An alternative and complementary classification used by the World Bank applies the criterion of per capita Gross National Product (GNP), as shown in Table 3.6. While the data do not extend to such centrally planned economies as the Russia and others who were not members of the World Bank group, they show the differences in population, area and average indices of food production per capita on the basis of per capita GNP. Among other things, Table 3.6 demonstrates the major discrepancies in per capita GNP in amount and annual growth rate between economy groups, with Sub-Saharan Africa and South Asia having very low levels on both counts. These data, together with the Food Production Index, show the parlous condition of the predominately agrarian Sub-Saharan Africa economies and, by inference, the impact of the 'green revolution' in East Asia, and, to a lesser extent, in South Asia. Table 3.7 considers the same economy groups from the standpoint of past agricultural growth and projected growth of population to the years 2007 and 2025.

Table 3.7 shows a projected population increase in the low-income economies from 2.8 to over 5.1 billion between 1987-2025. Within this group, Sub-Saharan Africa presents the frightening picuture of population almost tripling, from 443 million in 1987 to 1259 million in the context of annual average growth in agriculture, between 1980 and 1987 at 1.2 per cent and average index of food production per capita unchanged in 1985-87 from 1979-81 levels. Average annual growth rates in agriculture of low-income economies at 4 per cent overall were boosted by the performance of India and China at 5.1 per cent in contrast to other low-income economies at 1.9 per cent, with average indices of food production per capita at 119 and 106 respectively.

Table 3.6: Population, land area, GNP per capita and average index of food production per capita by major economy groups

Economy	Population[b] (millions)	Area[b] ('000s km²)	Gross National Product (GNP) per capita ($) 1987	Average annual growth rate (%) 1965-87	Average index of food production per capita (1979-81 = 100) 1985-87
Low-income	2823	37015	290	3.1	115
China and India	1866	12849	300	3.9	119
Other low-income	957	24166	280	1.5	106
Middle-income	1039	36118	1810	2.5	101
Lower-middle-income	610	16781	1200	2.2	101
Upper-middle-income	432	20272	2710	2.9	101
Low- and middle-income	3826	73133	700	2.7	111
including					
Sub-Saharan Africa	447	20999	330	0.6	100
East Asia	1513	14014	470	5.1	121
South Asia	1081	5158	290	1.8	109
Europe, M. East and N. Africa	391	11430	1940	2.5	105
Latin America and Caribbean	404	20306	1790	2.1	98
High-income	777	35757	14430	2.3	104
OECD members	746	31085	14670	2.3	103
Other[a]	31	2673	7880	3.5	134

[a] Other: Saudi Arabia, Israel, Singapore, Hong Kong, Kuwait, United Arab Emirates.

[b] Figures may not add due to rounding.

Table 3.7: Agriculture and population growth by major economy groups

Economy group	Population (millions)	Average annual growth of population (%)	Average annual growth rate in agriculture (%)	Average index of food production per capita (1979-81=100)	Projected population (millions)	
	mid-1987[a]	1980-87	1980-87	1985-87	2007	2025
Low-income of which	2824	2.0	4.0	115	3625	5161
China and India	1866	1.6	5.1	119	2279	2893
Other low-income	958	2.8	1.9	106	1346	2268
Middle-income of which	1038	2.2	2.5	101	1329	1862
Lower-middle-income	610	2.3	2.3	101	795	1145
Upper-middle-income	432	1.9	2.6	101	539	726
Low- and middle-income of which	3862	2.0	3.4	111	4954	7023
Sub-Saharan Africa	443	3.2	1.2	100	659	1259
East Asia	1513	1.5	5.9	121	1825	2261
South Asia	1081	2.3	1.4	109	1408	2004
Europe, M. East and N. Africa	390	2.1	–	105	505	743
Latin America and Caribbean	404	2.2	2.2	98	512	689
High-income of which	777	0.7	2.8	104	830	883
OECD members	747	0.6	2.6	103	787	814
Other[b]	31	3.0	10.1	134	42	69

[a] Slight differences from Table 3.6 and failure to add are due to rounding.

[b] Other: Saudi Arabia, Israel, Singapore, Hong Kong, Kuwait, United Arab Emirates.

Middle-income economies in 1980-87 demonstrated an average annual growth rate in agriculture slightly greater than projected population growth and the food production index, changing little between 1979-81 and 1985-87. The high-income economies, which are largely industry based, nevertheless demonstrated a 2.8 per cent annual agricultural growth rate from existing high levels, a slightly increased per capita food production index, but a projected annual population growth of only 0.7 per cent.

In spite of the lack of precision among data of this kind, several important conclusions may be drawn. First, projected population increases in low-and middle-income economies, from 3,862 million in 1987 to 7,023 million in 2025, will necessitate enormous increases in the productivity of both land and labour engaged in their agriculture. With a few exceptions, agriculture is a most important part of their economies, providing food and fibre for their populations and foreign exchange from export of industrial crops such as coffee, cocoa, palm products, rubber and cotton. Export sales are generally inadequate to permit significant import substitution for staple foods, hence agricultural development is vital for economic growth and enhanced well-being of their burgeoning populations.

Second, notwithstanding annual growth rates in agriculture well ahead of population in the high-income economies, exports of agricultural products outside the high-income group on a free market basis will continue to be severely constrained by the inability of the other groups of economies to pay full market prices. This prospect raises questions of international trading practices and food aid that are addressed later in this chapter.

Third, rapidly increasing populations in developing countries, by virtue of their demands for agricultural products, will inevitably place the natural resource base under continuing and increasing pressure. Additional land well be brought under cultivation and use of land best suited for extensive cropping and livestock management systems will be intensified. These developments will add to the existing loss of forests in such countries, to watershed degradation and floods, siltation of river systems, and other phenomena that always follow such environmental abuses.

Granted, improved technology already available and being developed as rapidly as financial and human resources permit will go some way to ameliorating this scenario. Higher-yielding crop varieties and breeds of livestock, and the management skills and imputes to go with them will certainly play an important part in enhancing the productivity of the agriculture of the countries under discussion. But their impact will be uneven. Differing ecological conditions demand different technologies and the fine tuning of such technologies within quite small areas, in response to variations in soil type for example. Further, as the lessons of the 'green revolution' have already shown, technology is only one part of the agricultural development package. Sustained government commitment, as exemplified by appropriate policy measures and the machinery for their effective implementation, is a *sine qua non* that, until recently, has not been sufficiently appreciated.

Fourth, and at the other end of the scale, increasing education, wealth and leisure are leading rapidly to the need for changes in the role of agriculture in the middle- and high-income economies. These role adjustments are deriving from such factors as increased public and political awareness of ecological consideration, demands for improved housing resulting in loss of prime agricultural land and, in the more densely-settled countries in eastern Europe particularly, strong, urban pressure for country-based recreation activities such as national parks for walking, equestrian facilities, lakes for inland watersports, in addition to organically-produced plant and animal products. These demands present challenging opportunities for new enterprises that farmers in high-income economies cannot afford to ignore.

Other Aspects

The other side of the picture also cannot be ignored, however. Increased intensification of agricultural practices involving heavy use of agricultural chemicals in crop and livestock enterprise have assisted in achieving the very high levels of productivity that characterise the industry in most high-income economies.

Table 3.8 compares the use of fertiliser between 1970 and 1986 in the major geographic regions with levels holding in OECD member countries. Taking fertiliser use as a proxy for intensification of agricultural practice, the table illustrates the increasing level of intensity of all regions. Noteworthy are the overall low levels of usage in Sub-Saharan Africa, the increased usage in East and South Asia and, particularly, the levels of application in East Asia in 1986 which was significantly above that in OECD countries. Increased Asian consumption can be ascribed largely to 'green revolution' technology involving irrigated double or triple cropping, exacerbated in East Asia by significant plantings of economic tree crops such as oil palm and rubber which typically are fertilised comparatively heavily at the establishment stage.

Table 3.8 Fertiliser consumption by geographic region

Geographic region	'00s g of NPK per ha of arable land	
	1970	1986
Sub-Saharan Africa	33	86
East Asia	367	1326
South Asia	114	586
Europe, M. East and N. Africa	475	960
Latin America and Caribbean	176	451
OECD members	995	1163

But these achievements can have undesirable side effects. Widespread increase in the nitrate content of rural and urban water supplies in the United Kingdom is one example of the effect that continuous heavy applications of nitrogenous fertilisers may have on light land. Problems in the Netherlands associated with the disposal of slurry deriving from intensive livestock operations is another. Increasing political pressure brought by environmental groups and enhanced public concern over the long-term impact of intensive agricultural practices on the national heritage are major factors that are changing the face of agriculture and hence its role and importance most immediately in the high-income economies but also, inevitably, wherever the industry exists.

In short, agriculture and farmers worldwide face enormous challenges if the world is to be adequately fed and clothed in the future. Past contributions of science and technology to agricultural development have not been trouble-free. The production increases permitted by new technologies together with policies designed to stimulate growth of production have led to large agricultural surpluses in the high-income economies of North America and western Europe. But, as already noted, the intensive production of crops and livestock involving a measure of indiscriminate chemical use and the cultivation of marginal land have resulted in soil degradation, ecological change and environmental pollution.

Soil degradation is particularly acute in low- and middle-income economies where population pressure has led and will continue to lead to reductions in the fallow periods of traditional cropping cycles. The consequences are serious, especially where farmers have moved, or been forced, into tropical rain forest or on to land marginal for traditional cropping because of low inherent fertility, uneven topography or unreliable rainfall.

Science and technology will continue to be central to production and sustainable practices. But they can only maintain and enhance their impact in an appropriate policy climate. Thus in developing countries, past research has tended to favour exportable cash crops, grains such as wheat and rice, and generally, the more favourable agro-ecological zones. Economic policies and development programmes have not been designed to meet the needs of resource-poor farmers and those living in marginal areas. Greater emphasis is required on these problems, particularly in the context of sustainable development. This implies, among other things, strengthened links between agricultural research, extension and farmers, with more attention to social equity, and recognition of the needs of women, who constitute a large proportion of farmers in developing countries, in the design of agricultural research programmes. Meeting these criteria requires that science and technology for agriculture must be based on a thorough understanding of:

- the physical and biological production environments;
- the genetic potential for increased productivity; and
- prevailing socio-economic circumstances.

It should be emphasised that, while technical issues are many and complex, they are matched in most developing countries by policy and institutional questions of comparable complexity. Extension services, input supply, output storage, marketing, and primary processing are just some of the complementary factors essential to the broad-based adoption of relevant agricultural research results.

International Trade, Food Aid

World trade in agricultural commodities expanded in the decades of the 1990s and 2000s and is likely to continue to expand in the foreseeable future. Reasons for the expansion include increased and diversified demand from consumers and expanded surpluses in major commodities in some areas coinciding with serious shortfalls elsewhere. These basic conditions were facilitated by expansion and development of marketing practices including trading in commodity futures which was itself assisted by computerisation and electronic message transfer. Government to government barter trade agreements involving such commodities as grain and oil also expanded substantially during the period. Physical aspects of trading have also benefited from the development of container-based and bulk transport facilities in exporting countries, together with improved handling facilities in major exporting and importing countries. Table 3.9 illustrates changes in world exports between 1965 and 1984.

Table 3.9: Growth of world exports, 1965-84
(annual percentage change in constant 1980 prices)

Exports	1965-70 average	1971-84 average
Agriculture	+3.21	+4.64
food	+2.66	+5.27
non-food	+4.33	+3.00
Metals	+9.65	+4.90
Fuel	+12.70	–3.25
Manufactures	+8.46	+4.78
Total	+9.32	+2.60

Note that food exports increased most rapidly of all itmes in annual percentage terms in the latter period. Other export commodities showed declining comparative growth, with fuel showing negative growth in the 15 years to 1984.

Movement of export shares of major agricultural commodity groups between the early 1960s and 1980s also makes interesting reading (see Table 3.10).

Table 3.10: Export shares of major agricultural groups (%)

Country	Beverage		Food		Raw materials		Total agriculture	
	1961-63	1982-84	1961-63	1982-84	1961-63	1982-84	1961-63	1982-84
Developing countries	98.1	94.9	44.8	34.2	69.2	65.3	63.1	48.4
Low-income countries	27.6	23.8	9.0	3.6	15.6	13.6	15.1	8.3
Africa	19.6	15.8	1.5	0.3	6.0	4.9	6.9	3.5
Asia	8.0	8.0	7.5	3.3	9.6	8.7	8.0	4.8
Middle-income oil exporters	17.1	17.6	6.5	3.3	33.9	24.7	14.8	8.8
Middle-income oil importers	53.4	53.5	29.3	27.3	19.7	27.0	33.3	31.3
Industrial market economies	1.7	4.7	46.2	62.7	23.5	24.0	30.5	47.9

Data are weighted by 1978-82 world export unit prices to permit cross-country comparisons. Beverage comprise coffee, coca and tea. Food comprises cereals, sugar, meat, poultry, dairy products, roots and tubers, pulses, fruits and vegetables. Raw materials comprise cotton, jute, rubber and tobacco.

Overall, developing countries have lost out to the industrial market economies in terms of export shares of total agriculture, from 63.1 per cent to 48.4 per cent and 30.5 per cent to 47.9 per cent respectively over the period in question. The 'food' group shows the greatest mobility, declining from 44.8 per cent to 34.2 per cent in developing countries and expanding from 46.2 per cent to 62.7 per cent in the industrial market economies. These changes have come about simultaneously with a slight overall decline, in real terms, of the prices of traded agricultural commodities between 1950 and 1984 as shown in Table 3.11.

The most important message of Table 3.11 is that over the 35-year period, 1950-84 the largest annual declines in the real prices of traded agricultural commodities have been for cereals, and fats and oils, groups that encompass the staple foods of the bulk of the world's population. In the light of increasing demand deriving from increasing world population these figures point to increasing efficiency in agricultural production by nations engaging in agricultural trade, with the industrial market economies in the forefront.

Table 3.11: Real growth of commodity prices (average annual percentage change)

Commodity	1950-59	1960-69	1970-79	1950-84
Total agriculture	-2.92	0.00	+0.1	-1.03
Beverages	-2.08	-1.26	+7.46	-1.13
Cereals	-3.84	+2.72	-1.31	-1.30
Fats and oils	-3.73	-0.73	-0.81	-1.29
Raw materials	-2.51	+0.50	-1.72	-1.08

Seen from the perspective of growth in export volumes, Table 3.12 shows the relative downturn in low-income economies' annual average export volume of primary goods (largely agricultural) between 1965-73 and 1986. In Sub-Saharan Africa average annual growth in total export volume and export of primary goods fell from just over 15 per cent to 1.1 per cent in each case. In the low-income economies generally, exports of manufactured goods grew very substantially but from a very small base. The high-income economies showed an average annual increase in export volume of primary goods from 8.9 per cent in 1965-73 to 10.2 per cent in 1986. The declining rate of their export of manufactured goods can be attributed to a general slowdown in the world's economy.

Table 3.12: Growth of export volume, 1965-88 by country group

Country group	Average annual change in export volume (%)					
	1965-73	1973-80	1980-85	1986	1987	1988
Low-income economies	9.6	2.3	1.5	7.0	4.3	6.8
manufactures	1.8	8.5	10.0	15.9	23.3	10.8
primary goods	11.2	1.1	-1.1	3.4	-4.0	4.6
Middle-income economies	3.9	4.4	5.3	5.2	7.3	7.2
manufactures	16.8	13.8	9.7	7.1	14.8	9.5
primary goods	2.4	1.2	2.5	3.7	1.2	6.1
High-income economies	9.9	4.8	1.8	4.8	5.9	8.8
manufactures	10.6	5.5	3.8	1.3	5.7	7.1
primary goods	8.9	3.5	-2.5	13.6	6.3	13.6
OECD members	9.4	5.4	3.3	3.6	6.4	7.8
manufactures	10.6	5.2	3.7	1.4	4.9	6.1
primary goods	6.7	5.9	2.2	10.2	10.6	13.6

International cooperation in agricultural trade is clearly accepted as an effective means of fostering economic growth in developing countries. But cooperation has been hampered by a continued failure to liberalise agricultural trade and by the declining and volatile agricultural terms of trade faced by some developing countries. Thus other means than unregulated commercial trade have been sought to serve the interests of such countries while simultaneously expanding trading opportunities for the industrial market economies. They include commodity agreements, schemes to compensate commodity producers for earning shortfalls and attempts to improve developing countries' access to the markets of industrial countries. At best these measures have had limited success partly because of the agricultural and trade policies followed in developed and developing countries.

Food Aid

Food aid may be defined as the provision of foodstuffs to low-income countries by governments or non-governmental organisations (NGOs) on highly concessional terms, or as gifts to help mitigate domestic supply shortages. Food aid may be provided in cases of acute food shortage occasioned by natural disasters such as drought or floods, or by man-made disasters such as war. It may also be provided on a more continuous basis to help alleviate situations where local agricultural capacity, perhaps weakened by inadequate and erratic rainfall or repeated flooding, and exacerbated by population pressure, is chronically unable to meet the population's basic nutritional needs; Bangladesh is a case in point. Shipment of food in the mid-1970s to African countries affected by the Saharan drought is an example of food aid to offset a natural disaster. Provision of food to Vietnam following the war in that country is an example of food aid to mitigate a man-made disaster. The need for and shipment of food aid to Ethiopia in the 1980s is an example where the impact of inappropriate agricultural policies and a civil war combined with excess population pressure on drought-prone land of marginal agricultural productivity led to recurrent acute shortages of food over wide areas of the country

Improved communications, the existence of large food surpluses in developed countries, and increased population pressure in some developing countries have all been factors in the growth of food aid from the middle to the latter part of the twentieth century. To these must also be added the political and moral initiatives from the populations of developed countries to help their less fortunate brethren elsewhere, to support NGOs and force action by their governments and international development agencies, in the provision of food aid.

The type of food aid supplied is conditioned by availability, ease of storage and shipment from suppliers, facility of handling and distribution by recipient agencies and, to an increasing extent, by the nutritional needs and preferences of the ultimate consumers. It tends to be dominated by cereals such as maize and wheat, both of which were markedly in surplus in the 1970s and 1980s. However, other commodities, including pulses, vegetable oils, sugar and dairy products such as butter, oil and dried skimmed milk powder, also feature. In the mid-1980s about one million tonnes of these commodities were provided as food aid.

The Food and Agriculture Organisation of the United Nations (FAO) indicates that between 1982-83 and 1987-88 shipments of cereals as food aid ranged between about 9 and 13 million tonnes per years, and that over this period these amounts approximated from 15 per cent to 23 per cent of the cereal imports of low-income food-deficit countries. Further evidence of the volume of food aid is provided by data from the FAO affiliated World Food Programme (WFP) in Table 3.13.

Table 3.13: WFP emergency operations approved annually, 1982-87

Year	No. of operations	No. of countries	WFP regular resources	International emergency food reserve (US $ millions)	Total
1982	68	37	19.7	172.2	193.2[a]
1983	68	36	45.0	155.3	200.3
1984	63	40	54.2	178.2	232.4
1985	44	32	43.9	181.0	224.9
1986	50	24	34.5	144.2	178.7
1987	79	35	50.9	220.4	271.3

[a] Includes US $ 1.3 million from the International Conference on Assistance to Refugees in Africa.

The volume and scope of the FAO/WFP affiliated International Emergency Food Reserve (IEFR) is illustrated by the fact that in 1987 twenty-three donors contributed 643,000 tonnes of cereals and over 58,000 tonnes of other commodities for distribution via WFP operations. But WFP is only one source of food aid. Others of major significance include the USA, the EC and its member countries, and Canada, all of which contribute to the overall development assistance programmes.

As has already been noted, many NGOs such as OXFAM, and a range of charities with religious affiliations, are also involved in food aid operations along with their other activities in needy countries. The size and scope of their work is perforce limited by the availability of resources but their effectiveness is arguably greater than large organisations because of less bureaucracy and their proximity to the grass-roots in donor and recipient countries.

Impressive though they are, the scope and volume of food aid should be viewed in the perspective of total external assistance to developing countries. Official development assistance (ODA) from all sources reported to the World Bank totalled nearly 31.8 billion US dollars in 1987 compared with 25.2 billion US dollars in 1981. External assistance to agriculture typically constitutes about 40 per cent of total ODA. Further, these figures refer to external assistance for agriculture as a whole and not merely food production. In absolute terms therefore food aid represents a very small proportion of total external assistance to developing countries. Its influence in political terms may be much more significant however. Thus, the governments and people of recipient countries might be inclined to look kindly on those who provided food aid in times of need. This may result in strengthened economic and cultural ties and all that these could entail.

The efficacy of many early food aid interventions suffered greatly from such design faults as provision of food unfamiliar to recipients and therefore not easily prepared, inadequate port handling and distribution mechanisms resulting in delays and high rates of spoilage, leading to diversion from target groups, and racketeering and price escalation. Further, provision of comparatively large volumes of food commodities free or at heavily subsidised prices led in many cases to the collapse of local markets for such home-grown produce as was available. This situation was commonly exacerbated by delays in distribution which resulted in food aid not arriving until the following harvest with disastrous economic consequences for local producers, particularly if there had been a good growing season. More recently food aid has been applied with greater attention to its capacity for short-term amelioration of human hardship together with longer-term economic benefits. The concept of payment with food for productive work is now widely applied such that beneficiaries receive food itmes equivalent to the wages they would otherwise expect, to create productive assets such as levelled plots for irrigation, soil and water conservation measures, shelter belts, and so on, in order to make better use of their environment. However, great care and considerable local knowledge are needed to ensure that such interventions, and indeed all food aid provided to rural communities, do not permanently damage indigenous food production and marketing systems. In this respect a clear understanding of the role of women is a prerequisite for success.

Major Crops

Most of the people of the world obtain the bulk of their energy needs from about 20 crops, while a number of others are cultivated for alternative nutritional or other purposes. The principal food crops comprise cereals, roots and tubers, pulses, oil-bearing plants and sugar cane and sugar beet. Also important are the beverage crops, fibre crops, rubber and a range of fruits and vegetables that provide nutritional supplements to staples.

Succeeding paragraphs touch on the major agricultural crop groups to provide an indication of their present relative importance, geographic distribution, productivity, and discernible trends in their popularity in different regions and cropping systems. Notwithstanding the comparatively small numbers of species cultivated by man, space does not permit other than superficial treatment of the subject.

The Cereals

Cereals are mankind's predominant source of energy. Annual cultivation of over 700 million hectares and producing 1,800 million tonnes of grain each year worldwide, they dwarf all other cultivated crops in area and production. Table 3.14 provides basic information on the world's major cereals, illustrating the predominance of *wheat*, at one-third of total area and about 30 per cent of total cereal production, with *rice* and *maize* second and third in importance. Among other features, the table shows the comparative stability of crop area harvested within the group in the period between 1979-81 and 1987 with all crops except rye showing a downturn of a few percentage points. Yields, on the other hand, are uniformly up, leading to increased production overall. Notwithstanding the very narrow time-frame, the table illustrates the underlying long-term trend towards increased productivity of major cereal in the mid- to late twentieth century.

Table 3.14: Global area, yield and production of cereals

Type cereals	Area harvested ('000 ha)		Yield (kg/ha)		Production ('000 t)	
	Average 1979-81	Average 1987	Average 1979-81	Average 1987	Average 1979-81	Average 1987
All cereals	719450	694000	2210	2597	1590000	1803000
Wheat	235000	221600	1887	2334	443600	517100
Rice paddy	143750	141500	2756	3283	396300	464500
Maize	126100	126000	3346	3636	421900	458000
Barely	81000	78100	1936	2327	156800	181700
Sorghum	44100	42700	1459	1472	64500	62800
Millet	40100	36300	693	763	27800	27700
Oats	25700	23500	1659	1840	42600	43200
Rye	15100	16300	1657	2098	25000	34100

Figures may not add due to rounding.

Asia dominates all other regions in terms of harvested area of the three major cereals with about 40 per cent of the world's wheat, nearly 90 per cent of its rice, and 30 per cent of the maize. The yield picture is quite different, however, with average Asian yields about half those achieved in Europe. European yields for wheat and rice are some 60 per cent and 20 per cent respectively more than those in North America, the next highest. But North America dominates the maize yield picture, 20 per cent above Europe and more than double the yields obtained in Asia. Nevertheless, average Asian yield of wheat and rice have risen significantly since the early 1970s increasing by 33 per cent (wheat) and 20 per cent (rice) between 1980 and 1987, demonstrating the continuing impact of the 'green revolution'.

Yield differences largely reflect the systems applied in different regions for cereal production. Thus, the generally high levels achieved in Europe and North America compared with other areas may be attributed to heavy use of fertilisers and other agricultural chemicals using seed of high genetic potential, combined in capital-intensive systems with relatively favourable natural growing conditions, as, for example, the mid-West corn belt of the USA. At the other extreme, African cereal production systems utilise few fertiliser or other agricultural chemicals, using largely unimproved seed, employing sub-optimal cultivation techniques frequently on inherently infertile soils and under difficult growing conditions. In contrast, much of the wheat and rice grown in Asia is irrigated, thus mitigating drought and flood hazards. Strains of planting material responsive to moderate fertiliser applications and irrigation, with a high grain: straw ratio are comparatively widely distributed. This combination results in Asian yield of wheat and rice roughly double and maize yields averaging half as much again as those in Africa.

Whereas wheat and rice are grown predominantly for human consumption worldwide, *maize, sorghum* and *millet* have differing end uses in different geographic regions. They are cultivated predominantly for human consumption in Africa and Asia (though maize is a vital component of developing poultry industries in these regions) in contrast with Europe and North America where the grain is typically fed to livestock. However, as incomes rise so dietary preferences change, and there is little doubt that coming decades will see a relative reduction in the amounts of these cereals grown for human nutrition worldwide. This relative reduction will be at least matched by increased production of wheat and rice, the preferred cereals. The pace of such changes will depend on many economic and biological factors: however, the development of gene manipulation and other techniques for use by plant breeders has already vastly accelerated man's capability to adjust plant characteristics and performance. Thus desirable features, such as drought- and flood-tolerance, pest- and disease-resistance can now be introduced more readily than hitherto, thereby broadening the geographic basis for crop cultivation and reducing the biological limitations on the spread of preferred cereals and other crops into the farming systems of many regions where they could not previously be grown. The increase in area of wheat harvested in Bangladesh from 430,000 hectares to nearly 600,000 hectares between 1979-81 and 1988 is an example of this phenomenon.

Roots and Tubers

While and tubers occupy less than 7 per cent of the area used for cereals worldwide they yield about one-third of the tonnage of cereals produced annually. Table 3.15 provides an indication of the global area, yield and production of the five root and tuber crops for which FAO collects national statistic. In common with other data in this section relating to staple foods in developing countries, their accuracy is open to some question for reasons such as mixed cropping practice, especially in Africa. This requires careful interpretation of relative shares of land use, yields and production of different crops grown simultaneously on the same piece of land. Notwithstanding these cautionary words, the data presented give as good an overall picture as possible of the current global situation of the crops in question.

The area of root and tubers harvested in the 1980s remained roughly constant with production up slightly between 1979-81 and 1987. *Potatoes*, the dominant crop in the group, are grown most widely in the Russia, some 6.9 million hectares, followed by Europe with 5.7 million hectares, and Asia (dominated by China) with 4 million hectares harvested. Yields range between a high of over 26 tonnes/ha in North America, to 19 tonnes in Europe, around 12 tonnes in Asia, slightly less in Russia and China, down to about 8.5 tonnes/ha in Africa. These figures reflect the relative overall productivity of the agriculture of all the regions concerned, with North America and Europe operating high input/high output systems under generally favourable soil and climatic conditions.

Cassava, the next most important in terms of harvested area is concentrated in tropical regions, particularly in Africa with 7.2 million hectares, followed by Asia with 3.8 million hectares, and South America (its region of origin) with some 2.5 million hectares. As for other crops, African cassava yields, at less than nine tonnes/ha are lower than elsewhere, with South America and Asia each obtaining yields of over 11.5 tonnes/ha. Agriculture (IITA) in Nigeria hold promise for widespread distribution of highly-productive, disease-resistant strains to farmers in major production areas.

Cassava is typically the starchy staple of the poor, frequently grown as a famine reserve crop because of its capability to yield under conditions of low soil fertility, and to be left in the ground for several months after it first matures until needed. However, it is also grown as an industrial crop for starch and to produce direct chips for stock feed. This latter end use has been developed notably in Thailand which has some 37

Table 3.15: Global area, yield and production of roots and tubers

Type of root/tuber	Area harvested ('000 ha)		Yield (kg/ha)		Production ('000 t)	
	Average 1979-81	Average 1987	Average 1979-81	Average 1987	Average 1979-81	Average 1987
All roots and tubers	46000	46200	12183	12625	560300	583800
Potatoes	19000	18000	14177	15708	269700	283800
Sweet potatoes	8900	9200	15133	14458	134200	132800
Cassava	13800	14500	8956	9305	123500	135100
Yams	2400	2500	9868	9361	23600	23500
Taro (coco yam)	1000	990	5517	4914	5500	4900

Figure may not add due to rounding.

per cent of the area harvested and produces about 43 per cent of the Asian output of the crop, largely for export to developed countries in the EC and elsewhere.

Sweet potatoes, at about 20 per cent, occupy a suprisingly large share of the world area of roots and tubers, yielding roughly the same weight of product as potatoes themselves. Asia, with, with over 80 per cent harvested area, is the predominant region and within Asia, China, at nearly 6 million hectares and 114 million tonnes, harvests about two-thirds of the world's sweet potato area and 85 per cent of the crop. China's substantial pig population consumes a significant proportion of the country's output.

Yams and taro are minor crops overall but occupy important segments of production systems in West Africa for human consumption. Grown in theorist and Guinea Savannah zones, the Nigerian and Cote d'Ivorie yam crops together accounts for 72 per cent and 78 per cent of the world's area and production respectively. Requiring heavy labour input in comparatively fertile soil, yam production has remained roughly static in recent years and is likely to decline as supplies of newly-cleared forest land diminish with growing population pressure.

Pulses

The group of grain legume crops classified as pulses contributes an important part of the calorie and protein intake of a large proportion of the population in developing countries. Table 3.16 presents the area, yield and production expansion of the pulse groups in the decade of the 1980s, with dry peas and lentils showing the greatest growth overall. Asia dominated the regional distribution with 56 per cent of the area and 52 per cent of world pulse production. India, with two-thirds of area harvested and one-half of production in the Asia regions, has slightly less than the area hearted in the rest if the world by only about one-half of world production excluding Asia.

Dry beans are grown most widely in Asia, with half the world's harvested area, followed by South America with about one-fifth. India has more than two-thirds and one-third, respectively, of the Asian and world area of dry beans harvested. Productivity overall is low, with Asia, at 465 kg/ha having the lowest regional yield, a level pulled down by India, where dry beans typically average around 300 kg/ha, half the European

Table 3.16: Global area, yield and production of pulses

Type of pulse	Area harvested ('000 ha)		Yield (kg/ha)		Production ('000 t)	
	Average 1979-81	Average 1987	Average 1979-81	Average 1987	Average 1979-81	Average 1987
All pulses	60333	67645	676	807	40807	54619
Dry beans	24298	25437	554	563	13482	14315
Dry broad beans	3685	3288	1162	1400	4284	4603
Dry peas	3439	9612	1144	1624	8453	15613
Chick peas	9530	10028	624	680	5971	6817
Lentils	1996	3233	596	792	1190	2560

Figure may not add due to rounding.

and one-third of levels attained in north Central America. Brazil dominates dry bean production, reflecting the importance of the crop in Brazilian diets.

Dry peas, the second most important pulse in production terms, dominate in area and production in the Russia with about half the world's total in both respects. Asia, with some 30 per cent of world area and production occupies second place; China accounting for over 80 per cent of the region's output. Productivity is highest in Europe with yields of 4,000 kg/ha treading sharply upwards. Yields in Russia also rose in the 1980s in contrast with other regions where they remaind roughly static or showed a slight downward trend over the same period. Grown mainly for human consumption, Table 3.16 shows clearly the upward movement of area, yield and production of the crop on a worldwide basis throughout the decade of the 1980s.

Chick peas, the third most important pulse overall in area and production, are centred mainly in the Asian region in general, and in India and Pakistan in particular. These countries together account for 85 per cent of harvested area, and 80 per cent of chick pea production worldwide. Yields have trended slightly upwards in the 1980s in contrast to area harvested which, in India and Pakistan fell from over 8 million hectares to about 6.5 million hectares in the period. Reasons for this decline may very between different localities but it can be attributed in part to increased double cropping of irrigated rice and wheat, and to a move towards groundnuts at the expense of chick peas in irrigated area. This latter trend has been fostered by government policy designed to increase vegetable oil production, especially in India, which fell increasingly into deficit for this commodity in the 1970s and early 1980s.

Over 60 per cent of the world's *dry broad beans* are grown in China where they constitute an important energy and protein component of the human diet. Yields of around 1300 kg/ha compare unfavourably with levels in some European countries such as France and Switzerland, which consistently achieve more than 3,000-3,500 kg/ha and where the crop is grown over comparatively small areas principally for stock feed.

Lentils, the least important of the pulses in area and production are another predominantly Asian crop. India again dominates the area harvested, with nearly 50 per cent of the global total, and about 30 per cent of production grown entirely for human consumption. Yields overall have shown a substantial increase in the 1980s, a move mirrored in Asia and particularly in India where they were raised on average by more than 40 per cent, in varying proportions due to improved planting material, cultural practices and pest control.

Oil Seeds

With the exception of linseed and castor beans, the oil seeds listed in Table 3.17 are important protein and energy sources in human diets worldwide. Linseed and castor beans on the other hand are grown for industrial end uses, primarily in the manufacture of paints and for lubrication purposes respectively; though linseed residues after oil extraction, in common with other oil seeds, make valuable stock feed by virtue of their high protein content.

The table illustrates the overall expansion of area and production of the major edible oil-seeds in the period 1979-81 to 1987. The area expansion of over 4 millions hectares, while considerable, is overshadowed by the production increase of 38 millions tonnes over this period, as a result of yield increases achieved from all the crops in question.

Soya bean leads the groups in all aspects shown, with the relatively modest area expansion overshadowed by a significant average yield increase on an existing high base level. The USA is the world's largest producer followed, in area and production, by Brazil and China. Whereas USA area and production declined in the 1980s, both Brazil and China registered substantial increases in both respects over the period. The Brazilian expansion is an excellent example of the conjunction of appropriate technology with government policy designed to encourage crop production. Thus, area planted in 1960 was 171,000 hectares, rising to over 720,000 hectares by 1968 and to over 10.5 million hectares some 20 years later. Production soared accordingly, from 205,000 tonnes in 1960 to 650,000 tonnes in 1968 and 18 million tonnes in 1988. Located principally in the centre-south of the country, soya bean production utilises capital-intensive production systems developed in the USA, and germplasm adjustment to Brazilian conditions from material produced by the International Soya bean Resource Base (INTOSY) based in the USA. The Brazilian

Table 3.17: Global area, yield and production of oil seeds

Type of oil/seed	Area harvested ('000 ha)		Yield (kg/ha)		Production ('000 t)	
	Average 1979-81	Average 1987	Average 1979-81	Average 1987	Average 1979-81	Average 1987
Soya bean	50500	52500	1701	1909	86000	100200
Cotton seed	34400	32200	972	1023	27200	30800
Groundnuts (in shell)	18500	999	1157	18500	21400	
Sunflower seed	12300	14500	1172	1426	14400	20600
Rapeseed	11500	15800	972	1425	11200	22500
Sesame seed	6400	6400	306	359	2000	2300
Linseed	5400	4300	450	571	2500	2500
Castor beans	1500	1500	577	601	800	900

Figures may not add due to rounding.

soya bean explosion was stimulated by pricing and credit policies aimed at placing the country in the front rank of world exporters of soya products. This objective was helped considerably by the development of productive wheat varieties for use in a cropping system designed to permit a smooth succession between the two crops. The concurrent expansion of wheat area and production, from 1.1 million to 3.5 million hectares and 0.7 to 5.4 million tonnes respectively between 1969 and 1988, resulted in a massive reduction in wheat imports, thereby saving valuable foreign exchange.

Cotton seed, occupying a position second to soya bean in area and production, is a valuable by-product of a crop grown primarily for its fibre. Seed production is greatest in Asia where China, India and Pakistan, in that order, are responsible for 90 per cent of total output. The USA failed to keep pace with China in the 1980s with production in the latter country moving from rough parity with the USA in 1979-81, at around 5 million tonnes, to 8.5 million tonnes in 1988, compared with a USA level of 5.4 million tonnes of cotton seed in that year. As a high-cost producer the USA continues to face strong competition from developing countries, especially in Asia and Africa. Further relaxation of centralised control of cropping in China in the early 1980s, as part of a policy to meet domestic demand for cotton fibre, went far to provide the stimulus for the production expansion noted above. This result, like that in Brazil for soya beans, illustrates clearly the receptivity of farmers from countries with totally different social and political backgrounds, to appropriate market signals.

Groundnuts, third in the area and production table, are grown most widely in India and China, followed by countries such as Senegal, Sudan, Nigeria, Cameroon and Malawi in Africa. The USA dominates production in the western hemisphere, followed by Argentina and Brazil. With the exception of most Africa and South American countries, production trended upwards in the 1980s reflecting increased domestic demand for vegetable oil in Asian counties and expanded export opportunities in the USA. The comparative decline in importance of groundnuts exports from African producers was due partly to unfavourable prices relative to competing domestic food crops, and inadequate research, extension, input supply (especially seed) and marketing services. Additionally, failure to respond adequately with control measures against the carcinogen aflatoxin, which develops rapidly in groundnuts at harvest, contributed somewhat to the African loss of export market share. In Brazil and Argentina groundnut area and production declined as a result of substitution by soya beans in cropping systems.

Sunflower seed production increased sharply in the 1980s due to significant overall yield enhancement derived from the widespread use of dwarf and hybrid planting material, especially in Europe, and to some increase in area planted. With about 4 million hectares and an annual production in excess of 6 million tonnes, the Russia harvests the largest area. However, European area and production expanded sharply over the period, with production more than doubling from 3 million to over 6.5 million tonnes, centred mainly in France, Romania and Spain. As it produces a high-quality cooking, low in unsaturated fatty acids, demand appears likely to remain strong for this product in the industrialised countries of the world.

Rapeseed is another oil crop that has seen substantial increase in yield, area and production over the decade of the 1980s. China and India account for Asia's pre-eminence as a producer but neither country matches the European rate of area expansion or productivity. Thus, while rapeseed harvested in Asia increased from 7.5 million to 9.5 million hectares with yields climbing from 700 to 900 kg/ha between 1979-81 and 1988, data for Europe over the same period showed an area increase from 1.5 million to 3 million hectares, with yields moving from 2,100 to around 2,700 kg/ha. This phenomenon is another example of farmers' responsiveness to an appropriate technology, coupled with favourable market signals, the latter provided to a large extent by the Common Agricultural Policy of the EC.

Sesame seed yields a high-quality cooking oil and also fulfils a minor confectionery role in many developed countries. India produces about one-quarter of the world's sesame seed, with China and Burma also important Asian growers. Africa, the only other significant regional producer, at about 600,000 tonnes from nearly 2 million hectares, has the crop occurring most commonly in Sudan and Nigeria, with and estimated 1.1 million and 250,000 hectares respectively. Productivity overall is low, rarely exceeding 1,000 kg/ha, and more commonly lying between 200 and 800 kg/ha in major areas of production. While strains with multilocular seed pods, and hence significantly greater yield potential, have been developed in recent years, shattering of the pods at harvest continues to result in the low yields that characterise the crop.

Oil palm is the major tree crop source of vegetable oil, producing around 9 million tonnes of edible palm oil worldwide in the late 1980s, up from 5 million tonnes at the beginning of the decade. A native of the West Africa rain forest, oil palm has been planted widely and with great success in similar ecologies in Asia, and in a few locations in Brazil. Substantial breeding and other research, notably in Malaysia, Cote d'Ivoire and Nigeria has resulted in the development of high-yielding strains within the dura, tenera and pisifera oil palm varieties. These are characterised respectively by the increasing proportion of palm oil to shell and kernel within the fruit. Palm kernels are the second important oil palm product yielding around 3 million tonnes worldwide in 1988. With a typical 50 per cent extraction rate, the higher greases of palm kernel oil are used in the manufacture of margarine and cooking fat. Lower grades go for the production of soap.

The latter part of the twentieth century has seen a major shift in the relative importance of Africa and Asia as palm oil and kernel producers. This has occurred largely as a result of the establishment of substabtional and well-managed plantings of highly productive strains, notably in Malaysia and Indonesia. With the exception of the Cote d'Ivoire, other historically important producers such as Nigeria and Zaire have done little recently to increase or improve their oil palm holdings. And in Nigeria particularly, domestic demand for palm oil, expanded by rising population, has contributed to the country's move from a substantial exporter to a net importer of palm oil. Recent advances in rapid vegetative propagation of high-yielding planting material hold promise for further expansion of the crop on smallholdings as well as commercial plantations, wherever it can be grown.

Coconuts are produced entirely in tropical regions and overwhelmingly in Asia, with about 30 million tonnes out of a world total of 36 million. Although eaten fresh as a vegetable, coconuts are more usually processed by stripping out the flesh from the hard shell and drying it to produce copra, a material containing 60-65 per cent of coconut oil that is used for cooking worldwide. According to FAO data, between 4 million and 5 million tonnes of copra were produced annually in the late 1980s, mainly in Indonesia and the Philippines.

The coconut palm occurs wild in groves in the pacific regions. It is also a common household tree in many tropical countries planted singly or as an integral part of traditional, mixed crop systems. Improved, commonly dwarf, planting material produced by hybridisation in several Asian countries and Cote d'Ivoire is increasingly used for household gardens and in the establishment of smallholder coconut groves and large plantations. These occur particularly in Indonesia and the Philippines, but smallholder and commercial planting programmes also operate in Malaysia, Thailand, Sri Lanka and India.

Sugar Cane and Sugar Beet

Sugar cane ideally requires an annual water supply around 1,600 mm with a distinct four to five months dry season for ripening in order to maximise sucrose production. Typically grown in tropical and subtropical countries, Brazil with over 4 million hectares, followed by India with 3.3 million hectares, Cuba with 1.3 million hectares and China with 1 million hectares contribute most in area terms to the world's 16.3 million hectares of cane. Other important producers include the USA, Mexico, Pakistan, the Philippines and Thailand. Yields overall have increased slightly in the decade of the 1980s, from 56 to 60 tonnes of cane per hectare, with countries such as Peru, Swaziland, Malawi and Ethiopia recording yields of 100 tonnes and over per hectare from modern, irrigated plantations. Global production, at nearly one billion tonnes of cane, contributed significantly to world supplies of raw sugar, estimated at 115 million tonnes in the late 1980s.

Sugar beet, the other source of mankind's sucrose, has remained roughly constant in area in the 1980s, around 8.7 million hectares. Yields, on the other hand, have increased overall from 30 to nearly 35 tonnes of beet per hectare resulting in annual global production of around 295 million tonnes towards the end of the decade. Production is centred largely in the temperate zone, with Europe and the Russia each harvesting about 3.4 million hectares annually. China and the USA are the other principle producers with areas harvested roughly equal, at about 530 million hectares each. Productivity is greatest in France, Europe's largest producer, at 66 tonnes of beet per hectare. The USSR and China achieve yields less than half this level, with Europe and the USA roughly equal at 43 tonnes/ha.

The Beverage Crops

Coffee is one of the crops for which an international commodity agreement (ICA) has been in existence for some years. Compared to others, the ICA for coffee has been the most successful in protecting growers' and consumers' interests by achieving a degree of stability in prices. The ICA for coffee has also promoted diversification schemes to reduce the proportion of marginal coffee area and thus enhance the overall productivity of the crop. The area of harvested coffee expanded from 10 to 11 million hectares in the 1980s, increasing from 2.5 million to 2.9 million hectares in Brazil, the world's largest producer. In Africa harvested area went from 3.5 million to 3.8 million hectares, with Cote d'Ivoir, Africa's dominant producer of the robusta type, registering an area increase from 1 million to 1.2 million hectares harvested as new planting came into bearing. Traditional producers of arabica-type coffee, such as Ethiopia, and Central American countries including Costa Rica, Guatemala and El Salvador, retained roughly constant areas partially due to civil unrest, and to strong competition from the robusta type in the growing market for instant coffee. In Asia where robusta is the predominant coffee, India, Indonesia and the Philippines together account for over 85 per cent of the region's coffee. All expanded their area harvested in the 1980s, led by Indonesia with an increase from 490,000 hectares to nearly 640,000 hectares.

World production of green coffee trended upwards overall but fluctuated through the 1980s in an annual range between 5.2 million and 6.2 million tonnes. Average yields lie between 470 kg and 560 kg of green beans per hectare, roughly the level achieved in Brazil. Highest yields, in the region of 1,500 kg/ha in Costa Rica, contrast with Cote d'Ivoire and Ethiopia where maximum average annual yields failed narrowly to reach 250 kg. Relative immaturity of the Ivorian crop and over-maturity of that in Ethiopia, together with extensive and low-input production systems account largely for these differences from the highly productive Costa Rican coffee industry.

Tea production, at around 2.5 million tonnes of made tea worldwide is less than half that of green coffee. Area, yield, and production all increased in the 1980s in Asia and Africa where most of the world's tea is grown on upland, acid soils, and under the high rainfall conditions that favour the crop. Asia, with the world's second largest area, but less than one-third of that harvested by China, produces over 650,000 tonnes annually compared with China's annual output of about 550,000 tonnes. Indian yields of made tea at 1,700 kg/ha are among the highest obtained by significant producers. Indonesia, with an estimated 90,000 hectares obtains yields of 1,600 kg/ha, but Kenya, by far Africa's biggest producer at around 160,000 tonnes from 84,000 hectares, overtops its Asian competitors, notably India and Sri Lanka, with yields of 1,900 kg/ha and more. The tea industry in India and Sri Lanka is based largely on industrial-type plantations established in colonial times, many of which are fully mature or ageing. Kenya, on the other hand, has built much of its thriving tea industry since the early 1960s on smoulder production via the quasi-commercial Kenya Tea Development Authority (KTDA). KTDA utilised the experience and reasserts finding from foreign-owned Kenyan tea estates with financing and management skills originally provided by the Commonwealth Development Corporation (CDC) and the World Bank. Since its establishment it has created an efficient and highly productive export-oriented industry, operated by Kenyan nationals and based on thousands of individual farmers' plots, that has increased its market share of made tea at the expense of long-established Asian competitors.

Cocoa owes much of its importance to demand for confectionery products based on chocolate. Largely confined to the wet tropics, cocoa displays a number of quality characteristics in its major varietal types. These range from the delicately flavoured criollos to the more robust foresteros. Amelonado cocoa which is extensively grown in West Africa belongs to this latter group. Recent years have seen the development of highly productive hybrid strains, and propagation by leaf cutting to provide populations of uniformly high genetic potential. Cocoa doses not lend itself well to large-scale industrial plantations because of its high labour demand for maintenance and harvesting. More typically it is grown by smallholders, or on small (5-10 hectares) commercial holding, especially in Africa which has two-thirds of the world's harvested area of cocoa and more than half of its production volume. Area harvested increased by about 850,000 hectares to nearly 5.6 million hectares in the 1980s. West Africa maintained its dominance at about 3.7 million hectares, but South America, led by Brazil with more than half of the region's 1.2 million hectares, expanded its harvested area by more than one-third over the decade. Productivity of South America cocoa declined from 525 kg/ha to 400 kg/has of direct beans in this period of contrast with West Africa where estimated average yields increased from 300 kg/ha to 350 kg/ha. This result can be attributed in part to the comparative immaturity of cocoa harvested in Brazil and Colombia in the late 1980s following from significant new plantings in those countries. Production increased in all cocoa growing areas by nearly one-third to over 2.2 million tonnes in the decade. West Africa maintained its dominance at 1.3 million tonnes, led by Cote d'Ivoire with more than half the volume. Government policy in Cote D'Ivoire has been to promote agricultural production for export. Commodities targets have included cocoa, to the extent of that country becoming the world's biggest exporter of cocoa beans in the 1970s with production levels exceeding 650,000 tonnes per annum in the mid- to late-1980s. A consequences has been a degree of over-production worldwide. This threw the ICA for cocoa into disarray and has resulted in substantial falls in cocoa prices in recent years.

Fibre Crops

Most of the world's five major fibre crops: cotton flax, juste, sisal and hemp, are produced in tropical and subtropical area. Natural fibres are under strong competition from synthetic products derived from petrochemicals, but cotton lint, by far the most important in terms of volume, more than held its own through the 1980s in contrast with hemp, jute and sisal which all lost ground to synthetics.

Cotton lint production rose from 14.4 million to 18 million tonnes in all regions with Asia showing the largest increase, from 5.5 million to 8.2 million tonnes, principally due to the massive upsurge in Chinese production already noted in respect of cotton seed. Egypt, a traditional producer of long-staple fibre registered a decline in the

1980s from 500,000 tonnes of lint, whereas neighbouring Sudan which also produces long-staple cotton under irrigated conditions lifted its production, albeit spasmodically, from an estimated 115,000 tonnes to 175,000 tonnes in 1987, a level that receded to 130,000 tonnes the following year. This unevenness illustrates the impact of irregularity in supply of inputs such as seed, fertiliser and insecticide on comparatively intensive production systems. The Sudan case also demonstrates the influence of the development of resistant strains of insect pests on crop yield. Build-up of white-fly strains resistant to established chemical control measures caused serious problems in the Sudan cotton industry in the early 1980s. Acceptable solution have been complicated by the high mammalian toxicity and related adverse environmental effects of new chemicals introduced to control the white-fly population. Such phenomena need constantly to be borne in mind in the design of pest management components of intensified cropping systems.

Distribution of *flax* worldwide is dominated by the Russia, with nearly 75 per cent of area harvested, followed by Europe. Led by France with Romania, Poland and Czechoslovakia the principal producers, at about 18 per cent. France leads in productivity with yields of 1300 kg/ha, at least twice those of the other major European producers, and almost four times the average levels in the Russia. Flax production trended upwards in both the Russia and Europe through the 1980s from 625,000 tonnes to 715,000 tonnes.

Hemp fibre is principally a product of China and India which share 40 per cent of area and over 50 per cent of global production, of around 220,000 tonnes. The crop is also grown extensively in the Russia and Romania, but yields in the Russia particularly, around 230 kg/ha, barely reach 20 per cent of those of the irrigated Chinese crop. The 1980s saw a 30 per cent reduction in area planted to hemp and a 10 per cent reduction in production, the domination being partially compensated by a market rise in productivity in China.

Jute, like hemp, is principally an Asian crop. Bangladesh, China, India and Thailand are together responsible for 90 per cent of the world's area and production of 2 million hectares and over 3 million tonnes respectively. Both declined in the 1980s with some compensation afforded by a global yield increase from 1,300 kg/ha to 1,500 kg/ha.

Sisal is grown principally in Brazil, Mexico, Tanzania and Kenya. Brazil produces over half the world's sisal, amounting to 370,000 tonnes, from about half the 575,000 hectares harvested. As in Mexico, yield in Brazil and Tanzania at around 600 kg/ha in the 1980s remained consistently below the estimated 1,100 kg/ha achieved in Kenya which combines good growing condition with effective plantation management of the crop.

Rubber

Natural rubber, although originating in Brazil, is grown almost exclusively in Asia. Its 93 per cent share of world production, of 4.7 millions tonnes, is shared largely between Malaysia, Indonesia and Thailand, all of which have major planting or replanting schemes of high-yielding clonal material for plantations and/or smallholders. Funding for these developments is by the governments concerned with assistance from international and regional development agencies. These investments, and others by lesser Asian producers, many of which were initiated in late 1960s, are reflected in the 23 per cent production increase registered in the 1980s decade. Rubber grown elsewhere, principally in Nigeria, Cote d'Ivoire, Cameroon and Brazil, tends to be less productive, either because it is less well established or managed, or susceptible to such diseases as South American leaf disease, as in Brazil. Given amelioration of such factors, there appears to be scope for expansion of the crop in these areas to meet domestic demand and to establish a modest share in export markets for the product.

Other Crops

FAO collects and publishes statistics on a wide range of other crops, including 15 vegetables, 19 fruits, 6 nuts, grapes, wine, hops and tobacco, which cannot be detailed here. However, they all form parts of production systems that vary in importance between and within different countries and regions. Indeed, given appropriate growing condition, farming and entrepreneurial skills, and market opportunities, most of these crops provide excellent returns on the capital and energy invested in them.

Major Livestock and Livestock Products

The principal domestic livestock may conveniently be divided into ruminants and non-ruminants, with cattle, sheep and goats dominating the former and pigs, poultry and equines largely constituting the latter groups. Principle animal products can be divided into meat, dairy, eggs and wool. This typology is followed in succeeding paragraphs after consideration of some important issues common to the livestock sector.

Livestock are complementary to many farming systems by providing draught power, converting crop residues into high-value meat or other products, and simultaneously producing organic manure to enhance subsequent crop production. However, in some parts of the world livestock compete with crops for scarce land resources. In such situations livestock population tends to exceed carrying capacity leading to a vicious spiral of declining productivity of land and livestock, environmental degradation, increasing poverty and social dislocation. Many parts of the arid and semi-arid zoned of Africa, and parts of Asia and the Middle East are affected by this set of circumstances. On the other hand, given appropriate management, livestock can utlise the productivity of land unsuitable for cropping, thereby reducing pressure on area of higher agricultural potential, as in the hill farm areas of the UK.

Livestock generally are inefficient converters of vegetable matter into energy and protein for human consumption. Their products therefore are more expensive per unit of nutrient than those from crops, though certain amino acids and vitamins, for example in the B_{12} complex, of key importance in human nutrition can be most economically obtained from animal products. It is also within the realms of possibility that widespread intensive livestock production systems will be developed based on the conversion of petro-chemical and like products by bacteria, yeasts and other micro-organisms into forms suitable for livestock feed and hence into human food. Such developments could therefore reduce the power of the 'inefficient converter' argument. Finally, there is an increasing body of evidence from animal geneticists and others that major breakthroughs in livestock productivity, utilising genetic manipulation, ova transplantation and similar techniques will be commonplace in the twenty-first century. These developments, together with more efficient feeding and management systems, can be expected increasingly to assist livestock to meet the future nutritional needs and dietary preference of mankind in an economic and sustainable manner.

Ruminants

Cattle stocks worldwide were estimated to have increased in the 1980s by some 50 million to 1.25 billion head. Perhaps significantly, cattle numbers declined only in North Central America, Europe and Oceania, regions predominated by the world's largest industrialised economies which are characterised by their highly-productive but comparatively high-cost livestock production systems. Africa, South America, the Russia and Asia, with 75 per cent of the world cattle numbers, kept largely under low-cost extensive conditions, expanded their herds by 8 per cent to over 940 million in this period. This has important implications, since these regions contain significant areas suffering from scarcity of crop land as a result of burgeoning populations, that could well lead increasingly to the situation described at the beginning of this section.

Buffaloes and Camels are valuable draught animals and sources of milk and meat in their respective habitats. Each increased in number in the 1980s, from 124 million to 137 million, and from 17 million to 19 million respectively. Their contrasting distribution in comparatively fertile but increasingly populous of climatically harsh and arid regions, also given cause for concern as to the extent to which these environments can continue to sustain them and their owners.

Sheep numbers worldwide remained roughly constant in the 1980s at 1.2 million. Asia maintained its dominant share of about 25 per cent followed by Oceania (dominated by Australia and New Zealand) with 229 million (19%). Africa, at 200 million, and Europe and the Russia, each with 140 million, also contributes heavily to the world's sheep population. As for other livestock types, the generic term covers a wide range of breeds each selected to make best use of the environmental conditions and consumer preferences in its native locality. Successful transfer and adaptation of breeds to foreign localities is exemplified by the case of the Merino in Australia, a breed that has changed substantially from the original breeding stock imported from

Europe in the nineteenth century. Unfortunately, no comparable productivity increases have occurred in the sheep of Africa or Asia that account for over 40 per cent of the world's sheep population.

This observation applies more broadly to the majority of the world's *goats*. Increasing from 465 to 520 million in the 1980s, especially in Asia and Africa where nearly 90 per cent are located, the species demonstrates a notable capacity to survive and multiply on crop residues and sparse natural vegetation. It is a valuable but irregular source of animal protein to the world's poor, that has seen few serious efforts to improve its genetic capabilities or management system in regions where it is most common. The goat is commonly excoriated for its adverse environmental impact by grazing and browsing habits that can involve terminal weakening of plants and soil structural damage by trampling that leads to wind and water erosion. While there is truth in such claims, goats play an important role in the economic lives of many millions of people that cannot currently be filled by other livestock species. In that respect, the comparative lack of focus on goats in research programmes is to be regretted.

Non-ruminants

Global *pig* stocks increased in the 1980s from 780 million to 823 million head with all regions except North America (heavily influenced by the USA) showing an expanding trend. The pig population in the USA declined steadily from 64 million to 43 million compared with an equally steady increase in Europe, from 173 million to 190 million over the decade. These different trends from regions having essentially similar intensive, industrial-style production systems and dietary habits are more sharply contrasted in terms of slaughter statistics. Thus, in the USA, numbers slaughtered from 93 million to 89 million (4%) compared with a 32 per cent drop in herd numbers. In Europe slaughtering increased by 9 per cent, from 237 million to 259 million in line with the herd increase over the period. Elsewhere, notably in Asia with a pig population increasing by 39 per cent to 505 million, of which some 83 per cent are in China, slaughterings fell behind herd expansion showing no more than a 10 per cent growth over the decade. These figures are a reflection of growth in human population demanding a larger volume of pigmeat and generally less intensive pig production systems, which require longer to bring pigs to preferred slaughter weight, operated in Asia compared to Europe.

In many of the world's industrial market economies of Europe and North America pig production systems based on precise genetic tailoring of desired slaughter types together with equally precise and economically efficient feeding practices are now the norm. These compare with use of the pig as a converter of crop residues and kitchen waste in the backyard systems of much of Asia. Africa with yet more extensive systems, in which pigs are kept as relatively uncontrolled scavengers, typically illustrates the opposite extreme from those practised in industrial market economies.

The popularity of pigs as sources of meat in the world's richest countries could well be enhanced by concerns over the impact of methane production of ruminants on damage to the ozone layer, and hence on global warming. The scientific debate is likely to continue for some years, fuelled by the cattle and sheep lobby. On the other hand, as already noted, and beyond any doubt, parts of Europe are already suffering environmental degradation from inadequate handling of waste products (no doubt containing methane and other toxic products) from intensive pig production systems.

Poultry numbers, especially in developing countries, are subject to even wider margins of tolerance than other livestock. FAO estimated a world population of over 10 million that increased by 3 million in the 1980s. Asia, with 4 billion, and Africa together account for almost half the world's poultry, followed by North America with about 20 per cent, and Europe and the Russia with 13 per cent and 11 per cent respectively. All regions have registered increased numbers, but Asia with an 83 per cent increase, strongly influenced by China where poultry numbers (generally accurately recorded) rose from 860 million to 1850 million and North America with a 39 per cent expansion, are clearly the dominant influence.

The poultry industry displays many similarities to that for pigs. At one end of the spectrum are the highly-productive, intensive, scientifically-designed and industrially-operated broiler systems that dominate Europe and North America but which have also spread widely into South America and parts of Asia and Africa. At the other extreme are the backyard scavengers almost totally unmanaged, in terms of control of breeding, diets, pests, or diseases that are irregularly and occasionally slaughtered, and whose eggs are rarely collected from human consumption. In common with other intensive livestock production systems, those for poultry demonstrate potential environmental and human health hazards. Salmonella infestation in eggs is one of the latter that pre-occupied UK poultry producers as well as consumers in the late 1980s. Fuelled by opportunist politicians, animal rights activists and supporters of the 'green' movement, issues of this kind are likely to continue to be vigorously pursed in the wealthy economies. While some may argue their irrelevance to the overall picture, such instances provide pointers to the partially obscured costs of intensive production systems that need to be carefully taken into account in their design and operation.

Equines, comprising horses, mules and asses, all increased slightly in the 1980s from 64 million to 65 million horses, from 14 million to 15.5 million mules, and 38 million to 41.5 million asses. Such increases in the world's poorer countries are a reflection of population growth and the increased demand for draught power on farms and for rural transport. Countries with policies and the means to mechanise their agriculture, such as China, Turkey and Poland, registered declining numbers of equines. The UK and USA each showed a reverse trend with horse numbers increasing from 147,000 to 189,000 and from 10 million to 11 million respectively. This phenomenon, attributable to increasing wealth and the use of horses for recreational purposes, is one that deserves consideration by farmers with ready access to urban markets for the service.

Animal Products

Meat and meat products constitute an important part of human diets with consumption increasing roughly in line with disposable income. Table 3.18 illustrates movements in average numbers slaughtered and meat production from the principal animal sources worldwide in the 1980s.

Within the species, pork is clearly the most popular form of animal protein but poultry meat showed a 39 per cent production increase over the decade compared with 25 per cent for pork, and 11 per cent for beef, the world's second largest meat source. Growth rates depend on a number of factors of differing importance in different areas but production costs and dietary preference are of obvious significance. Increased preoccupation in Europe and the USA with environmental questions, animal rights and the health implications of different types of meat in human diets can be expected to exert even greater future influence in these area than in those with less disposable income and lower meat consumption levels.

Asia, as the world's most population region, produces and consumes more meat, especially pork, and at a more rapidly expanding rate, than elsewhere. Thus, in the 1980s, meat production expanded by 56 per cent in Asia from 26.9 million to 41.9 million tonnes. Europe, the world's largest meat producer at 42.9 million tonnes in 1988, increased production by barely 10 per cent in the same period reflecting the more stable population size, the relative saturation of the domestic meat market and the limited export opportunities for the product.

Table 3.18: Global annual average slaughtering and meat production

Source	Animals slaughtered (millions)		Meat production (million tonnes)	
	1979-81	1988	1979-81	1988
Beef and veal	223.3	236.5	44.1	48.1
Mutton and lamb	385.2	442.0	5.7	6.6
Goat	159.9	197.5	1.8	2.4
Pork	735.8	870.5	51.5	64.4
Poultry	–	–	26.5	36.9
Other	–	–	4.3	4.4
Total meat production	–	—	133.9	136.5

Dairy products from cattle, comprising milk and its derivatives, including butter and cheese, are obtained from a global dairy herd estimated at 220 million in 1988. Dairy cow numbers are greatest in Asia with 54 million, followed by Europe with 46 million. Productivity is greatest in the USA, which averages almost 6,500 kg per cow annually compared with Europe at 3,750 kg, Asia at 900 kg, and Africa at 480 kg.

Nationally, the range is from Israel with 8,500 kg compared with the UK at around 5,000 kg and Ghana at 55 kg. Each reflects the nature of the production systems employed. Thus in Israel with strictly limited land resources which are heavily dependent on irrigation, it is based on high-cost, capital-intensive practices. Milk production in the UK employs a mixture of these techniques with lower-cost grazing and related feed sources in a large national herd with a broader range of genetic potential. In the Ghana case, as the figures suggest, recorded milk production represents the minor surplus beyond that utilised by suckling calves, with little or no attention paid to systematic breeding or feeding regimes.

Dairy production in the EC and USA is characterised by a complex web of subsidies and other mechanisms designed to influence production. The productivity of the dairy industry in these areas has resulted in periodic bouts of over-production leading to the build-up of stock that have been disposed of by give-away schemes to low-income groups locally, and as food-aid shipments to needy countries elsewhere. While the political logic underlying the activities is quite clear, their economic underpinning is less secure, as mentioned earlier. However, the development of the Indian dairy industry and others owes something to the free provision of milk power as food aid to the government agencies concerned, on condition that proceeds from the distribution of the reconstituted product were utilised to promote the domestic dairy industry, for which financial and technical assistance were also provided.

Dairy products from other species such as goats and sheep are traditionally important in diets in some developing countries. They also fill specialised markets in the world's richest economies where the output from dairy goat breeds, particularly, adds to the wide variety of bovine dairy products available.

Egg production in the 1980s rose from an estimated 27 million to 35 million tonnes worldwide. Relative growth rates were greatest in Asia, from 7.5 to 13.2 million tonnes, and only Oceania, influences heavily by Australia and New Zealand, registered a contraction in the volume of eggs produced. As noted earlier, reasons for the expansion can be attributed to wide adoption of intensive production systems that are sufficiently flexible for their less sophisticated versions to be employed successfully in developing countries having less reliable supplies and support resources, and lesser management skills than those available in regions such as Europe or the USA. The poultry industry in general, and egg production in particular, provides a good example of the successful transfer and modification of modern agricultural technology from developed to developing countries, a process that has by no means been universally successful for all technologies. Reasons for successful transfer include the fact that the technology is relatively scale-neutral, in that it is capable of rapid expansion in line with market demand, given the availability of capital and increasing management skills.

Table 3.19: Global sheep numbers and greasy wool production 1988

	Numbers (millions)	Greasy wool production ('000 t)
World	1173	3123
Africa	200	224
North Central America	19	46
South America	110	310
Asia	332	486
Europe	142	311
Oceania	229	1270
Russia	141	476

Wool, amounting to over 3 million tonnes of the greasy (unwashed) product, is an important output from the world's 1.2 million sheep. Wool-bearing sheep are confined mainly to the temperate zones with tropical regions characterised by a hairy sheep population adapted to the higher temperatures and other conditions that militated against the development of wool production characteristic in the course of ovine evolution. Table 3.19, which shows sheep numbers and volume of wool production by major geographic region in 1988, illustrates this point and the successful colonisation of Oceania by wool-bearing sheep since their introduction over 200 years ago.

The table points out the importance and high productivity of Oceania, essentially Australia and New Zealand, in wool production, and the comparative insignificance of the product in North Central America. Whereas sheep numbers and wool production have increased substantially in Oceania over the years, they have declined in North Central America influenced largely by a continuing fall in sheep numbers in the USA, from over 20 million in 1960 to 10 million in 1988. The phenomenon is due in part to a decline in competitiveness with cattle, in economic terms, exacerbated by government policy provisions favouring cattle relative to sheep.

Land Use

Mankind's use of the land resource can be divided into half a dozen overlapping categories. Table 3.1 shows five divisions which can be used to define land usage in order to provide the bases from which to present an account of the world's major agricultural production systems. The relative importance of the categories described briefly below will vary between countries and regions in direct relation to population density and the level of economic activity in the country or region concerned.

Thus land may be used for urban purposes to provide housing, factories, roads, and sites for other services for urban-based economic activities. Urban land is subsumed under the category 'Other' in Table 3.1. So also is land used for *recreational* purposes, principally by urban dwellers. This may include urban parks, sports grounds and the like, as well as national parks, wholly or partially closed to agriculture, and footpaths and bridle-ways that frequently pass through land that is being actively farmed.

Forestry is a third major land-use category, sometimes created by tree planting, or as existing, albeit declining, natural tree cover that represents the ecological climax or sub-climax vegetation of the region concerned. Overlaps include forest land used for recreational purposes as in national parks, for watershed protection in hilly areas, for extensive livestock raising, as in the declining forest savannah in Africa, and for sustaining the world's few remaining populations of forest-dwelling hunters and gathers, principally in the equatorial forest areas of South America, Asia and Africa.

Permanent pasture, a fourth major class of land use principally for extensive livestock production, may be climax vegetation in the world's more arid areas, or derived from original forest cover following clearing or burning by indigenous hunters and/or transhumant populations. Complementary and increasingly competing usage can include nature reserves for tourism, hunting, or species protection (as in sub-Saharan Africa) and encroaching crop farmers. Permanent pasture generally is not subject to close agricultural management practice. Vegetative cover is normally naturally-occurring; artificial fertilisers are not usually applied; and productivity is more a function of nature's bounty that man's ingenuity.

Permanent crops, the fifth category of land use, are examples of relatively intensive management practice. Thus, for example, oil palm, coffee, cocoa, tea, coconuts and sugar cane each demand particular skills for their establishment, maintenance and harvest. Overlapping usage is limited but can include watershed protection, intercropping with annuals and, occasionally, livestock production concurrently as with coconuts in some Asian countries.

Arable crops, the sixth category, include land used for annuals or short-term plants, such as leys, to produce human food directly or via livestock kept under relatively intensive management conditions. It is the land-use category upon which mankind directly depends for the bulk of its sustenance. Because of its specialised nature there is little overlap with other forms of usage. However, in the world's richest economies, recreation for urban populations is becoming more important.

Succeeding paragraphs examine some of the major agricultural production systems employed in world agriculture. Scope for variations within each is considerable because of the interplay between overlapping biological, economic, social and environmental factors as they affect the individual farmers concerned.

Biological factors include the limitations imposed by soil, climate and topography on choice of crop of animal species. Incidence of pests and/or diseases are other biological factors which, until the advent of cheap and effective control measure, whether through use of resistant genetic material or chemicals, frequently governed, and often still limit, crop and livestock cycles on farm land.

Economic factors such as cost of production, and returns to labour and land also play a key role in determining choice of production system. Their interaction with biological factors is exemplified by, say, the effect of soil-borne disease on crop yield. While control measures may exist, the extent of their utilisation, or indeed the decision whether or not to grow the crop in question, will depend on the anticipated return from continuing to produce it with or without the disease control measures. Crop rotations offer good examples for the interplay between biological and economic factors. Thus the benefits of sheep in the Norfolk four-course rotation included their manure and the consolidating effect of them being folded on light land, both of which served to improve yields, and thus economic returns, from succeeding crops.

Social factors apply most in less-well-developed countries. However, consumer dietary or other preference for particular crops or livestock apply across the board. Other preferences expressed by subsistence farmers may involve traditions associated with male or female responsibilities for production operations, such as land clearing (usually by mean), weeding (usually by women), and so on. The first rice varieties introduced to Asia at the beginning of the 'green revolution' prove a good example of social and economic factors affecting the uptake of higher yielding material. In many areas the variety IR8 was not readily acceptable, partly because of perceived inferior grain quality to those transitionally grown, and partly because the dwarf habit and higher silica content in the straw needed to reduce lodging, reduced the volume, feeding value and palatability to the work oxen and buffaloes which were dependent on straw for a significant part of their feed. Slight changes in food preparation practices and some fine tuning in varietal straw and grain characteristics to take account of these economic and social factors, resulted in ready acceptance of the new planting material and widespread changes in planting preference from traditional to modern varieties.

The latter part of the twentieth century has seen a marked increase in the importance of economic factors on crop and livestock production systems in developed and some middle-income developing countries. This has arisen largely because of the substatinal increase in the range of biological tools available to farmers, and to their enhanced economic awareness and preoccupation with profit maximisation. Government manipulation of prices in such countries, as mentioned earlier, has resulted, for example, in the large upsurge of oilseed rape in many traditional grain areas of Europe. The Brazilian soya bean/wheat explosion is another case. However, the closing decade of the century saw another factor limiting economic considerations in production systems.

In the developed countries there will continue to be dependence on specialised *capital-intensive crop production* systems based on a handful of annual cereals, oilseeds, pulses, vegetables, or fodder crops, grown in flexible sequences for human consumption directly or industry through livestock. The components of these systems will continue to be dictated by biological, social and economic considerations but with rather more account taken of environmental facts than was the case in the 1980s. Pressure on land and profits will inevitably require continued dependence on intensive livestock production systems based on high-yielding fodder crops, protein-rich supplements and other dietary components designed to utilise fully the genetic potential of the animals concerned, whatever the species. Fertility enhancement and biological manipulations to increase inherent productivity are likely to be subject to close scrutiny because of possible adverse effects on consumers.

In developing countries *rain-fed production* systems also tend to be based on a narrow species range, reflecting needs and market opportunities. Most are labour-rather than capital-intensive; some rely heavily on the land's inherent fertility with perhaps a temporary boost derived from clearing and burning of existing vegetation. Dependence on organic manure varies, but chemical fertiliser use is limited. Where individual land-holding rights are comparatively tenuous, land improvement incentives are weak. System, variously known as 'shifting cultivation', 'slash and burn', or 'bush falling' are typically practised in Africa. They are breaking down rapidly as population pressure on available land increases, leading to rapid reduction of the bush fallow, from 25 or 30 years frequently to no more than 3 or 4 years between cropping sequences. Fertility regeneration suffers in consequence, leading to a fast-moving downward land productivity spiral, soil degradation and increasing poverty.

System practised are frequently complex with risk aversion high on the agenda. In many areas, mixed cropping is the norm as an insurance against drought, flood, pests and diseases, and to spread peak demands for labour through the farming year. Thus in the Northern Guinea Zone in Nigeria, millet may be planted in anticipation of the first rains. Shortly after their onset and germination of the millet, sorghum is interplanted, followed thereafter by cowpeas, and/or groundnuts also interplanted, to obtain optimal ground cover, avoid overshading, and maximise utilisation of sparse and erratic rainfall that rarely exceeds 700 mm per annum over a period of 120 days. In rainforest areas, equally if not more complex, cropping systems are followed in order to utilise inherent soil fertility, nutrient recycling by leaf fall, and available sunlight. Exemplified in the Kande gardens of Sri Lanka these multi-storey systems appear commonly in similar ecological conditions elsewhere in Asia, as well as in Africa. Unlike the slash and burn systems, the Kande garden is largely self-sustaining, provided population pressure does not reduce land availability below critical levels for particular localities.

Irrigation in temperate and sub-tropical countries has proven increasingly popular as a means of safeguarding crop water requirements and enhancing yield. Large-scale sprinkler equipment dependent on pumping that minimises investment in land levelling, albeit at higher recurrent cost than gravity schemes, is a feature of many production systems on light lands in the EC and the USA. (However, much of the irrigated land in the southwestern states of the USA is irrigated by gravity.) In arid and semi-arid areas the timely provision of irrigation water frequently makes the difference between acceptable returns and crop failure. Economic returns are greatest from high-value crops such as fruits and vegetables, as in Israel which employs sophisticated water-efficient drip irrigation techniques. In oil-rich Arabian countries irrigation of cereals and fodder crops of low intrinsic economic value is common-place, thanks to heavy government subsidies on petro-chemical incomes, and justified on national pre-occupations with self-sufficiency in cereal and dairy products.

Irrigated cropping systems are most widespread in Asia with over 60 per cent of the world's total of 227 million hectares under irrigation. Frequently based on rice and/or wheat, many Asian irrigated cropping systems seek to produce two or more crops per calendar year with rice planted to take advantage of the rainy season followed by another, quick-maturing crop, wheat, or perhaps cotton, and if time allows, a short-season pulse or oilseed such as chick peas, gram or groundnuts to round off the cycle. With the advent of productive, early-maturing cereal varieties, cropping intensity is largely governed by the availability and control of water supplies. The latter can be a serious limitant on many of the state-controlled gravity schemes that abound in China, India, Pakistan, Indonesia and the Philippines, which all tend to suffer from inadequate maintenance and less than fully-competent management, particularly at the lower levels. These problems are less acute in the many thousands of privately or publicly owned tube wells that rely on pumping from underground water supplies. Such schemes are dependent, however, on efficient pump operation. Further, in many areas, inadequate attention to drainage in the design of irrigation schemes generally, together with over-watering, has led to major problems of salinisation and loss of irrigable area pending expensive de-salting operation. Most Asian irrigated cropping systems are based on large numbers of small farmers each responsible for one or two hectares, cultivating by hand or animal draught and largely dependent upon family labour. In Europe and the Americas the converse tends to happen. Holdings are much large (50 hectares and up), production is by capital-intensive means with labour demand generally seasonal for harvesting high-value fruit or vegetable crops.

Perennial crop production systems are concerned primarily with orchard crops in temperate and sub-tropical regions, and with other economic tree crops such as oil palm, coconuts, beverages and rubber, as well as sugar cane, predominantly in the tropics. They are characterised by their relatively long duration, up to 40 years of economic life, for some orchard and tree crops. They normally only involve one species grown for its economic return, plus one or two others grown in association to provide ground cover or shade. Organisation may be based on private or publicly owned plantations in contiguous blocks often several hundred hectares in extent. Alternatively, it may consist of a group of smallholders each with a food crop farm but also tending plots established round a nucleus plantation by a company that provides them with production, marketing and processing services. Both styles are favoured for oil palm and rubber, particularly in Malaysia and Indonesia. But other crops are also produced similarly: thus coffee and coca plantations are common in Brazil, as are tea plantations in India. Kenya's tea industry, however, is based largely on smallholders serviced by KTDA as noted earlier. In this case, as also commonly in other African countries, smallholders operate separate small-scale crop and livestock production enterprises to cover subsistence needs. This is typical with coffee and cocoa in much of Africa with primary processing and marketing either carried out by individual farmers or at scattered facilities operated, frequently inefficiently, by co-operatives or government agencies. Productivity and incomes tend to suffer because of inadequate services to provide inputs, extension advice or marketing. Further, because the tree crops are a charge on family resources, they tend to come second to subsistence production in the allocation of labour and purchased inputs.

Mixed Crop and Livestock Systems

Many crop production systems incorporate livestock for draught and to utilise by-products such as straw to produce organic manure, and to that extent they are mixed. Conversely, some livestock production systems are designed round home-grown feed and fodder crops, as for example the small dairy farms that were a feature of Pennines' agriculture in the mid-twentieth century.

Perhaps the ultimate mixed system is the association developed in Asia between rice cultivation, duck rearing and fish farming. In this case, ducks eat weeds and insect pests and manure rice paddies that also serve as sites for raising carp, and eggs are sold as a 'cash crop'. Less exotic and more venerable is the interdependence between sheep, roots and grain crops upon which the Norfolk four-course rotation was built, and from which many variations have evolved. The symbiotic relationship between the transhumant Fulani and hoe farmers of West Africa is yet another example. In this case, depending on circumstance, farmers either pay the Fulani, or the Fulani pay the farmers, to fold their livestock on crop land after harvest. They consume the surplus millet straw, cowpea stover, and groundnut haulms, manuring the land and treading in the material that is not eaten, to the benefit of both parties. Unfortunately, pressure on land in these areas is rapidly resulting in the system's breakdown as grazers encroach on crop land during the growing season and farmers clear and plant land formerly used for grazing.

Livestock Production Systems

Notwithstanding the overlaps that exist between crop and livestock production systems there remain several of the latter that are comparatively free-standing. They exist in different forms in low-, middle- and high-income economies and in different climatic zones in response to such factors as land-use capability and market demand. Some are very land-extensive, other are essentially industrial in that land is only needed to provide housing and ancillary structures.

Ruminant livestock with equines are of enormous economic value because of their ability to convert natural vegetation, as well as planted feed and fodder crops, into products and power for use by mankind. This capability is utlised in a number of systems. Perhaps the most extensive is that characterised by *transhumance or nomadism*. Evolving in areas climatically unsuited to sustained crop farming, generally because of inadequate precipitation, transhumance is commonest in Africa and Asia. Grazing and water are the factors governing the movement of animals which typically may include cattle, sheep and goats, with horses and/or camels. Most transhumant have traditional grazing rights shared between family, clan and tribal units. Water rights are jealously preserves and guarded, with water points, wells and oases are clearly owned by particular groups. Livestock are the most tangible capital assets, sold to meet cash requirements, exchanged for food or presented as dowries. Stock numbers are maintained in line with herdsmen's estimates of the carrying capacity of available land and water supplies, taking account of erratic rainfall regimes, outbreaks of disease and other likely hazards, underlain by attempts to maximise their herd numbers and share of common resources at the expense of each other, to the detriment of all. Margins for error in these estimates have declined progressively in many areas as increasing populations have led to the spread of crop farmers into land hitherto only used for grazing. This phenomenon was thrown into sharp relief in Africa suring he drought years of the 1970s; it is also apparent in Asia, with the construction of major irrigation schemes in traditional grazing areas in north-west China for example, and in the Middle East where the plight of the Kurds exemplifies pressures on transhumant by settled populations in Iran and Iraq.

Past efforts by development agencies to assist governments to improve the production of their transhumant populations have been generally unsuccessful, mainly due to a lack of understanding of the interaction between the technical and social problems involved. It is now generally accepted that the solution lines in adjustments in traditional practice to improve productivity within the broad framework of evolving social organisation of the population concerned.

Ranching is another form of extensive livestock production on land unsuited to rain-fed crops. It differs from transhumance in that its practitioners operate in the monetary economy and have a permanent home base, normally located centrally within the land used for grazing by the ranch animals. It is a western concept based on ownership of land and water rights within clearly demarcated boundaries. While practised successfully in the Americas, Australia and parts of southern Africa, it has usually failed when transplanted elsewhere, primarily for social reasons. In recent years, ranching has complemented more intensive operations by producing young stock for fattening in feed-lots and subsequent slaughter. A great benefits of ranching is that land ownership provides the incentives and collateral to finance management practices for range improvement, such as aerial seeding and fertiliser (usually phosphate) application, controlled burning and additional water points, all of which lead to increased productivity.

Hill farming in Europe is another form of ranching usually practised on a smaller scale than in Australia or the Americas and most commonly with sheep, though cattle are locally important. Productivity of hill farms is typically limited by topography, soil type and rainfall, on land unsuited to annual crop production. Lambs and calves, typically produced from natural grazing, are sold for fattening as part of mixed crop and stock enterprise mentioned earlier on an industrial feed-lot. Land ownership and subsidy provisions over the years have provided incentives to invest in measures to enhance land and flock productivity, but profit margins are typically narrow and, overall, hill farmers tend to obtain slim returns on their investments.

More *intensive livestock* production systems based on grazing include those for meat and milk production that utilise planted mixed grass and legume lays, and break crops, such as kale and other fodder's offered to livestock of high productive potential. Typically of the developed countries having high land values and high-cost livestock, the systems are designed to optimise area and animal productivity. Division of fields into paddocks by electric fencing in sized selected to reduce unnecessary movement by grazing animals and rotation of carefully fertilised grazing plots to capture feed at its most nutritious, are among the management practices employed to this end. The intensity of such systems can most readily be gauged by the level of management attention given and the volume of investment made in the land on which they are applied.

The top of the intensity scale is represented by production systems that provide livestock with their feed *in situ*, whether as broiler chickens, housed pigs, feed-lot cattle or permanently-housed dairy herds. The three former can and often do function on the basis of 100 per cent bought-in feed and represent the ultimate in factory farming. This phenomenon is less common with dairy cattle with the herd frequently receiving the bulky part of its diet from home-grown fresh or preserved forage, topped up by high-nutrient supplements that may not be purchased off the farm. In all cases specialised and skilled management is a prerequisite for success. Skills needed include detailed knowledge of nutrient requirement, preventive animal health measures as well as the indefinable stockmanship qualities essential for animal well-being and good performance.

Conclusion

Lack of space has seriously limited the coverage given to the foregoing review of world agriculture. Nevertheless one message in clear if mankind is to be fed and clothed at anything approaching adequate standards. The inexorable increases in human population at least over the next five decades underlines the importance of agriculture as the industry most directly concerned with sustaining human life. Contemporaneous movements towards democratic forms of government will result in more freedom for individuals. Therefore, new, as never before, priority must be given to the design and operation of production systems that are efficient, economic, equitable and above all, environmentally-sustainable wherever they are applied. This is the challenge facing world agriculture today and in the foreseeable future.

❑❑❑

4

The Principles of Crop Production

Introduction

The primary objective of crop production is to obtain as large a quantity as possible of a desired product per unit of the most limiting resource, consistent with the further objective in cash economies of having the largest financial return. An important subsidiary objective of crop production should also be to produce crops in ways that are conducive to the maintenance of natural species and a visually appealing landscape. The desired product may be some part of a crop, such as a seed or tuber, or it may be most of he above-ground part of a crop, such as grass for animal feed. Where only part of a crop forms the main product, the less important fractions are known as by-products. The desired product does not have to be a food for man or feed for livestock, but could be a raw material for industry such as timber,,biomass for fuel, fibre for clothing or chemicals for the pharmaceutical industry. Where the desired product is a food, it must take into account the way in which the quality of the product in ways that may be, or perceived to be, injurious to health.

Output is expressed usually as the weight of some specified plant product per unit area of land, this is usually called the crop yield, and that this refers to a yield per unit area is often implicitly assumed. It is, of course, perfectly valid to discuss crop production in terms of yield per unit of some other resources, such as labour, water or energy, depending upon which factor is regarded as being the most limiting. The importance of water in many areas of the world, or of the contribution of fossil fuels to crop production in capital-intensive agricultural systems, serves to remind us of the importance of using the most appropriate ratios when discussing crop production.

In this chapter, discussion is centred on the more conventional use of the term yield as the weight of produce per unit area, partly because land is usually the most limiting resource, which is not unconnected with the fact that light, the ultimate limit to crop growth, is received by crops on an area basis.

The Determination of Yield per Unit Area

The yield of useful product, sometimes referred to as 'economic' or 'commercial' yield, is only a fraction of the total organic matter synthesised by the crop, the total yield of which is often called the 'biological' yield. Biological yield is difficult to estimate accurately as it must include the root system, and roots are every difficult to recover from the soil, where they may extend to depths of 1-2 m. For this reason, the terms 'recoverable biological yield' may be used, which is restricted to all the plant material contained in the above-ground portion of plants, together with any below-grounded storage organs such as tubers and storage roots.

Farmers also commonly refer to economic yield in terms of the fresh weight of the product, such as tonnes of tubers, or grain, or of turnips. This poses problems when discussing the biological principles determining yield, since the water content of these commodities may differ substantially. In seeds harvested at maturity, the water content is usually low, often between 12-16 per cent, while vegetables, tubers and fleshy fruits may contain 70-80 per cent water. It is therefore convenient when comparing the growth of different crops, or the effects of the environment on growth, to discuss yield in terms of dry weight.

All the components of a plant necessary for growth, development and maintenance can be derived from about 17 elements, of which carbon, hydrogen and oxygen, obtained from carbon dioxide and water, make up the bulk. The remaining elements are conventionally referred to as mineral elements and are normally absorbed by plant roots from the soil. The most important exception to this is nitrogen, which may be assimilated from the air by leguminous plants in symbiotic association with root nodule bacteria (*Rhizobium* spp.). Although plants may absorb many additional elements, only this relatively limited number have been proved to be essential. To be considered essential, the absence of the element must directly cause abnormal growth and premature death. The effect must be specific to the element in question and can be prevented or corrected only by supplying this element. The element must be directly involved in the mutrition of the plant, quite apart from its possible effects in correcting some unfavourable microbiological or chemical condition of the soil or other cultural medium (Wild and Jones 1988). The elements are required in vastly different quantities, and they are therefore conventionally subdivided into major (macro) or minor (trace) elements (Table 4.1).

Table 4.1: Concentrations of mineral nutrient elements in plant material at level considered adequate

Element	Concentration in dry matter	Relative number of atoms with respect to molybdenum
	(ppm)	
Molybdenum	0.1	1
Copper	6	100
Zinc	20	300
Manganese	50	1000
Iron	100	2000
Boron	20	2000
Chlorine	100	3000
	(%)	
Sulphur	0.1	30000
Phosphorus	0.2	60000
Magnesium	0.2	80000
Calcium	0.5	125000
Potassium	1.0	250000
Nitrogen	1.5	1000000

Source: Epstein.

Since animals are directly or ultimately dependent upon crops for their food supply, it should be noted that in addition to a common requirement for the essential elements listed in Table 4.1, animals require nickel, selenium, chromium, iodine, fluorine, tin, silicon, vanadium and arsenic.

The combination of carbon dioxide and water, which provides the starting point for all the compounds found in plants, require a source of energy, which is provided by visible light (380-750 mm). Light energy is absorbed by pigments known as chlorophyll, which give plants their green colour. Photosynthesis is the process by which plants synthesise organic compounds from the inorganic raw materials mentioned, using the energy from sunlight to fuel the reaction, as shown in the following equation:

$$6CO_2 + 6H_2O \rightarrow \text{light energy} \rightarrow C_6H_{12}O_6 + 6O_2$$

In this process, light energy is converted into chemical energy, in the form of carbohydrates, proteins and other plant constituents, and oxygen is evolved as a waste product. The manipulation of crops in order to maximise the utilisation of light dominates all other issues in crop production and consciously or unconsciously dictates both the way crops are grown and their genetic improvement.

The Relationship between Light Interception and Yield

The total amount of radiation intercepted by crops can be estimated by the use of solarimeters, instruments which convert the radiation which falls upon them into an electric current, the voltage of which can readily be measured. By placing a solarimeter above a crop and one at ground level beneath the crop canopy, the amount of radiation intercepted by the levels of the crop can be measured, either over a specific part of the life cycle of the crop, or over its entire life span. Radiation receipts are usually expressed in terms of joules per unit area, usually per m^2.

In Fig. 4.1 (a), the total dry matter yields of sugar beet, potatoes, barely and apples have been plotted against the total amount of radiation intercepted. It is clear that these very contrasting crops converted intercepted total radiation into dry matter at approximately the same rate, i.e., 1.4 g/MJ. Approximately 50 per cent of the incoming radiation is photosythetically active and if dry matter yield is plotted against photosythetically-active radiation (PAR), then the conversion efficiency will be approximately twice as large.

In Fig. 4.1 (b), the total dry matter yields of sugar beet and potato crops grown with different varieties and times of planting and in different years, have been plotted against the total amount of radiation intercepted. With the exception of 1990, which was a dry growing season, total dry matter yield is again shown to be closet related to intercepted radiation. from which it may be deduced that treatments were affecting yield mainly by affecting the size of the crop canopy and its ability to intercept radiation.

By using this sort of analysis of the growth of crops, it can be seen that the total dry matter yield of a crop will depend upon (a) the total amount of radiation intercepted by the crop canopy, and (b) the efficiency with which intercepted radiation is converted into dry matter. Since the farmer is not necessarily interested in total dry matter yield, but in some fraction of it, he will also be concerned with (c) the way dry matter is distributed between the various parts of a plant.

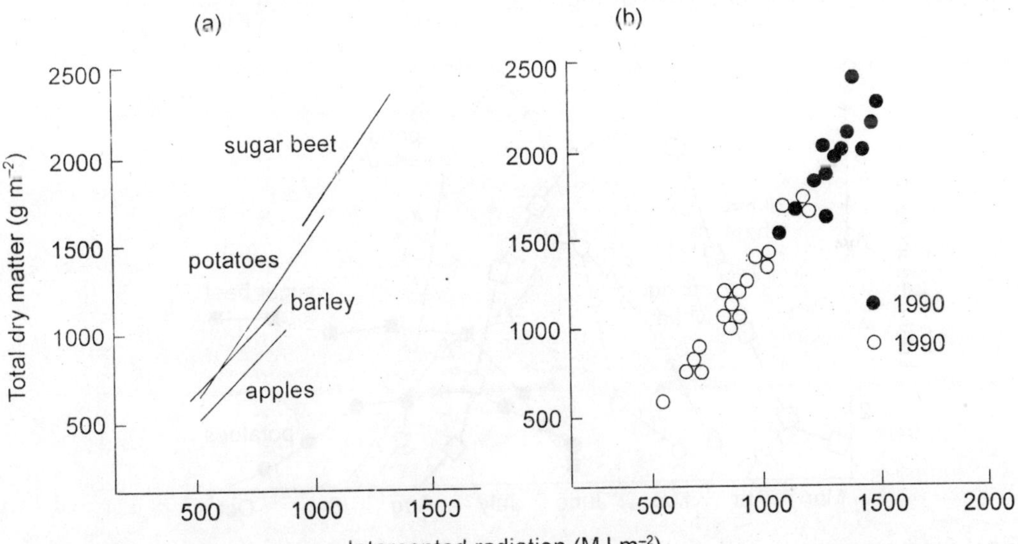

Fig. 4.1: Relation between total dry matter yield at harvest and radiation intercepted by foliage throughout the growing season. Closed symbols = sugar beet; open symbols = potatoes.

The Total Amount of Radiation Intercepted by Crops

This will depend upon the pattern of light receipts in a particular locality and season, which is less variable than most climatic parameters, and upon the development of the crop canopy—the above-ground light intercepting structures of the crop, principally the leaves. A useful measure of leaves, which is relevant to their ability to intercept light, is the leaf area index, *L*, defined as the leaf area per unit area of land. It expresses the number of complete layers of leaves displayed by the crop, although of course leaves do not form unbroken layers one above the other. Leaves occur in different shapes and sizes, and are held at different angles and may occur on long or short stems—all of which affect the 'structure' or 'architecture' of the crop canopy. In the cereal crop, leaf sheaths are included in *L*, and it should be borne in mind that other structures such as the ears of cereals, the pods of peas or the siliquas of oil-seed rape, are important photosynthetic organs. The typical development of *L* with time for various crops commonly grown in Britain is shown in Fig. 4.2. Under British conditions it would seem that values of *L* between 4-7 are required to intercept most of the incident radiation. A typical relationship between *L* and the proportion of the incident radiation intercepted by a crop is shown in Fig. 4.3.

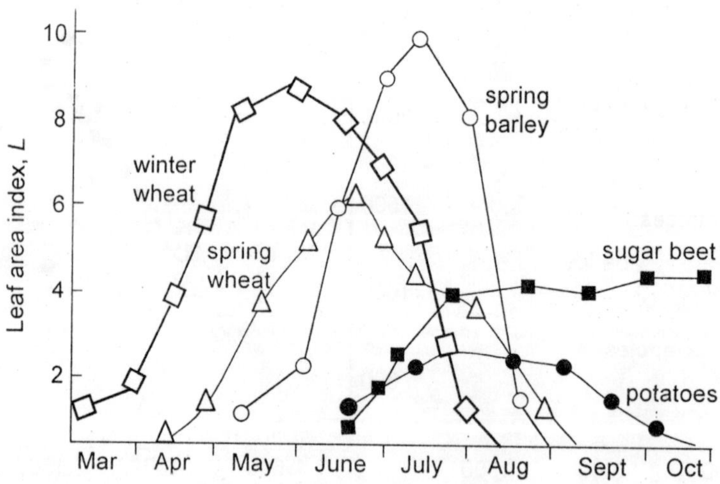

Fig. 4.2: Changes with time in the leaf index, *L*, of different crops grown in Britain. (Source: adapted from Watson).

Leaf area index is usually obtained by harvesting a known area of a crop, weighing the levels and, for a small sample of these obtaining the weight of a known area of leaves. From this can be calculated the area of leaves per unit leaf weight (the specific leaf area), which when multiplied by the total weight of leaves gives the area of levels

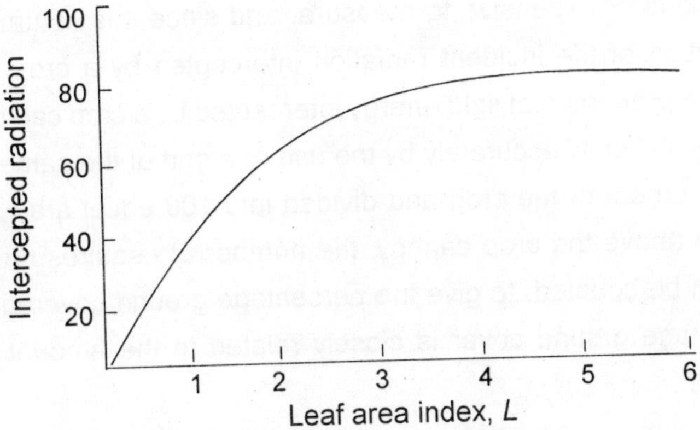

Fig. 4.3: Relation between intercepted radiation and leaf area index. (Source: Tostevin).

applicable to the harvested area. The ratio between these is the leaf area index, *L*. Dry weights are normally used, since fresh leaves are constantly losing water and changing weight during the measuring process. An instrument is now available which is capable of measuring leaf area index and leaf angle non-destructively, but it is expensive. In Fig. 4.4, the total light energy receipts, the development of *L* in a sugar beet crop and the resultant amount of light intercepted are shown to illustrate these relationships.

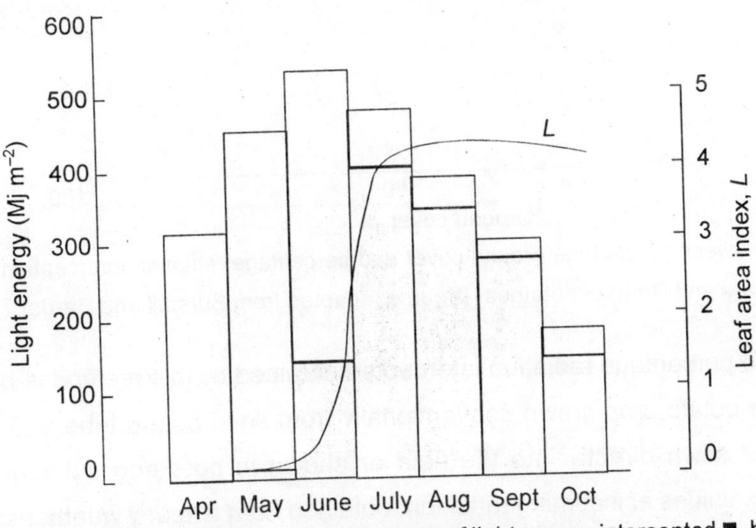

Fig. 4.4: Light energy receipts □ and the amount of light energy intercepted ■ by a sugar beet crop at Broom's Barn Experimental Station (Suffolk), and the development of leaf area index, *L*. (Source: Jaggard and Scott).

L is either laborious or expensive to measure, and since it is commonly used to estimate the proportion of the incident radiation intercepted by a crop canopy, it is useful to note that the proportion of light energy intercepted by a crop can be estimated cheaply, easily and sufficiently accurately by the use of a grid of the same dimensions as the spatial arrangement of the crop and divided into 100 equal areas. By placing the grid immediately above the crop canopy, the number of esquires more than half filled with leaves can be counted, to give the percentage ground covered by the crop canopy. The percentage ground cover is closely related to the amount of radiation intercepted (Fig. 4.5).

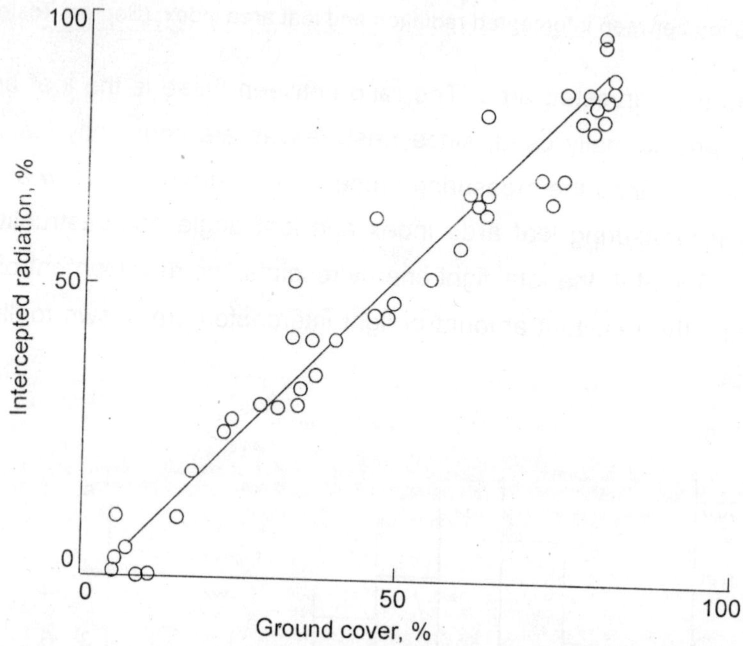

Fig. 4.5: Relation between percentage ground cover and percentage radiation interception in two varieties (King Edward and Cara) of potatoes. (Source: adapted from Burstall and Harris).

In Fig. 4.6, the percentage radiation intercepted obtained by this method is plotted against time for a potato crop grown conventionally from seed potato tubers, or from 'true' seed, either sown directly into the field or shown in pots and subsequently transplanted. Daily values of incoming radiation, obtained from a nearly weather station were converted into amounts of radiation intercepted by the crop canopy, using the data shown in Fig. 4.6.

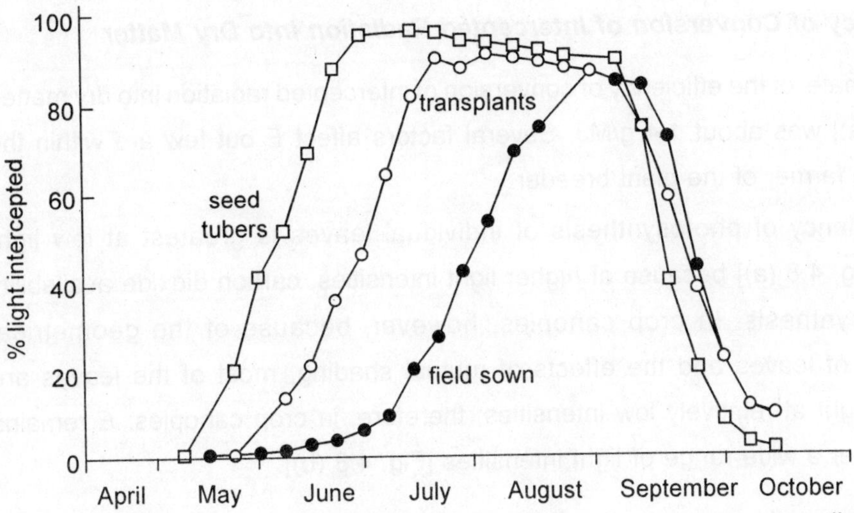

Fig. 4.6: The effect of method of propagating the potato crop on the percentage light interception. (Source: data in Shakya).

Values obtained for tuber dry matter yield at several harvests during he growing period have been plotted against the amount of radiation intercepted, and the results are shown in Fig. 4.7.

A knowledge of the temporal patterns of light energy receipts, the development of *L* (or percentage ground cover) and the resultant interception of radiation by plants is fundamental to an understanding of the biological basis of crop production.

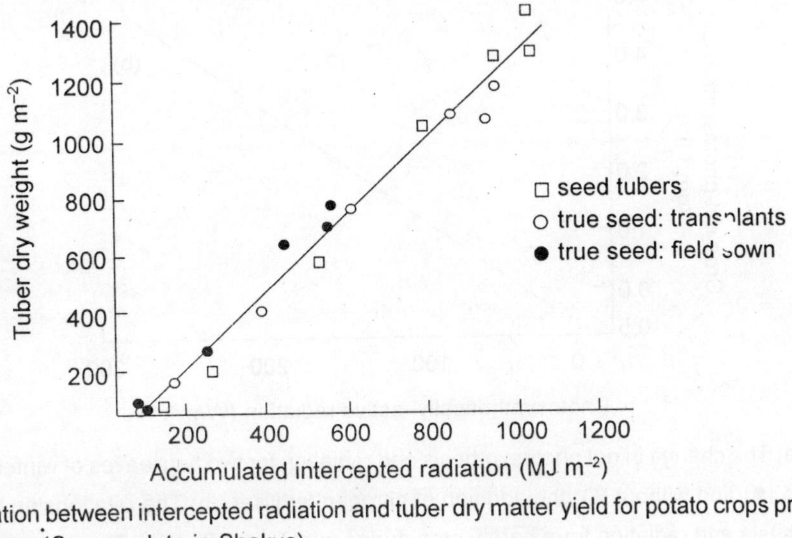

Fig. 4.7: Relation between intercepted radiation and tuber dry matter yield for potato crops propagated in various ways. (*Source*: data in Shakya).

The Efficiency of Conversion of Intercepted Radiation into Dry Matter

One estimate of the efficiency of conversion of intercepted radiation into dry matter, *E* [Fig. 4.1 (a)] was about 1.4 g/MJ. Several factors affect *E* but few are within the control or the farmer of the plant breeder.

The efficiency of photosynthesis of individual leaves is greatest at low light intensities [Fig. 4.8 (a)] because at higher light intensities, carbon dioxide availability limits photosynthesis. In crop canopies, however, because of the geometrical arrangement of leaves and the effects of mutual shading, most of the leaves are intercepting light at relatively low intensities; therefore, in crop canopies, *E* remains unaffected over a wide range of light intensities [Fig. 4.8 (b)].

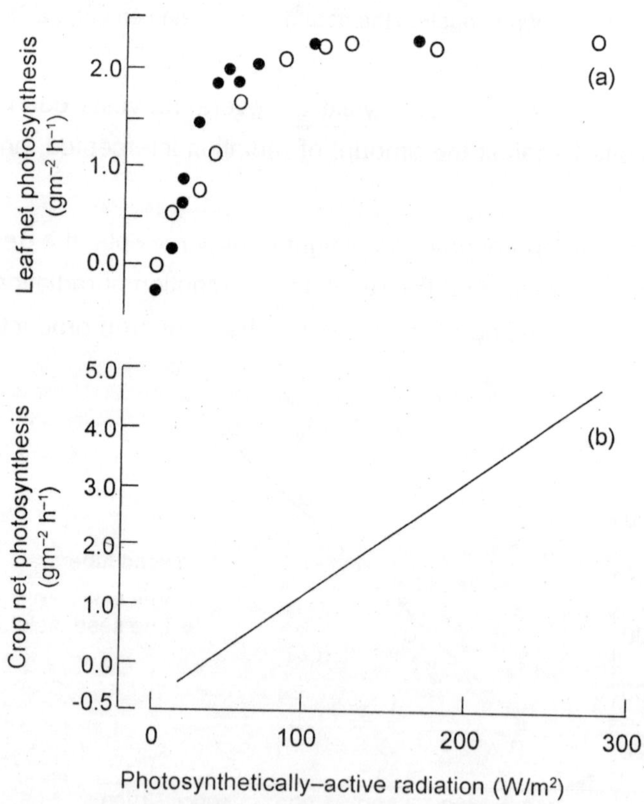

Fig. 4.8: (a) The change in net photosynthesis and radiation for two flag leaves of winter wheat from a crop with (●) and without (O) the addition of nitrogen fertiliser. (b) The relationship between net photosynthesis and radiation for a barely crop during mid-June. (*Source*: Biscoe and Gallagher).

There are important differences between species in the efficiency with which light is utilised. The most efficient group contains many of the high-yielding tropical and sub-tropical graminaceous species such as sugar cane, maize and sorghum. These are referred to as C_4 plants, as the first stable compounds formed during photosynthesis are acids containing 4 carbon atoms. The less photosythetically efficient plants are known as C_3 plants, as the comparable first stable compound contains 3 carbon atoms. A major difference between these groups is the greater efficiency with which C_4 plants utilise carbon dioxide. However, this greater efficiency is evident only at high temperatures and light intensities; therefore under temperate conditions these differences will be minimal and offset by the higher threshold temperatures required for the growth of C_4 plants, which restricts the period over which such plants can intercept radiation.

Plant breeding has had little or no effect on increasing *E* in plants and there is evidence that the photosynthetic rate has fallen during the evolution of wheat (Evans).

The photosynthetic efficiency of individual leaves tends to decline with age: *E* will therefore be affected by the ability of a crop to replace old leaves with new ones. For some crop species (e.g. wheat), this ability is limited by the termination of the leaves which can develop. This type of growth is described as 'determinate' and can be contrasted with crops such as peas and beans in which flowers develop in the axils of leaves and new leaves can be produced throughout the entire life-cycle: this is known as 'indeterminate' growth.

While carbon dioxide is being taken up and oxygen released in the process of photosynthesis, the reverse process, known as respiration, is taking place. *E* therefore represents the net effect of photosynthesis and respiration. Respiration accounts for a loss of some 30-50 per cent of the dry matter produced by photosynthesis and it is needed to release energy for the synthesis of new compounds required for growth and for the maintenance of the existing plant structure, the latter requirement increasing as the crop grows. Respiration rate is strongly affected by temperature, doubling for every 10°C increase: temperature, however, will only affect maintenance respiration as growth responds in a similar way to a rise in temperature.

Water stress is the most important environmental factor affecting *E* [see Fig. 4.1 (b)]. The perforations in leaves (stomata), through which carbon dioxide diffuses to reach the chloroplasts, close in response to water stress, reducing the supply of carbon dioxide and hence the rate of photosynthesis.

The amount of water lost by plants (transpiration) is related closely to the amount of radiation intercepted by them, hence transpiration is often closely related to yield. There may even be a closer relationship than that between intercepted radiation and yield, as the amount of transpiration takes into account the reduction of *E* associated with the build-up of water stress in crops (see Penman).

There is some evidence that the capacity of a crop to accept the products of photosynthesis, or the 'sink' capacity, may limit photosynthesis, as also may be ability of a plant to transport the products of photosynthesis from the 'source' (e.g. leaves) to the 'sink' (e.g. storage organ, such as grain, root or tuber). A limitation to cereal yield, for example, may be the number of grain sites in which assimilates can be deposited. The potential number of these sites, however, is determined before grain filling has started, primarily by how much radiation has been intercepted prior to anthesis. On the other hand, it has been shown that the photosynthetic capacity of potatoes is not limited by the size of the tuber 'sink', but is by the size of the photosynthetic 'source' of assimilates (Fig. 4.9).

Distribution of Dry Matter

The proportion of the total recoverable biological yield formed by the economic yield is often referred to as the harvest index, *HI*, thus: economic yield = recoverable biological yield × harvest index. In cases where there are linear relationships between total yield and economic yield with accumulated intercepted radiation (*IR*), then the ratio of *E* estimated from economic yield and of *E* estimated for total yield, shows how the crop partitions its dry matter into economic yield, and is reffered to as the *partition coefficient*. An example of this is given from the sugar beet crop in Table 4.2 which also shows the application of nitrogen fertiliser markedly reduced the proportion of assimilate used in the production of sugar.

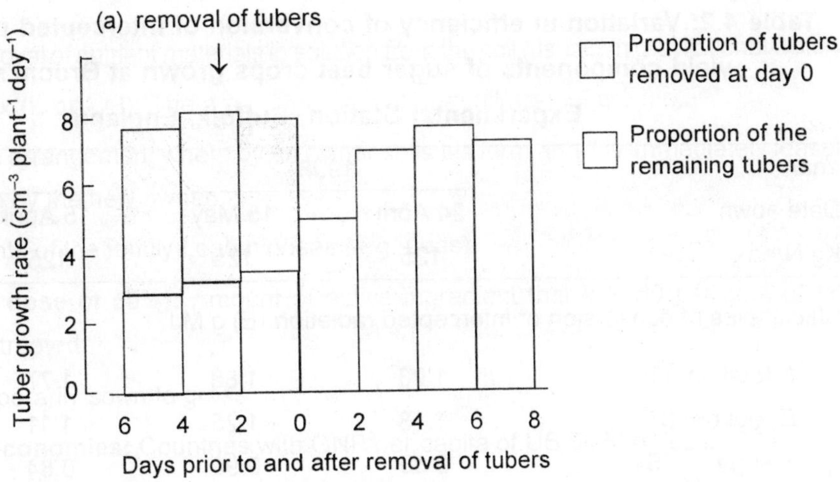

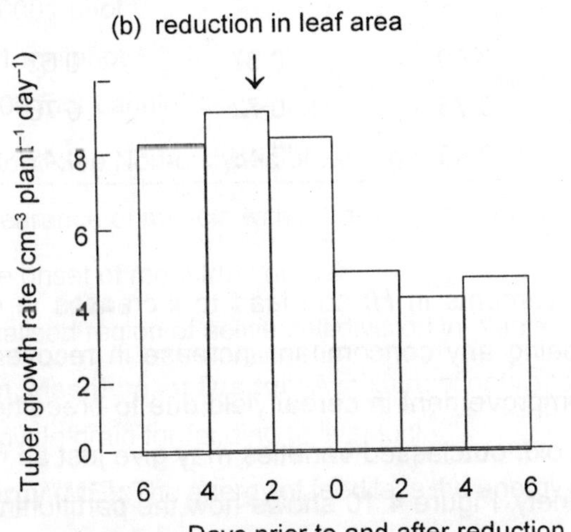

Fig. 4.9: (a) Tuber growth rates per plant as affected by removal of tubers. (b) Changes in tuber growth rate per plant in response to a 50 per cent reduction in leaf area. (*Source*: adapted from Engles and Marschner).

HI varies considerably between species: for example, maincrop potatoes may have a *HI* of about 85 per cent, modern winter wheat varieties 40-50 per cent and oil-seed rape about 15-25 per cent. Remember that these calculations are based on dry matter yields, and in the case of oil-seed rape, due allowance must be made for the high energy content of the dry matter of the seed, due to its oil content.

Table 4.2: Variation in efficiency of conversion of intercepted radiation into yield components of sugar beet crops grown at Broom's Barn Experimental Station, Suffolk, England

Year	1998		1999	
Date sown	24 April	15 May	15 April	15 April
Kg N/ha	125	125	125	0
Efficiencies of conversion of intercepted radiation (E) g MJ^{-1}				
E total dm (T)	1.98	1.88	1.77	1.52
E root dm (R)	1.18	1.25	1.11	1.31
E sugar (S)	0.89	0.90	0.84	1.00
Partition ratios				
E_R/E_T	0.60	0.67	0.63	0.86
E_S/E_R	0.75	0.72	0.76	0.76
E_S/E_T	0.45	0.48	0.48	0.66

Source: Milford *et al.*

It is obvious that improvements in *HI* can lead to increases in economic yield without there necessarily being any concomitant increase in recoverable biological yield. Indeed, much of the improvement in cereal yield due to breeding is attributable to improvements in *HI*, and old, outclassed varieties may give just as much biological yield as a new improves variety. Figure 4.10 shows how the partitioning of dry matter to tubers varied in two contrasting varieties of potatoes grown with a wide range of nitrogen inputs.

Changes in the environment or in the way in which crops are grown may have important effects on *HI* through the way crops grow and develop. Growth may be defined as an increase in size, and is usually measured in terms of dry weight (although it might be noted that, in seedlings, growth is better defined as an increase in fresh weight, as it is largest related to the uptake of water). Development may be defined as progress towards maturity. Some of these changes are abrupt and easy to recognise, such as germination, flowering and senescence, and clearly alter the growth pattern of a plant. Other changes such as those taking place on the growing point of a cereal plant, which determine whether leaves or flowering parts are being initiated, can be seen only by the microscopic examination of growing apices.

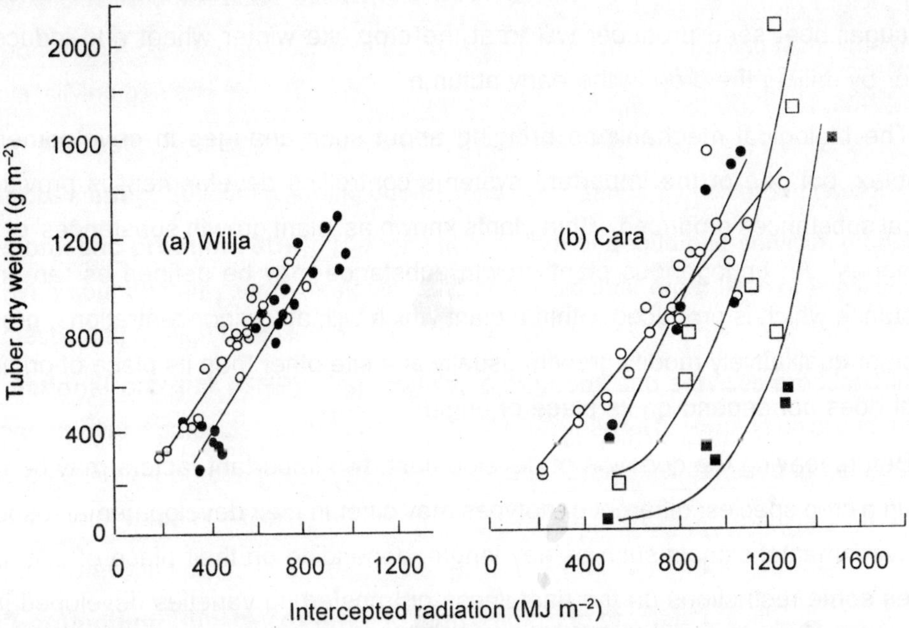

Fig. 4.10: Relationships between intercepted radiation and tuber dry matter yield at different levels of fertiliser nitrogen in two varieties of potatoes (a) Wilja, o = 0 – 70, ● = 150 – 750 kg N ha^{-1} and (b) Cara, o = 0 – 70, ● = 150, □ = 340, ■ = 750 kg N ha^{-1}.

It is evident, even to the casual observer, that plants grow and develop in an orderly manner. Developmental changes are under the genetic control of the plant, but they require triggering by signals picked up from the environment.

For example, the latest 'safe' drilling date of most winter wheat varieties in Britain is the middle to the end of February; winter wheat drilled later may remain vegetative and not produce any flowers and hence grain, thus there will be no economic yield. On the other hand, sugar beet drilled too early in the spring will be induced to flower in the same growing season and the plants are said to 'bolt'. In this case, flowering is undesirable and will reduce the *HI* and have other undesirable effects, such as increasing harvesting difficulties and giving rise to potential 'weed' beet problems in subsequent crops through shed seed. These crops are in effect biennials, and the environmental cues needed to induce flowering are periods of low temperature and/or short days. The choice of appropriate drilling dates enables the farmer to treat these crops as annuals or biennials for the particular component of yield required. Of course, the sugar beet seed producer will treat the crop like winter wheat and induce it to flower by drilling the crop in the early autumn.

The biological mechanisms bringing about such changes in development are complex, but one of the important systems controlling development is provided by helical substances produced within plants known as 'plant growth substances' or 'plant hormones'. An endogenous plant growth substance may be defined as "an organic substance which is produced within a plant which will, at low concentrations, promote, inhibit or qualitatively modify growth, usually at a site other than its place of origin". Its effect does not depend on its place of origin.

Before leaving the question of development, two important factors may be noted. Within a crop species, different genotypes may differ in their developmental responses to environmental signals such as day-length, depending on their place of origin. This places some restrictions on the usefulness of transferring varieties developed in one latitude for use in areas further north or south, where, for example, the day-lengths are different. A historic example is provided by the potato, where the crop, adapted to low latitudes in the Andes of South America, had short-day requirements for tuberisation and tubered very late in the long summer days of temperate latitudes. It was nearly 200 years before the crop became sufficiently well adapted to these conditions to make any important contribution to food supplies.

The second point is that endogenous plant-growth regulators, or chemicals closely related to them, can be manufactured and applied to crops in ways which may affect their growth and development to the farmer's advantage. One such chemical is Chlormequat which is widely used to shorten straw and decrease the risk of lodging in cereals. A considerable research effort has been put into the development of such chemicals for use in agriculture and horticulture.

Changes in the crop environment which affect growth rather than development may also affect the harvest index: for example, heavy rates of fertiliser application (see Fig. 4.10) and high plant densities often decrease *HI*, although the absolute increase in growth achieved by such practices may outweigh their effects on yield through any reduction in *HI*. The approximate economic balance has to be determined empirically in field experiments.

Potential Crop Yield

The basic principle on which crop production is based is that of the large scale conversion of light energy into the chemical energy contained in crops. This principle can be used as a basis for action and is useful in estimating the potential yield of a crop in a particular environment with which comparisons of actual yield achieved can be made. Such estimates are useful to show how big a discrepancy there might be between the best yields, or between average yields and the potential yield, to elucidate the reasons for such discrepancies and by implication to find out what scope there is for improvement.

The simplest calculation is (a) to take the total amount of energy which falls on an area of land over a year, and (b) to divide this by the energy contained in known unit of dry matter per hectare (ha). In Britain (a) is about 3.3 GJ/m^2 and (b) is about 17.5kJ/g (Monteith). This would give rise to a yield of 1890 t dry matter/ha. The average total crop dry matter yields of three important crops which are grown in Britain have been estimated and are presented in Table 4.3. On the basis of this calculation, less than 1 per cent of the radiant energy is transformed into crop dry matter, and the process of converting radiant energy into crop dry matter cannot be regarded as being very efficient!

Table 4.3: Average economic and recoverable biological yields of three major crops and the proportion of the total annual light energy their yields represent

Crop	Component	Economic yield Fresh wt (t/ha)	Economic yield Dry wt (t/ha)	HI (%)	Recoverable biological yield (t/ha)	Yield as % of 1890 t/ha: Economic yield	Yield as % of 1890 t/ha: Biological yield
Potatoes	Tubers	40	8.0	85	9.4	0.42	0.49
Sugar beet	Sugar	6.4	6.4	50	12.8	0.34	0.34
Winter wheat	Grain	6.7	5.7	40	14.2	0.30	0.75

Before considering how much of this inefficiency can be attributed to the shortcomings of the farmer, it should be made clear that much of it is unavoidable.

Only 50 per cent of the incident radiation is photosythetically active and, of this 50 per cent, some 15 per cent is reflected from or transmitted through the leaf. The maximum biological efficiency with which photosythetically-active radiation (PAR) can be converted into glucose is 22.5 per cent and this is only the case at low light intensities; finally, an allowance must be made for the loss of dry matter due to respiration. The effect of these losses on the conversion efficiency of radiation into crop dry matter in a C_3 plant is given in Table 4.4, where it is shown that about 5 per cent of the photosyntetically-active radiation (or 2.5% of the radiation) could be converted into crop dry matter.

Table 4.4: An account of the conversion of photosythetically-active radiation (PAR) into stored energy in crop dry matter by a C_3 crop

	Relative yield
(1) Photosynthetically-active radiation (PAR)[a]	100
(2) The crop absorbs only 85 per cent of PAR because of reflection and transmission	85
(3) Maximum efficiency of conversion of PAR into glucose, 22.5 per cent	19
(4) Less 40 per cent to allow for leaves in the crop canopy receiving light at a higher radiant flux	11
(5) Less 30 per cent to allow for photo-respiration	8
(6) Less 40 per cent to allow for dark respiration, i.e., 5 units of energy stored in the crop dry matter	5

[a] About 50 per cent of the total radiation.

Source: Heath and Roberts.

Given that the total amount of radiation received per day in June is about 17 MJ m^{-2} and that 2.5 per cent of this is converted into crop dry matter with an energy content of 17.5 KJ/g, then the rate of crop growth would be 243 kg/ha^{-1} m^{-2} day^{-1}. Short-term maximum growth rates of many crops in north-west European condition reach about

200 kg ha^{-1} day^{-1} (Fig. 4.11), which agrees well with this calculation if allowance is made for the fact that in the estimates of crop growth rates in the field, roots are not included.

Since crops convert intercepted radiation into dry matter at a rate of about 1.4 kg/ MJ total radiation, we can estimate that a crop in midsummer receiving 17 MJ of total radiation m^{-2} day^{-1} would give rise to an increase in yield of $17 \times 1.4 = 23.8$ g m^{-2} day^{-1} or 238 kg ha^{-1} day^{-1}, which again agrees well with the foregoing estimates and measurements.

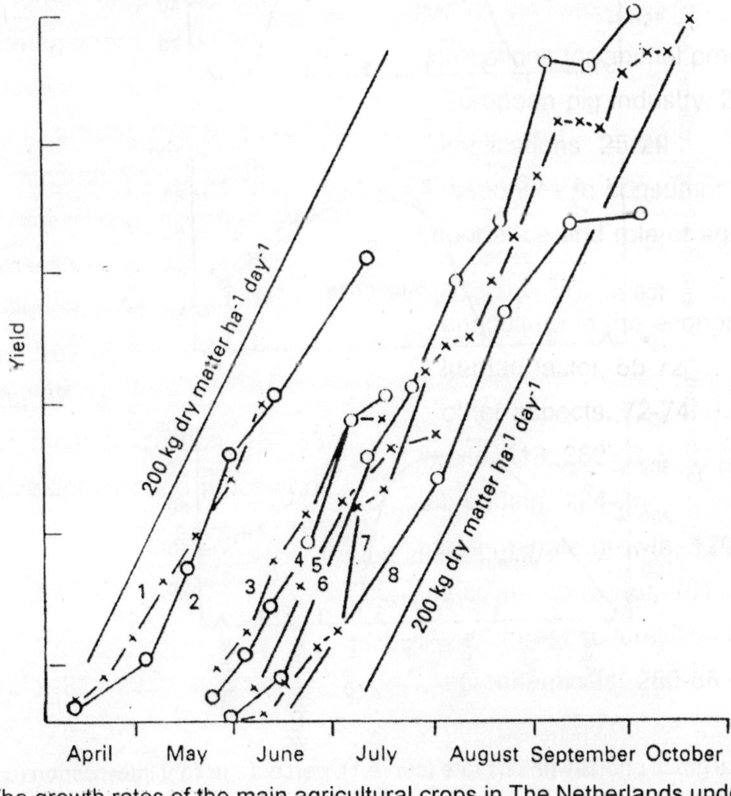

Fig. 4.11 The growth rates of the main agricultural crops in The Netherlands under (near-) optimal growing conditions, compared with a growth curve corresponding to 200 kg ha^{-1} day^{-1} (a) grass; (2) wheat; (a) oats and peas; (4) peas; (5) barely; (6) potatoes; (7) sugar beet; (8) maize. (*Source*: Wit *et al.*).

It is possible to estimate the potential yield of a crop if we know the pattern of radiation receipts over the year, the pattern of leaf development (i.e., the percentage of radiation intercepted by the crop) and the proportion of dry matter yield found in the economically useful part of the crop.

In Fig. 4.12, diagrams are shown of three crops of commercial importance in western Europe: potatoes, sugar beet and winter wheat (Sibma).

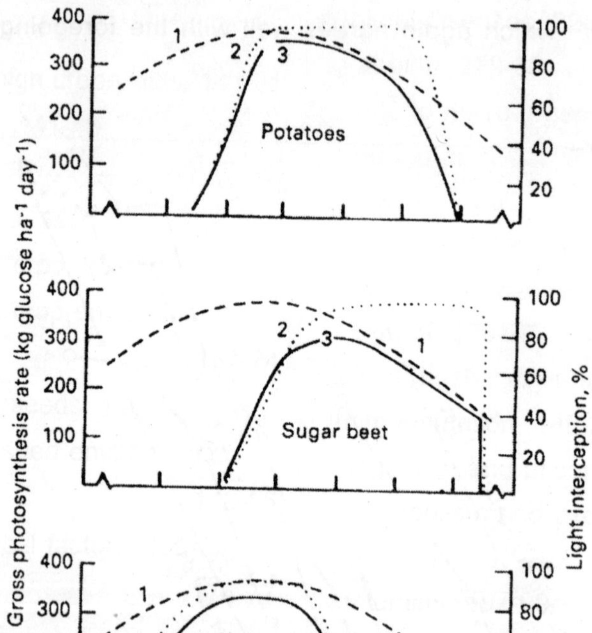

Fig. 4.12: Potential gross photosynthesis rate (curve 1), percentage light interception (curve 2) and calculated actual gross photosynthesis rate (curve 3) for potatoes, sugar beet and wheat. (**Source:** Sibma).

In each diagram, line 1 is the gross production rate (dry matter/ha) for a green, closed crop canopy optimally supplied with water and minerals, which is determined by the amount of radiation and how efficiently this is converted into dry matter. Curve

2 shows the percentage of light intercepted by the three crops and curve 3 shows the total dry matter in kg ha^{-1} day^{-1}. As this line represents total dry matter, allowance must be made for respiration, loss of dead leaves, and the *HI* in estimating economic yield. Assuming a respiration loss of 25 per cent of dry matter, the calculated potential yields of these crops, for radiation received at latitude 52°N is 96 t of tubers, 77 t of sugar beet (approximately 12 t of sugar), and 11 t of wheat ha^{-1} year^{-1}. These compare with average yields of 40 t tubers, 40 t sugar beet and 6.75 t of wheat ha^{-1}.

In north-west Europe, it is apparent from Fig. 4.12 that crop canopies are present for less than half the year, and that a full crop canopy, capable of intercepting all the incident radiation, is maintained for an even shorter period. This is because temperatures are normally too low to permit crops to be grown over a large part of the year, as can be seen in Fig. 4.13. Even when crops such as winter wheat, or grass, do cover the ground during the winter, radiation receipts (Fig. 4.13) are too low to permit much growth to occur.

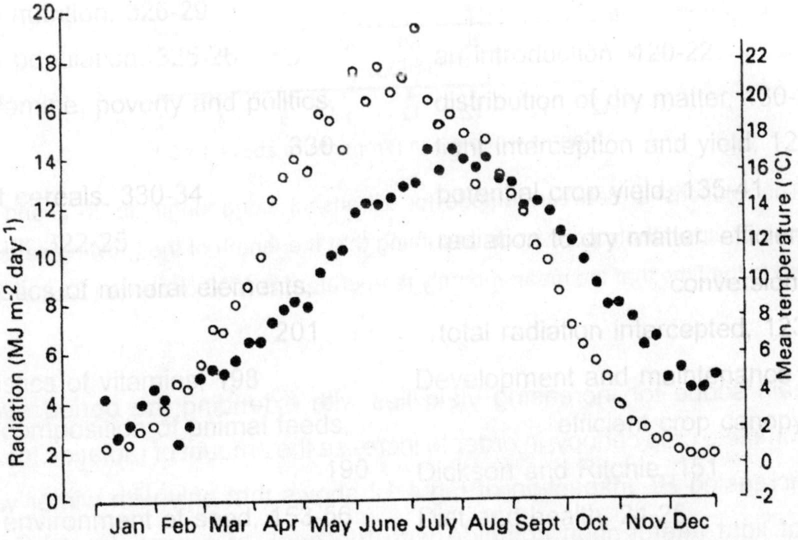

Fig. 4.13: Mean daily values of incident radiation (o) and temperature (●) over the period 1990-99 measured at Reading University Farm, Sonning.

For these reasons it would seem apparent that yield potential should increase from north to south as the potential length of the growing season becomes less and less restricted by temperature, and the intensity of radiation increases. In Fig. 4.14, potential gross photosynthetic production per annum is plotted against latitude and shows that potential crop production in the sub-tropical latitudes is more than double that of the latitudes in which Britain lies.

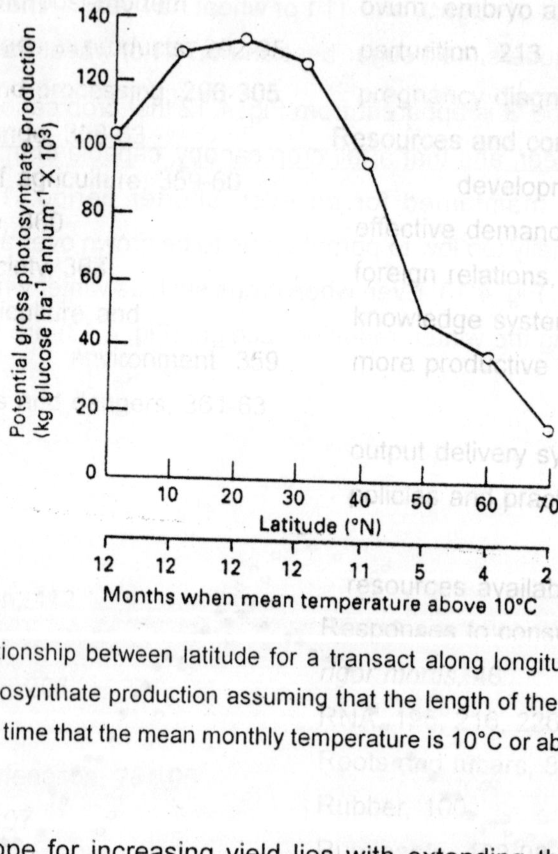

Fig. 4.14: The relationship between latitude for a transact along longitude 20°E and estimated potential gross photosynthate production assuming that the length of the growing season can be approximated to the time that the mean monthly temperature is 10°C or above (*Source*: Heath and Roberts).

The main scope for increasing yield lies with extending the period over which there is a complete crop canopy, in order to increase the amount of radiation intercepted, and with increasing *HI*. Inspection of Fig.4.12 shows that only with winter wheat did the peak of light interception coincide with the peak of potential production (i.e., midsummer), but with the potato crop there would seem to be scope for intercepting more radiation earlier in the growing season, and even more so for the sugar beet crop, where there is scarcely any radiation intercepted before the beginning of June. It has been estimated that if all the radiation in April, May and June were to be intercepted then the potential yield of sugar would be about 20 t ha-1. This would require the crop to be sown in the autumn, which in turn would require some method of preventing the crop from running to seed in the following summer; a potential prospect for the development of a growth regulator which would counter the environmental signals which govern the development of the crop.

These considerations of potential crop yields are useful both to the practising farmer and to those engaged in typing to raise yield potential, whether through techniques aimed at modifying the plant environment or the genotype. In many farming systems, particularly those encountered in countries without strong market economies, the actual rates of production may be severely restricted by the lack of cash to purchase inputs which permit farmers to supply factors (e.g., fertilisers) which are limiting leaf growth. This is shown clearly in Fig. 4.15 (Haverkort), which shows the potential yield of potatoes grown at different elevations in Rwanda, in Central Africa.

The actual yield levels are every low, but could be doubled by the use of low-cost inputs such as mulches, and the use of improved varieties and healthy propagation material. These yields may be described as 'aspired' yields, capable of being attained at low cost. 'Attainable' yields could double the yield of aspired yields but would require costly inputs. Yields could be increased somewhat beyond this by the application of the best existing knowledge, for example, on research stations, and it can be seen that these may be very close to the yield potential of the crop, set by the climatic variables.

The Practical Achievement of High Crop Yields

The strategy for achieving yields that are close to a crop's potential, using economically justifiable methods of production, is relatively simple. It is to establish a crop canopy capable of intercepting most of the incident radiation, as rapidly as possible; to maintain that canopy, assimilating efficiently, for as long as possible (or desirable); and, where yield is a fraction of the recoverable biological yield, to maximise *HI*.

The means by which the farmer may seek to implement this strategy are:

(a) the genetic material at his disposal (choice of species and variety),

(b) the manipulation of this material (e.g., choice of method of propagation, sowing date, plant density),

(c) the modification of the crop environment (e.g., applying fertilisers),

(d) the control of 'negative' biological agents (weeds, pests and diseases).

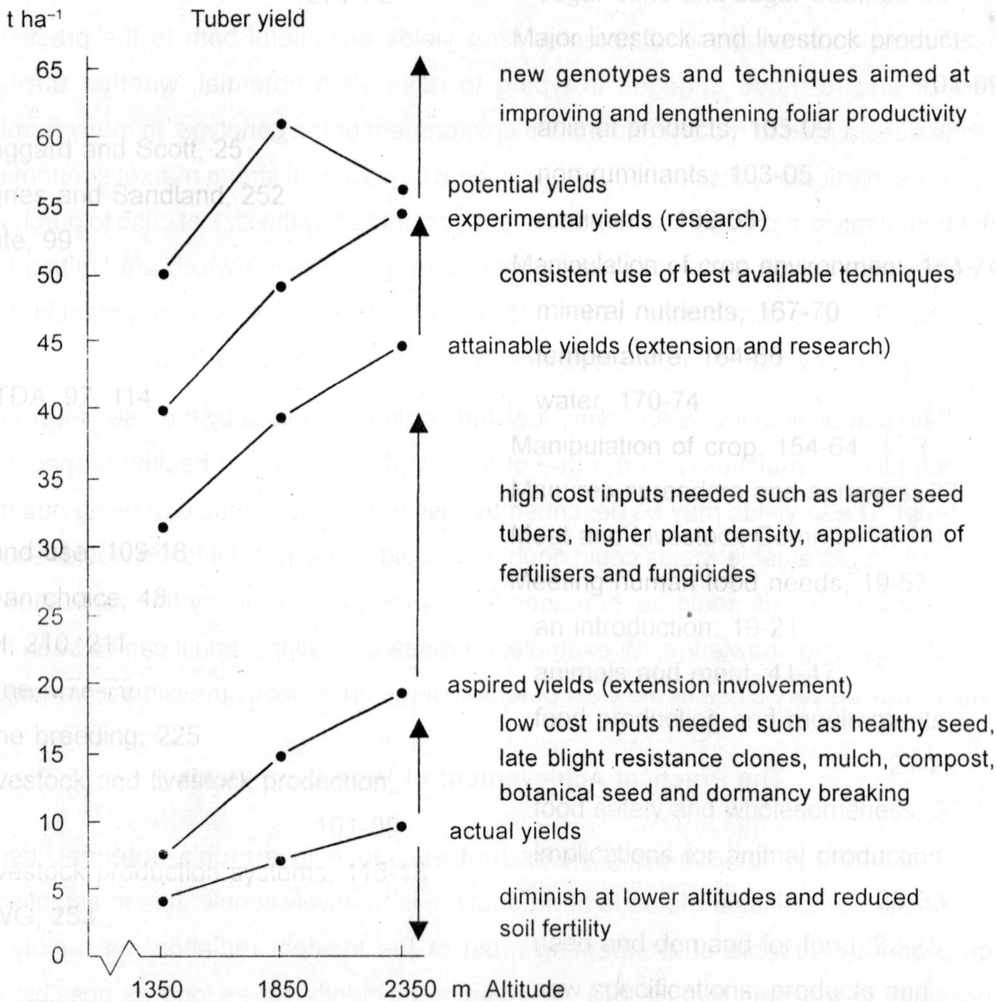

Fig. 4.15: Yield gaps in Central Africa and means of closing them. (Source: Haverkort).

In addition to attaining yields as close to the potential yield as his resources permit, the crop must be harvested and in most cases the produce stored, either for a few days, or in some cases months, although seldom by farmers beyond a year. As the farmer's returns are determined by the quantity and quality of the product sold, he must pay due attention to the efficiency of harvesting and storage, and to any effects his production practices may have on them and on the quality of the produce.

In Fig. 4.16 an attempt has been made to summarise the strategy in diagrammatic form. The resources or 'tools' available to the farmer can be directly related to the

(a) Strategy

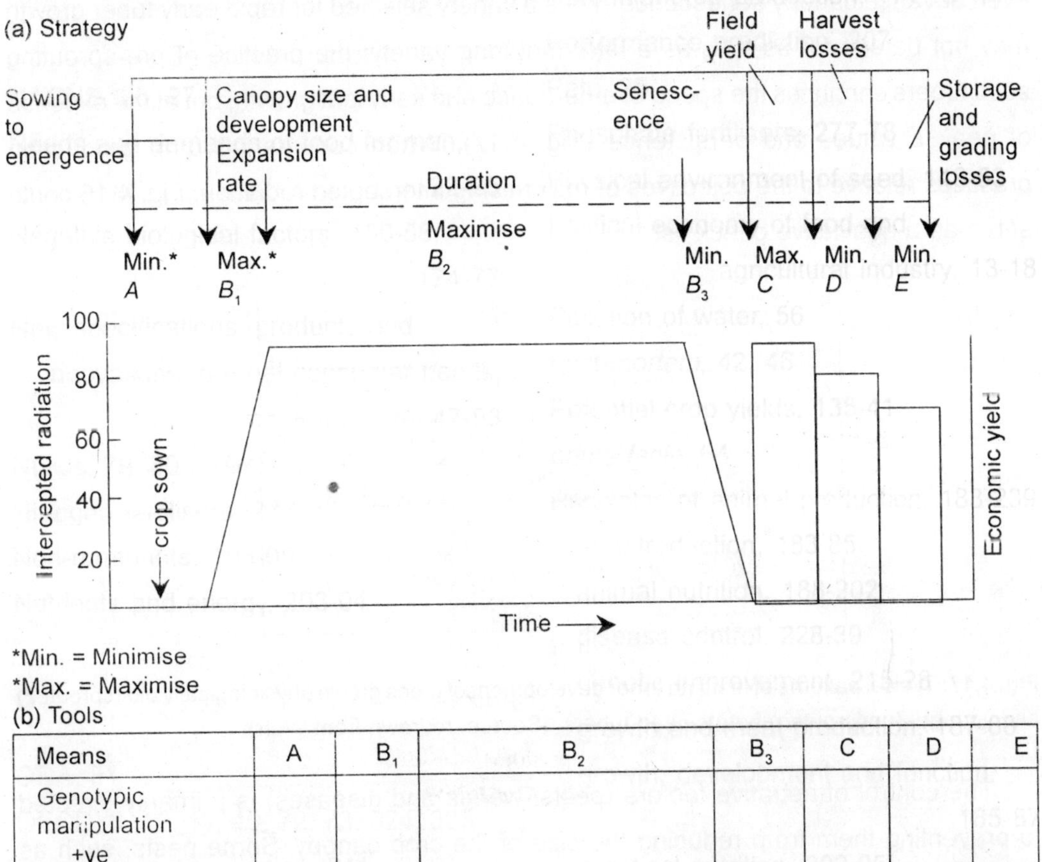

*Min. = Minimise
*Max. = Maximise

(b) Tools

Means	A	B_1	B_2	B_3	C	D	E
Genotypic manipulation +ve –ve							
Plant manipulation +ve –ve							
Environment manipulation +ve –ve							

Fig. 4.16: A schematic diagram of a strategy for attaining high yields of an arable crop.

strategy of maximising leaf growth, and the point that needs to be emphasised here, is that practices which may have positive effects on canopy size at one stage of growth may have negative effects at others. For example, promoting the speed of emergence of the crop above ground and the speed at which the ground is covered by leaves, could reduce the maximum size of the crop canopy (i.e., leaf area index) and many even advance canopy senescence. Thus a variety selected for rapid early tuber growth may not persist for as long as a later-maturing variety; the practice of pre-sprouting seed tubers enhances the speed of emergence and early leaf growth, but at the expense of peak *L* values and persistence (Fig. 4.17). When considering crop production practices relative to the objectives of maximising intercepted radiation, Fig. 4.16 could serve as a qualitative check list.

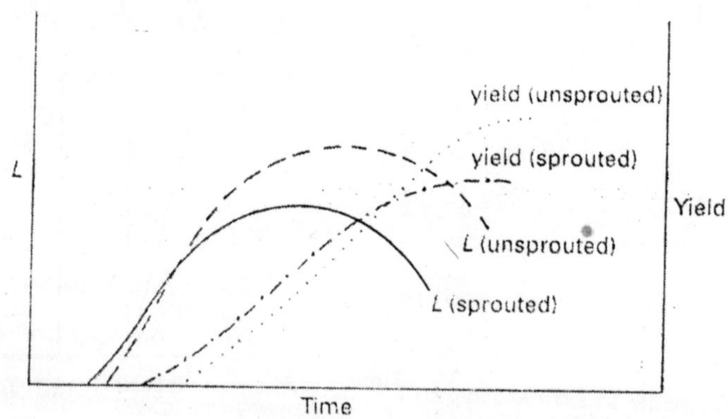

Fig. 4.17: Typical patterns of leaf and tuber development of crops grown physiologically old (sprouted) and younger (unsprouted) potato seed tubers. (Source: redrawn from Wurr).

The control of negative factors (pests, weeds and diseases) is primarily directed to preventing them from reducing the size of the crop canopy. Some pests, such as the Colorado beetle (*Leptinotarsa decemlienata*, say), reduce the leaf area of potatoes by the simple expedient of eating the leaves. Pests such as the potato cyst nematode (*Globoders rostochiensis*) reduce the leaf area of the potato crop just as effectively (Fig. 4.18), but do so indirectly by attacking the root system and depriving the plant of mineral nutrients and water.

All the means of production, as indicated in Fig. 4.16, are involved in their control. Thus varieties may be selected which have been bred and selected for resistance to specific pests or diseases, and large part of the world of the plant breeder is devoted to the production of disease/pest-resistant varities. An example of the use of plant manipulation in this context, consists of the possibility of varying the time of sowing in order, for example, to 'escape' pests. The frit (*Oscinella frit* L.) is an important pest of oats, and greatest damage is done to spring oat seedlings attacked before the four-leaf stage by the first generation of larvae. If the plant is somewhat older, and has 'tillered' (tiller means a stem developed in the axil of a leaf), then it is more resistant. Thus sowing date is an important method of controlling the harmful effects of this pest (Table 4.5).

Table 4.5: Relation between frit fly attack and sowing date of oats

Sowing date	Tillers attacked (%)
February 1	0
February 20	Trace
March 1	Trace
March 15	Trace
March 23	10
March 31	20
April 14	32

Source: Pethybridge, 1924, quoted in Jones and Jones.

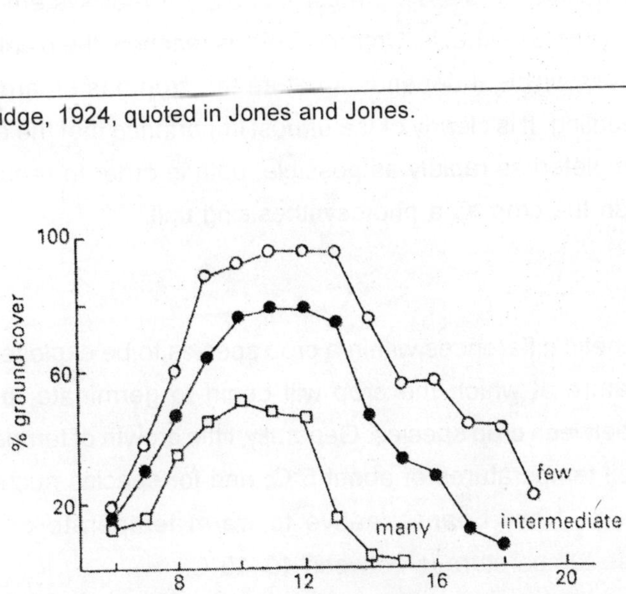

Fig. 4.18: Changes with time in ground cover of Pentland Dell potato plants infested with few, intermediate or many potato cyst-nematodes. (Source: redrawn from Evans and Trudgill) .

It is clear from Fig. 4.16 that it is possible to discuss the practical attainment of high crop yields either by considering the means of production and their effect on the objectives of maximising *IR, HI* and quality, or by considering the objectives and discussing the means by which they may be realised. The latter approach would seem more logical and has been adopted in the following discussion.

Establishment of the Crop

Before any light can be intercepted, crops established from seed must be sown, the seeds must germinate and leaves appear above the soil surface. This is usually described as the establishment phase and it is a very vulnerable period in the life-cycle of a crop. The seed must be in the correct physiological state to permit germination, while germination and seedling growth must not be inhibited by the physical and chemical state of the soil surrounding the seed. In mechanised agriculture, soil conditions must also be suitable for the passage of machinery required to produce a 'suitable' seedbed and permit drilling. Since seeds provide a welcome food supply for animals, birds, insects, fungi and other organisms, protection from them may be required. After germination the seedling must establish a root system and begin to intercept radiation and photosynthesis. Until this point is reached, the plant is dependent upon the seed reserves which, if dependent before the crop has emerged, will result in the death of the seedling. It is clearly of the utmost importance that the establishment phase should be completed as rapidly as possible, both in order to reduce the risk of death and to establish the crop as a photosynthesising unit.

Genotypic Effects

There may be genetic differences within a crop species to be exploited with regard to the base temperature at which the crop will begin to germinate, but the major differences here are between crop species. Generally, little growth of temperate species takes place below soil temperatures of about 5°C, and for species such as maize or phaseolus beans (e.g., Navy beans), native to warm temperate or sub-tropical environment, the base temperature is between 10–15°C.

Manipulation of the Seed

The germination capacity of the seeds sown is clearly an important determinant of how many seeds may be established. Legislation lays down minimum standards for germination and tests are carried out which indicate the potential capacity of the seed to germinate under ideal conditions. Conditions in the seedbed are often not ideal, and there is some evidence that different seed lots, all of acceptable germination standards, may perform differently in the field, especially under adverse conditions. The ability of a seed lot to perform well under such circumstances is termed 'seed vigour' and is defined as "a physiological property determined by the genotype and modified by the environment, which governs the ability of a seed to produce a seedling rapidly in soil and the extent to which that seed tolerates a range of environmental factors" (Perry).

Treating seed in order to speed up the rate and uniformity of emergence is a possibility which may be attractive with species which are slow to germinate and emerge above ground. These treatments include supplying the seed with water, but with insufficient to allow the seed to germinate. The seed is then dried and sown in the normal way. This procedure is said to initiate the first phase of germination, which includes the mobilisation of enzymes that make the food reserves available to the embryo and also assists in the repair of damaged cells. The process may be taken further, by soaking seeds in aerated water and allowing fermentation to take place. Such seeds would easily be damaged by conventional drills and are protected by a viscous gel in a process known as 'fluid drilling'. The techniques is not normally practised with commonly-grown farm crops.

Seed may also be selected on the basis of size: in general, larger, heavier seeds are more likely to be viable than small, light ones, which are often removed by sieving or other means.

The conditions under which seed is harvested may also affect the subsequent performance of the crop: for example, sugar beet harvested under cool damp conditions is more likely to 'bolt' or run to seed in the following year than seed harvested under warm, dry conditions. There is some disagreement about the optimum moisture content at which to harvest cereals for maximum seed quality, but this would seem to lie between the moisture content at physiological maturity, i.e., when the seed attains its maximum dry weight, and harvest maturity, when the seed has a sufficiently low moisture content to allow it to be harvested mechanically.

The storage environment of the seed has important consequences for seed viability. In general, the higher the temperature and moisture content at which seed is held, the faster the seed will deteriorate: hence seed should be stored under cool, dry conditions. An exception to this is the so-called 'recalcitrant' seeds, typified by large-seeded leguminous crops, in which higher moisture content are required to maintain viability under long-term storage. These problems are particularly important with respect to the long-term storage of seed in 'gene-banks'.

In British farming, seeds are the normal means by which crops are propagated, the exception to this being the potato crop, which is propagated vegetatively from tubers. With this crop, it is much easier to speed up development after planting, because growth of the tuber can be encouraged in the store before planting; this is a standard practice when rapid establishment is required for plants grown for an early market. In the tropics many important crops such as sugar cane, yams and cassava are grown by vegetative propagation. One of the major disadvantages of this method of progression is the increased potential for the transmission of diseases to the subsequent crop; interestingly and paradoxically, one means of combating this in the potato crop is to propagate the crop from 'true' seed. This technique is proving to be economically attractive in environments where disease-free tubers cannot be produced and where the cost of importing and transporting healthy seed tubers is very high.

Modification of the Seed Environment

The Physical Environment of the Seed

The positive requirements of the seed for germination are water, air and a suitable temperature. Soil temperature is largely outside the farmer's control, and depends primarily upon latitude, altitude and aspect. Sowing date will therefore largely dictate the soil temperature conditions experienced by the seed. In general, farm crops are not valuable enough to merit the expense of protecting the crops artificially; an exception to this is the early potato crop in Britain, where the premium paid for earliness makes it economic to grow early potato crops under plastic film 'mulches', the increase in soil temperature advancing the date of harvest by up to 10 days. In tropical conditions soil temperatures may be too high for optimal establishment of the crop. Here soil temperatures may be reduced by growing sensitive crops, such as potatoes, in the shade of tall crops such as maize prior to the latter being harvested.

Under given climatic condition, the supply of water and air to the seed are related to the physical condition of the soil in the vicinity of the seed. The basic problems, which will vary in severity according to the soil type and climate, are to ensure that the seed is provided with sufficient air and water, and to minimise the risk of erosion due to rainfall or wind, which may remove both soil and seed. The soil is normally devoid of vegetation at the time of sowing, since in temperate climates crops are normally grown as monocultures, and every effort is made by the farmer to provide a seedbed from weeds.

Before considering what steps might be taken to optimise the physical environment of the seed, it is important to be aware of the complexity of the soil environment, which may change rapidly both horizontally and vertically and with time. The soil can be broken up physically into inorganic constituents of various size particles, the proportions of which, together with the organic matter, determine the soil texture (Table 4.6). These particles, together with the soil organic matter, may be further combined into aggregates which give the soil its structure. Between the particles and aggregates of particles are pore spaces of varying size: some of which are small enough to hold water by surface tension against the pull of gravity; above a certain size, water cannot be held and the spaced will be filled with air. Superimposed on this 'fine' structure are fissures and channels which arise from many causes, such as shrinkage and expansion on drying and wetting, channels formed when root decay, and worm channels.

Owing to the structural characteristics just described, the soil can supply both oxygen and water to the seed and growing plant. It is logical to assume that there is an 'ideal' combination of particles and their arrangement into aggregates which will give rise to the most favourable air and water environment for the germinating seed. It has, for example (Russell), been stated that the ideal seedbed should consist of soil crumbs nor much finer than 0.5-1.0 mm and not much coarser than 5-6 mm in a fairly firm packing. Finer crumbs would block the coarser pores needed for drainage and coarser crumbs would give rise to 'open' seedbeds in which the seed may be in poor contact with films of water and the soil would also be prone to dry out in the vicinity of the seed; it is particularly important that a germinating seed should not run short of water and become desiccated.

Table 4.6: Classification of soil particle sizes

Soil fraction	Size (microns)
Gravel	2000–20000
Coarse sand	500–2000
Medium sand	200–500
Fine sand	100–200
Very fine sand	20–100
Silt	2–20
Clay	Less than 2

Pore size is also important with respect to root penetration. If roots have to exert even small pressures to enlarge the pores into which they are growing, their growth is considerably reduced. The majority of roots exceed 60 μm in diameter and the existence of a sufficient number of continuous pores into which roots can freely enter is an important requirement for their growth.

No simple solution can be given to the problem of achieving an optimum arrangement, assuming this to be known, of soil particles and aggregates or of how to maintain the stability of the soil structure. Traditionally, cultivations, especially mouldboard ploughing, have been used to achieve alterations in soil structure, while the stability of the soil structure has been favoured by the maintenance of organic matter through the application of organic manures and the operation of ley farming systems where grass crops of several years' duration alternate with arable crops.

Recent years have seen marked changes in attitudes towards cultivations particularly with regard to the basic cultivation process of ploughing. Reduced, minimal or zero tillage methods have received considerable attention, not only in order to reduce the cost associated with numerous tillage operations, but also with a view to improving the soil environment for the emerging and emerged crop.

In Table 4.7, the effects of various cultivation systems on labour, energy requirements and yields of winter wheat are given for two soil types. They indicate that under the conditions of the experiment, yields were relatively insensitive to the various cultivation systems, but that these could have a profound influence on the time they involve, on the energy required to perform them and on cost. Because of the reduction in time, some systems increase the flexibility in the choice of planting date, which may have implications for crop yield.

Table 4.7: The effects of cultivation systems on labour, energy requirements and yields of winter wheat

Cultivation system	man hours Ha^{-1}	energy (MJ ha^{-1})	yield (t ha^{-1})
Plough (20 cm): disc or cultivator: drill	3.5[a] (2.2)	330 (179)	5.92 (4.91)
Chisel plough (13 cm) × disc or cultivator: drill	3.4 (2.3)	310 (210)	5.78 (4.81)
Plough (20 cm): combined cultivator and drill	3.0 (1.9)	340 (178)	5.82 (4.76)
Shallow plough (10cm): combined cultivator and drill	1.6 (1.2)	193 (104)	5.71 (4.65)
Sprayer: direct drill	0.5 (0.6)	27 (36)	5.62 (4.53)

[a] Clay loam (silty loam)

Source: Bullen.

These cultivation systems involve the conventional use of farm equipment subsequent to the establishment of the crop, for example, the application of fertiliser, sprays and harvesting equipment. Recent experiments on a clay loam soil in south-east Scotland have investigated the effects on the yield of barley, potatoes and grass silage of zero compared with conventional traffic systems (Campbell and McGregor; Douglas and Crawford; Dickson and Ritchie). Meaned over periods of about three years, zero traffic increased potato yield by 13 per cent, barely yield by 6 per cent and grass silage yield by 15–25 per cent in the years following the initial establishment of the grass. Indirect benefits included a reduction in clouding when potatoes were harvested, and a reduction in the draught force required for cultivations after harvested. On the grassland, the recovery of nitrogen was increased from 50 per cent to 70 per cent.

Waterlogging is the prime cause of anaerobic soil conditions, which occur when the rate at which oxygen enters the soil from the atmosphere is less than that at which it is used in the respiratory processes of plant root, bacteria fungi and other soil organisms. Because oxygen diffuses some 10,000 times faster in the gas phase than in solution, filling the pore spaces with water has a major influence on the availability of oxygen for respiration. Lack of oxygen is not, however, the only injury plants suffer in anaerobic conditions for many other harmful changes, both chemical and physical, contribute to plant damage. For example, ethylene gas, an endogenous plant growth regulator is formed, which can induce biological effects at very low concentrations such as the stunting of roots and premature senescence. Ethylene will persist in poorly-drained soils in the field for weeks at levels high enough to have appreciable effects on root development. It is possible that the ethylene content of many poorly-drained soils is a more common source of damage to crop growth than lack of oxygen or a high concentration of carbon dioxide in the soil.

The injury caused by transient waterlogging can vary considerably with stage of growth, but the germinating seed is particularly prone to injury. With shoots above ground, the possibility of transfer of oxygen through the shoots to the roots exists, through the development of intercellular air spaces in the cortex of roots (aerenchyma), particularly well developed in rice and aquatic plant species.

Drainage is obviously important in humid climates when, for large parts of the year, rainfall may exceed evaporation (see Fig. 4.30). In Britain, for example, with spring-sown crops when the soil is bare at the time of drilling and also near to field capacity (i.e., the soil is holding all the water it can against the pull of gravity), little water is lost from the soil by evaporation once the top few centimetres of the soil are dry, and none is lost by transpiration, since there is no leaf cover. Rainfall will therefore fill up the large air spaces if for some reason, such as the presence of a clay subsoil, the water in excess of field capacity cannot drain away.

Remedies for such problems lie in draining the land artificially by means of an underground systems of pipes, often combined with the formation of continuous channels in the clay by means of the aptly named mole-plough. Drainage problems associated with high water table, as in river valleys, or on land below sea-level as in the Dutch Polders, or parts of the Fens, present more difficult drainage problems and may need expensive pumping equipment to keep the surface soil drained.

Soil erosion is particularly likely to occur when there is no vegetative cover of the soil surface. Water erosion can occur on sloping land, particularly under the impact of intense rainfall, and is of frequent occurrence in the tropics. Under the impact of raindrops, soil aggregates are broken down, the pores become blocked, the rate of infiltration of the rain into the soil drops dramatically and water then flows over the soil surface, taking with it particles of soil. While the erosion of soil by water is continually taking place in temperate countries such as Britain, the rainfall intensity is such that the effects are rarely dramatic and its serious incidence is localised in time and space. Solution to the problem are sought by ensuring that no large area of bare soil is exposed to rainfall, and that the velocity of water running over the soil surface is reduced, by techniques such as cultivating and cropping along contour lines.

Factors predisposing soil to wind erosions are the soil texture, fine sands (100-200 μm) and organic soil aggregates (up to 1 mm) being particularly at risk, and the absence of any wind breaks or vegetative cover. Solutions include the provision of wind-breaks and the retention or provision of vegetative trash on the surface soil.

The Chemical Environment of the Seed

The only chemical required for germination is water, and this has been discussed in the previous section. However, shortly after germination, given an adequate supply of water, the major limitation to growth will be the nutrient supply – although the reserves in the seed (or vegetative propagation unit) are likely to be sufficient to provide the emerging seedling with sufficient nutrients for the shoot to appear above the ground. Of the utmost importance is the pH of the soil.

A neutral soil has a pH of 7, soils are said to be increasingly acid as the pH falls below this value and increasingly alkaline as the pH rises above it. The pH scale is logarithmic and a decrease from 7 to 6 implies a tenfold and from 7 to 5 a hundredfold increase in acidity. Over a wide range, probably from pH 4 to 8, the harmful effects of acidity are indirect and are mainly related to the effect of pH on the concentration of different ions in the soil solution and hence their availability to the plant. A high aluminium content is the main cause of incidence of trace element problems—all except Mo tending to become less available to crop as the pH rises.

Crop species and even varieties vary considerably in their tolerance to soil pH, and can be roughly classified into three groups: those which only grow well on alkaline soils (calcicoles), those adapted to and which grow well only on acid soils (calcifuges) and those that tolerate a wide range of soil pH values. In Table 4.8, the soil pH below which growth may be restricted on mineral soils is indicated. These are only rough guides because of the varietal differences in tolerance alluded to. A slightly acid soil (pH 6.5) is probably best for a wide range of crops.

Acidity generally arises when soils are deficient in calcium. This is removed from soils by leaching and in crop and animal products and is therefore affected by the farming system and by the amount of drainage. Losses by leaching are more severe on soils where there is little negatively-charged colloidal material (i.e., sands as opposed to clays) to bind on to the positively-charged calcium ions. It is corrected by the addition of calcium carbonate (lime).

Problems arising from alkaline soil conditions are relatively rare in temperate agriculture, but may occur when land has been flooded with sea water. Under hot, arid conditions, soluble salts accumulate in the surface of soils whenever the ground water comes within a few feet of the surface. During dry periods, the surface of the soil is covered with a salt crust, which dissolves in the soil water each time it is wetted. The salts are usually the sulphates and chlorides of sodium and calcium, though occasionally they may be nitrates; the soils are light in colour and the pH, while high, is usually below 8.5. Under some conditions, sodium carbonate may accumulate and this raises the pH to 9 or 10. If sodium carbonate is predominant the humic matter in the soil may be dispersed and the soil assumes a black colour.

Table 4.8: Soil pH below which growth may be restricted on mineral soils

Arable crops		Grasses and legumes	
4.9	potato	7.4	Clover, wild white
	Rye		Fescues
5.3	Oats		Ryegrass
5.4	Kale	5.3	Cocksfoot
	Linseed		Timothy
	Swede	5.6	Clover, white
	Turnip	5.7	Clover, Alisike
	Mustard	5.9	Clover, red
5.5	Maize		Vetches
	Wheat	6.1	Trefoil
5.6	Rape	6.2	Lucerne
5.8	Mangel		Sainfoin
5.9	Barely		
	Sugar beet		
	Pea		
6.0	Bean		

Source: MAFF.

Saline soils may arise naturally or through faulty irrigation and in arid regions salt control on irrigated soil is essential. The irrigation water always contains dissolved salts. The salt content of irrigation water and soil solution may be specified by its electrical conductivity and used as a guide to the frequency with which additional water must be added to leach the salts down the profile and as a guide to which crops to grow, as some are more tolerant of alkaline conditions than others. The harmful effects of soil alkalinity may be due to the specific toxicity of the ions concerned or to the soil solution – which can affect nutrient uptake, transpiration, photosynethesis and respiration. The general consequence of such effects is to give dwarf, stunted plants and dull bluish-green leaves, which are often coated with a waxy deposit. The amount of soluble borates is important as boron is toxic to plants above certain minimum levels. Irrigation water containing more then 0.3 ppm of boron must be treated with caution, and 2-4 ppm restricts cropping to boron-tolerant crops such as sugar beet, lucerne and some brassicas.

Because of the importance of leaching with excess irrigation water in controlling salt concentrations in arid land irrigation, it is essential that irrigation schemes should be complemented by adequate drainage facilities. Failure to ensure this may result in crop production having to cease because of salinity problems.

It is the seedling stage that is most vulnerable to the effects of soil salinity. Some of the harmful effects at this stage can be avoided by drawing the land into ridges having flat sides either coming to an apex or having a flat top. Salts accumulate in the apex, or the centre of the flat top as this is where the wetting front of irrigation applied in the furrows will meet. Seeds planted on the sides or the edge of a flat-topped ridge will be able to develop in soil with a salt content close to that of the irrigation water.

In temperate regions, problems of salt concentration on the development of seedlings are most likely to occur from the use of fertilisers applied to the seedbed. Phosphates present no problem as they are relatively insoluble in the soil solution and so have little influence on its osmotic pressure. Salts carrying nitrogen, potassium or sodium (the last applied in the form of salt for sugar beet, for which sodium is beneficial, although not essential) may, if the levels applied are high and the rainfall after drilling low, have harmful effects on germination. In general the timing and placement of fertilisers can be manipulated to avoid harmful effects of these chemicals on emergence.

Another hazard for the germinating seed may arise from the use of soil-acting herbicides to control weeds in the early stages of growth. Faulty spraying, or environmental factors which reduce the tolerance of the crop or increase the concentration in the vicinity of the germinating seed, may have adverse effects on germination.

Negative Biological Factors

The seed or germinating seedling may be attacked by a wide range of birds, small mammals, and fungal and bacterial diseases. One strategy relevant to this stage of growth is to treat the seed with a fungicidal and/or pesticidal seed dressing: occasionally the seed may be dressed with chemicals with the object of repelling birds.

The other major biological hazard at this stage is provided by the potential competition from unsown plants, or weeds. Many arable soils contain a vast reservoir of weed seeds, often of the order of 5,000-10,000 viable weed seeds per m^2, but not all of these seeds will germinate at any one time. Figure 4.19 illustrates some of the factors influencing the total population of viable weed seeds in the soil. In addition to seeds, some weeds may persist and regenerate from vegetative parts of the plant, for example, from underground stems (rhizomes) in the case of couch grass (*Agropyron repens*).

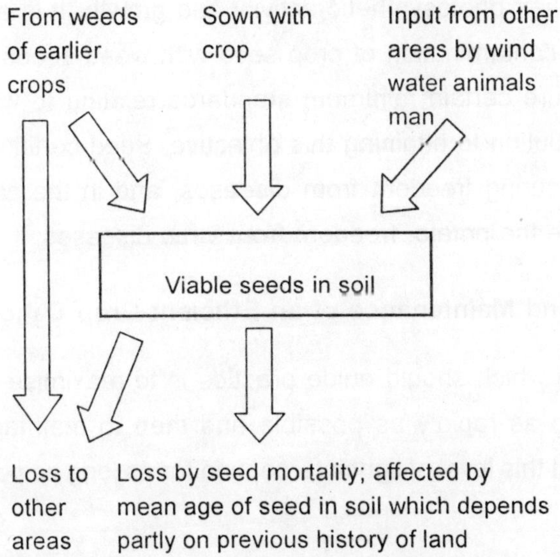

Fig. 4.19: Diagram illustrating some of the factors influencing the total population of viable seeds in the soil. (*Source*: Hill).

Before the fairly recent influx of a galaxy of chemical herbicides, permitting the selective control of weeds in crops, horse- or tractor-drawn hoes provided the only major, rapid method of achieving selective weed control in the early stages of crop growth. The introduction of the seed drill which enabled crops to be grown in rows and mechanically hoed, was a major step in the evolution of modern agriculture, and was due recognition of the general principle that unless crops are kept practically free from weed competition, particularly in the early stages of growth, yields may be expected to be drastically reduced. With the increasing interest in farming without the use of agrochemical (fertilisers and pesticides), or reducing their use, there has been a revival of interest in traditional methods of tilting the balance of competition between crop and weeds in favour of the crop. These include techniques such as the creation of 'false' seed beds, in which a seed bed prepared, but sowing of the crop delayed until a flush of weeds has germinated and been mechanically destroyed. Increasing the plant density of the crop (seed rate) and, in the case of cereals, growing taller more profusely tillering vanities may also alter the balance of competition in the crop's favour. In essence, the basic principle is to adopt strategies of manipulating the crop which will ensure that its canopy develops faster than the growth of weeds, thus shading them and reducing their photosynthetic activity and growth. It is also of the greatest importance to avoid contamination of crop seed with weed seeds. Seed certification schemes which ensure certain minimum standards relating to weed contamination make a major contribution to attaining this objective. Seed certification schemes are also important in ensuring freedom from diseases, and in the case of vegetatively propagated crops like the potato, freedom from virus diseases.

The Development and Maintenance of an Efficient Crop Canopy

A major principle which should guide practice is to maximise the interception of radiation by the crop as rapidly as possible and then to maintain the leaf canopy operating efficiently at this level of light interception for as long as possible or desirable (see Fig. 4.16).

Genotype

The choice of crop species and to a lesser extent the choice of cultivar will have marked effects on these objectives and will also interact with the other techniques at the disposal of the farmer. For example, leguminous crops (e.g., peas, beans and lucerne) will in general be unresponsive to additions of nitrogen fertiliser, due to their symbiotic association with bacteria which enable them to fix nitrogen from the atmosphere. Crops such as lucerne are far less affected by drought than others, such as grass and potatoes. In adverse conditions, such as the semi-arid tropics where crops are often short of water and plant nutrients, at appears that growing two or more crops together (intercropping) often makes better use of the limited environmental resources than growing crops in pure stands. Figure 4.2 illustrated the canopy development with time of some contrasting crops.

Manipulation of the Crop

The major factor of importance here is the manipulation of plant numbers and their spatial arrangement. Plant density, sometimes referred to as plant population, is the number of plants per unit area of land, and their spatial arrangement concerns the geometrical relationships between the plant units.

The greater the number of plant per unit area and the more evenly they are spaced, then the greater the speed with which the ground surface will be covered by leaves and complete interception of the incident radiation obtained (Fig. 4.20). However, this does not mean that the optimal plant densities are those which result in the most rapid attainment of complete light interception, or that the optional plant spacing approximate to square planting patterns, for the reasons discussed below.

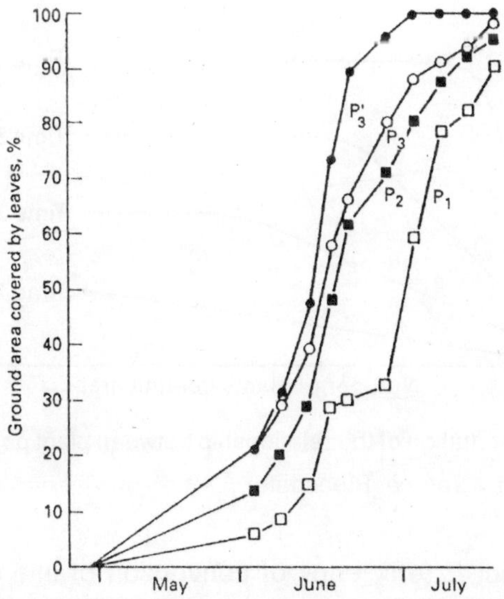

Fig. 4.20: The effect of plant density and spatial arrangement on the proportion of the soil surface covered by leaves when viewed directly from above, (□ = 32000 plants/ha 56 cm rows, ■ 86000 plants/ha 56 cm rows, O 133000 plants/ha 56 cm rows, ● 141000 plants/ha 26 cm rows.) (*Source*: Harris).

Figure. 4.21 shows the relationship between total biological yield and plant density at successive intervals of time. At the first harvest (time 1), taken so soon after the crop has emerged that the individual plants have been unaffected by competition (e.g., for light) with their neighbours, the yield will be directly proportional to plant density. As the plants grow they begin to compete with their neighbours and this will begin first at the highest density which leads to mutual shading between plants. This reduces the quantity of light received by an individual plant and its growth rate is therefore reduced. Competition is delayed between plants grown at wider spacings and the individual plants will continue to grow at a faster rate for some time. Thus the plant density at which maximum biological yield is attained gets smaller as time progresses. Note that the maximum biological yield foes not usually fall at high plant densities, for when complete light interception has been attained, the reduced growth of individual plants is compensated by the increase in their number. Crop growth rate, as opposed to the growth rate of individual plants, therefore reaches a constant which is primarily determined by the levels of incident radiation.

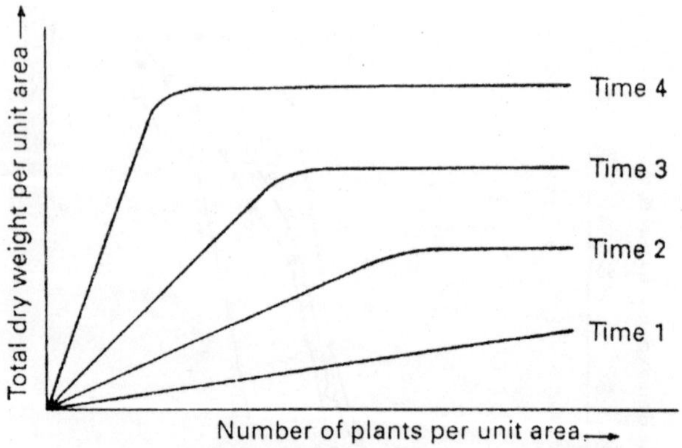

Fig. 4.21: Diagrammatic representation of the relationship between plant population and total weight per unit area on four occasions. (*Source*: Bleasdale).

Plant density may affect *E* (efficiency of conversion of intercepted radiation into dry matter) by altering the structure of the crop canopy, although this effect is probably less important than the indirect effect of increasing plant density on the rate of water loss from the soil, which increases the possibility of plant water stress reducing *E*.

In terms of economic yield, plant density has two important effects. In the first place, increasing interplant competition may affect the distribution of dry weight in the crop. In response to shading, stems tend to grow longer and leaves larger and thinner– mechanisms which enable the crop to intercept more light. This tends to result in a reduction in the supply of assimilate to storage organs such as seeds. Thus it may be expected that where yield is the result of reproductory growth, it will decline when plant density exceeds an optimum. Secondly, it will be appreciated from what has been said, that the maximum biological yield may be attained from the collective contribution made by many rather small individual plants. However, in certain crops, the size of the individual plant is important in terms of its saleability and hence economic yield. Figure 4.22 shows the effect of plant density on the yield of tubers of differing size. Since different sizes are required for particular market outlets, e.g., large tubers for baking, and small tubers for canning, control of plant density is an important means by which the yield of a particular size grade can be maximised. In a crop such as sugar beet, where the size of roots is of no intrinsic market importance, excessive plant densities contribute to a loss of yield through a reduction in average root size increasing harvesting losses (Fig. 4.23).

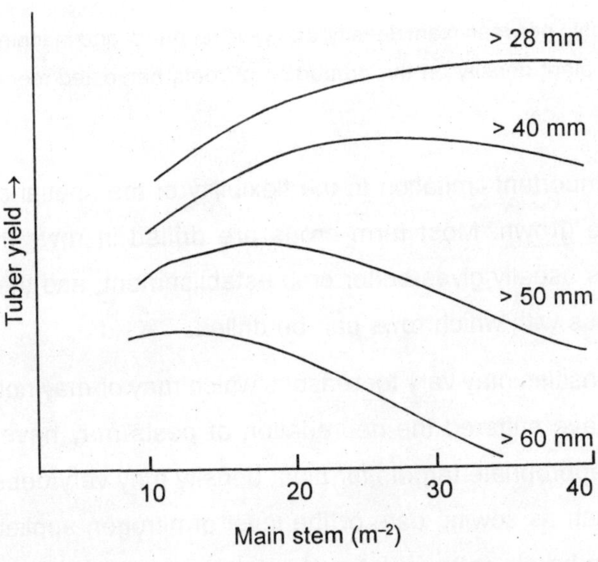

Fig. 4.22: Relationship between plant density in the potato crop and the yield of tubers of different size grades. (Source: Zaag van der).

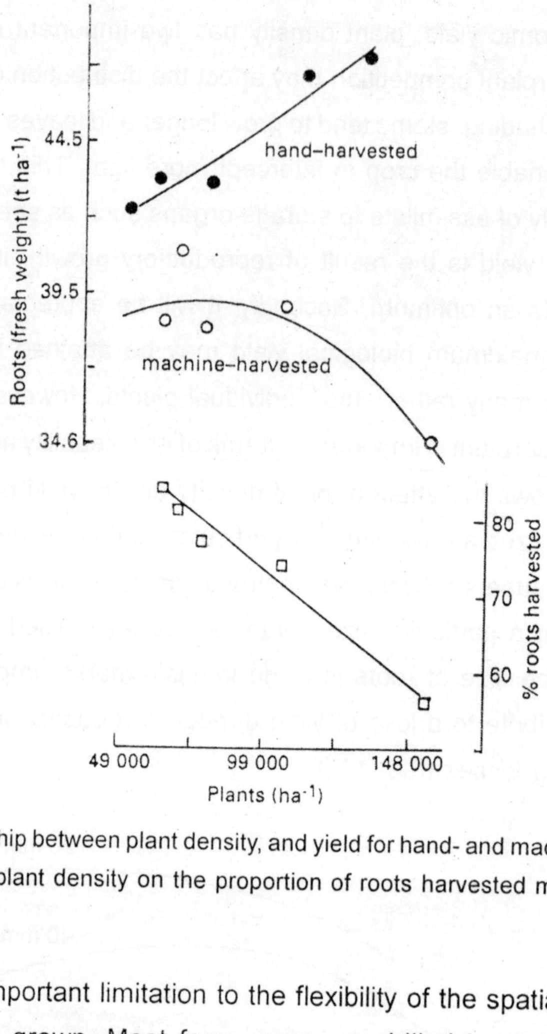

Fig. 4.23: The relationship between plant density, and yield for hand- and machine-harvested sugar beet and the effect of plant density on the proportion of roots harvested mechanically. (Source: redrawn from Harris).

There may be important limitation to the flexibility of the spatial arrangements at which crops can be grown. Most farm crops are drilled in rows in preference to broadcasting, as this usually gives better crop establishment, and there are practical limits to the closeness with which rows can be drilled.

Optimal plant densities may vary for reasons which may or may not be anticipated. Thus crops which have suffered the depredation of pests may have higher optimal sowing rates, while appropriate targets for plant density may vary depending on other agronomic inputs such as sowing date or the level of nitrogen applied. The effect of sowing date of spring barely on the relationship between grain yield and plant density illustrates this point (Fig. 4.24).

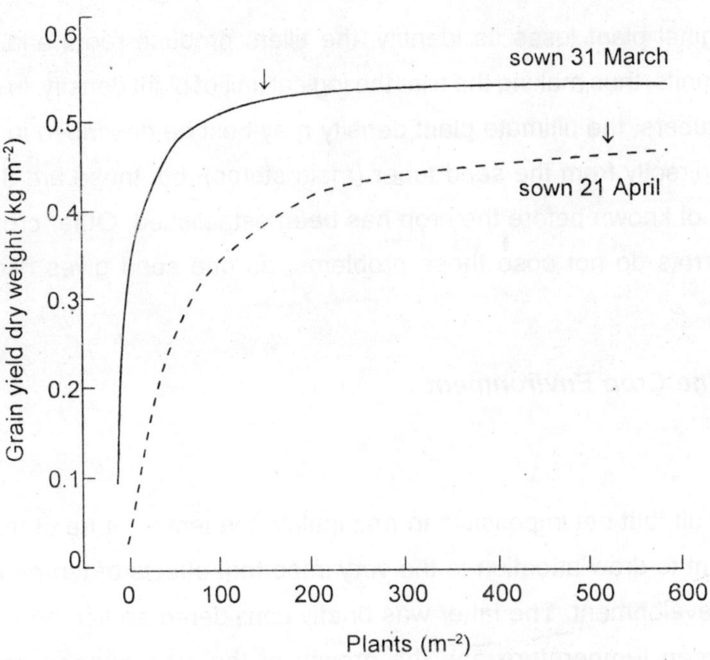

Fig. 4.24: The effect of sowing date on target (↓) plant density in spring barely variety Doublet grown at Reading University Farm, Sonning. (*Source*: Ellis and Ghassemi-Golezani).

For any given plant density, the wider the row the smaller the distance between plants within the row: the geometrical arrangement is often expressed by dividing the longest distance between plants by the shortest, the resulting ratio being referred to as rectangularity. Thus plants arranged on a square planting pattern would have a rectangularity of 1, while a crop (e.g., sugar beet) grown on 50 cm rows with a mean spacing of 25 cm between plants within the row would have a rectangularity of 2. Some crops can tolerate a fairly high rectangularity: for example, cereal crops have been sown traditionally on rows 18 cm apart, with 300 plants m^{-2} this would give a mean spacing of 1.85 cm and a rectangularity of 9.7; decreasing the row spacing to 10 cm and therefore the rectangularity to 3, would in this crop be expected to increase yield by a small amount.

Before leaving the subject of plant density, it should be pointed out that the unit in which density is described may give rise to some difficulties. For example, in cereals, as tillers (plants originating from buds developed in the axils of the lower leaves) are produced, the original plant loses its identity, the tillers produce roots and become independent plant units, thus making the tiller the logical unit of plant density. In potatoes grown from seed tubers, the ultimate plant density may best be described in terms of the stems arising directly from the seed tuber (main stems), but these are difficult to estimate and are not known before the crop has been established. Other crops, such as turnips and carrots do not pose these problems, as one seed gives rise to one plant.

Manipulation of the Crop Environment

Temperature

While it is difficult, but not impossible to manipulate the temperature of field-sown crops, it is important to draw attention to the very important effects of temperature on crop growth and development. The latter was briefly considered earlier, and here the relationships between temperature and the growth of the crop canopy are briefly discussed, using the sugar beet crop as an example. Milford *et al.* examined leaf area index (*L*) of crops grown on different sites in Britain or with different levels of nitrogen or times of sowing. The development of *L* could be described as a function of thermal time. Since leaf growth does not take place (or takes place infinitely slowly) below temperatures of 3°C, this temperature is reffered to as the base temperature for leaf growth. If, for example, the mean daily temperature ((maximum + minimum)/2) was 8°C then the thermal time would be 8 – 3 = 5°C. This is considered to be a more useful unit of 'time' than chronological time. Supposing, over a period of one week from the time of sowing, the temperatures calculated in the above way were 5, 8, 6, 5, 7, 7, 9°C, then the thermal time experienced would be 47 day-degrees. If, on the other hand, the mean temperature on each day was less than 3°C, then the thermal time would be zero. In each case, the chronological time would have been 7 days. It is not difficult to appreciate that the growth of the seedling would be better related to thermal than to chronological time. Milford *et al.* described the development of the sugar beet canopy (*L*) from the time the crop was sown as a function of thermal time using three characteristic: D_E, an estimate of the thermal time required for crop establishment (i.e. from sowing to the emergence of the crop above ground), *pL* the thermal rate of development of L, and D_L, the thermal time of the duration of increase in L. This model of leaf development is shown in Fig. 4.25. The variation in thermal time required for crop establishment (arbitrarily taken to be the time from sowing to reaching L = 0.2) varied between 381°C and 510°C-days. The reasons for this variation were not known, but could be due to a shortage of water or excess nitrogen (see section of crop establishment). While, within a particular experiment, thermal time accounted for much of the variation in the rate of expansion of the canopy, there were large differences between sites and seasons (varying from 0.53 to 0.97 units of L per 100°C-days). The thermal time interval required for leaf canopies to expand to the maximum value of L varied from 360 to 750°C-days. Since the supply of nitrogen is a major factor governing

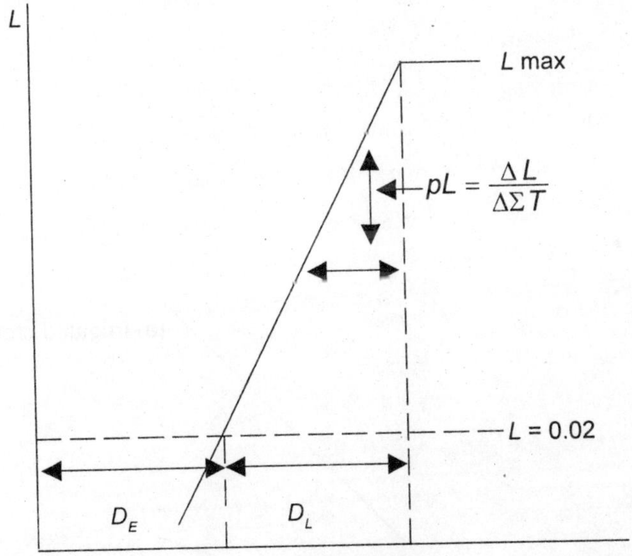

Fig. 4.25: Model to derive the thermal attributes of the expansion of leaf area index; pL = thermal rate of expansion of L, D_E = thermal time interval required to establish a rapidly expanding leaf canopy, and D_L = thermal duration of expansion of L (**Source:** redrawn from Milford *et al.*).

the rate of leaf expansion, as will be discussed, it was not surprising that a strong correlation was found between L and the percentage of N in the leaf lamina: a change in the mean N concentration from 3.9 per cent to 4.8 per cent was associated with a doubling of the thermal rate of expansion. Potassium and sodium (a beneficial nutrient for sugar beet) also had some effect on the relationship. On soils where water was not limiting leaf growth, the concentration of nitrogen, potassium and sodium accounted for 87 per cent of the variation in the relationship between leaf expansions and thermal time. In these conditions the rate of canopy expansion is largely related to temperature and the supply of fertilisers. On soils where water was limiting, the rates of leaf expansion were considerably reduced at comparable nitrogen contents in the leaf lamina (Fig. 4.26). Therefore on soils subject to drought (i.e., sandy soils) or in dry years, water is probably the main factor determining early leaf growth rates.

Thus the components of the environment which have the most marked effects on the development of the crop canopy up to the point of maximum intercepted radiation are temperature, water and mineral nutrients.

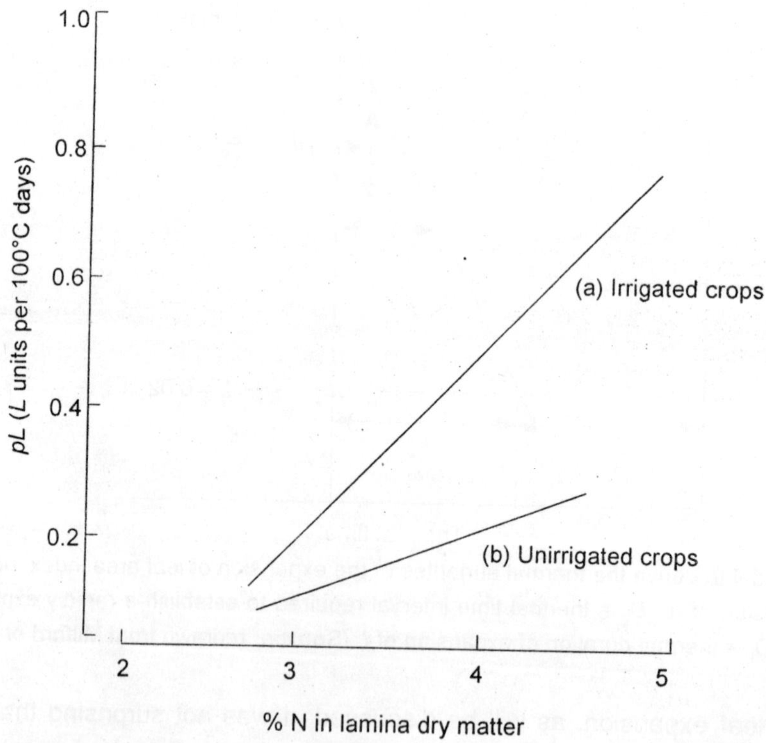

Fig. 4.26: The relationship between the thermal rate of expansion of *L* (*pL*) and the percentage of N in lamina dry matter for (*a*) irrigated and (*b*) unirrigated crops. (*Source*: redrawn from Milford *et al.*).

Mineral Nutrients

A shortage (or unavailability) of one or more essential mineral elements in the soil is one of the most frequent reasons for the failure of crops to develop sufficient leaf area fully to intercept all the incident radiation. As described earlier, carbon, hydrogen and oxygen are taken by plants from carbon dioxide and oxygen in the air and water from the soil. The remaining essential elements are taken up primarily from minerals in the soil. The most notable exception is the fixation of atmospheric nitrogen in the root nodules of leguminous plants via the root nodule bacteria (*Rhizobium* spp), while plants can absorb nutrients through their leaves, notably sulphur in the form of sulphur dioxide.

Figure 4.27 shows the effect of three major nutrients on the development of *L* in potatoes grown in an experiment at Rothamsted Experimental Station. Clearly this soil was fairly well supplied with phosphorus and potassium but not with nitrogen; supplying the latter as fertiliser would be expected to have large effects on the amount of radiation intercepted and on yield. Major effects of mineral nutrients on *E* have not been demonstrated.

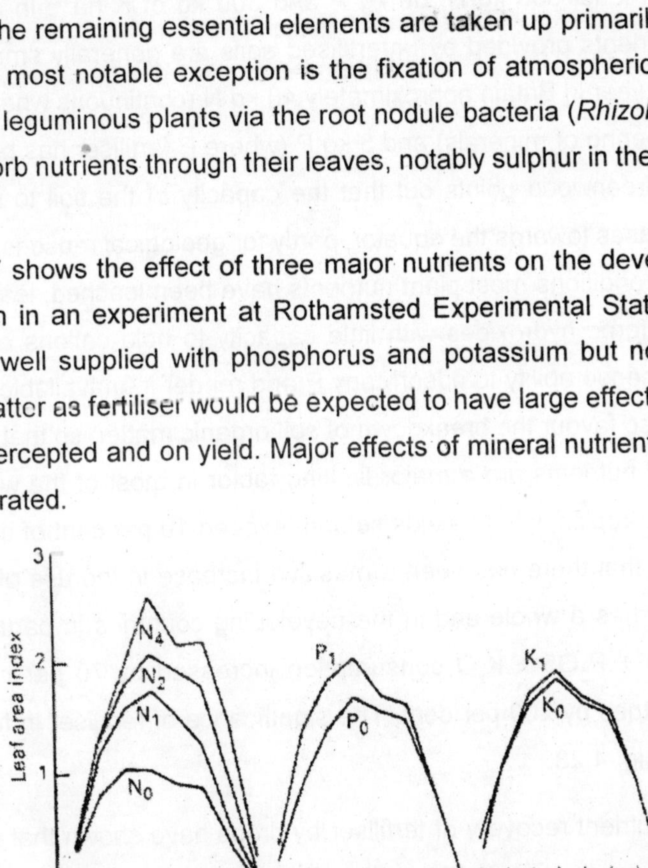

Fig. 4.27: The effects of N, P and K on changes with time in the leaf area index of potatoes, variety King Edward. (N_0 = 0, N_1 = 94, N_2 = 188, N_4 = 376 kg N ha^{-1}; P_0 ⇒0, P_1 = 156 kg P ha^{-1}; K_0 = 0, K_1 = 82 kg K ha^{-1}) (Source: redrawn from Dyson and Watson).

It has been pointed out (Greenwood), that the dry matter of crops grown with ample nutrients seldom contains less than 1.5 per cent N, 0.3 per cent P and 1.5 per cent K, the major nutrients which most commonly limit yield. In temperate regions the maximum potential dry matter yield is about 20 t ha^{-1} in the 100-150 days available for growth and about 30 t ha^{-1} in warmer climates because of the longer rowing season and the preponderance of photosynthetically more efficient crop species (C_4 plants). It is therefore arguable that if maximum potential yields are to be achieved, then crops must contain at least 300 kg N, 60 kg P and 300 kg of K ha^{-1}. In comparison, the amounts of nutrients provided by unfertilised soils are generally small, for example, per hectare per year in Britain approximately 40 kg N (continuous wheat), 10-100 kg K (from the weathering of minerals) and 5 kg P (where P fertiliser has been withheld for many years). Greenwood points out that the capacity of the soil to supply nutrients generally decreases towards the equator, partly for geological reasons, partly because under hot, wet conditions most plant nutrients have been leached, leaving a matrix of aluminium and ferric hydroxides with little capacity to hold cations against leaching and with an immense ability to adsorb any P and render it unavailable to plants. High temperatures also favour the breakdown of soil organic matter, so that it contains less N. Thus mineral nutrients are a major limiting factor in most of the world's soils, but acutely so in the tropics, where yields seldom exceed 10 per cent of the maximum. It is not surprising that there has been a massive increase in the use of fertiliser since 1950 in the world as a whole and in the developing countries in particular. Between 1966-77, total N + P_2O5 + K_2O consumption increased by 70 per cent and in the developing countries by 200 per cent. The significance of fertiliser in food production can be seen in Fig. 4.28.

Studies on nutrient recovery of fertiliser by crops have shown that while the need for more fertiliser use is overwhelming, there is an equal need for using more efficiently the fertiliser which are applied. Less than 50 per cent of the N, 15 per cent of the P and 5-80 per cent of the K applied for each crop is recovered in it (Greenwood). This nutrient loss is not only wasteful, but potentially damaging to the environment (e.g., through the eutrophication of water) and possibly to health (e.g., through the increase in the nitrate concentration in drinking water from nitrates leached from the soil in the drainage water).

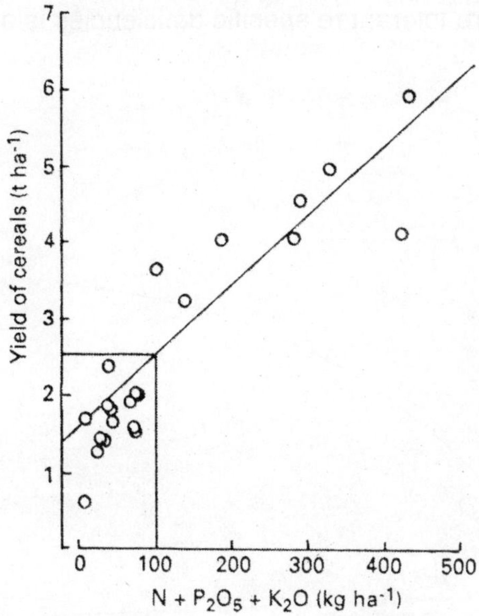

Fig. 4.28: Average yield of cereals for countries with a population of more than 35 million people plotted against the total application of fertiliser (N + P_2O_5 + K_2O) per hectare. The total population of countries within the rectangle is 2509 million, outside the rectangle 630 million. (Source: redrawn from Greenwood).

The response of total dry matter yield to inputs of fertilisers generally takes the form of a diminishing response curve. There have been few reports linking the supply of fertiliser to the amount of radiation intercepted by crops: in Fig. 4.29, however, an example is given to show that successive increments of fertiliser nitrogen had progressively smaller effects on *IR*, which would account for this form of response. However, economic yield may be depressed by high levels of fertiliser through adverse effects on the distribution of dry matter.

While crops need to absorb very small quantities of the trace elements to permit normal growth, the absence, or more usually the unavailability to the plant, of these elements restricts crop yields over large parts of the world. Because deficiencies of trace elements may be due to the unavailability of the elements in the soil, the correct formulation of the nutrient and the appropriate method of application are extremely important when attempting to control trace element deficiencies. Different crops and even varieties of specific crops vary in the degree to which they are affected and breeding varieties which are tolerant to specific deficiencies is an important means of combating the problem.

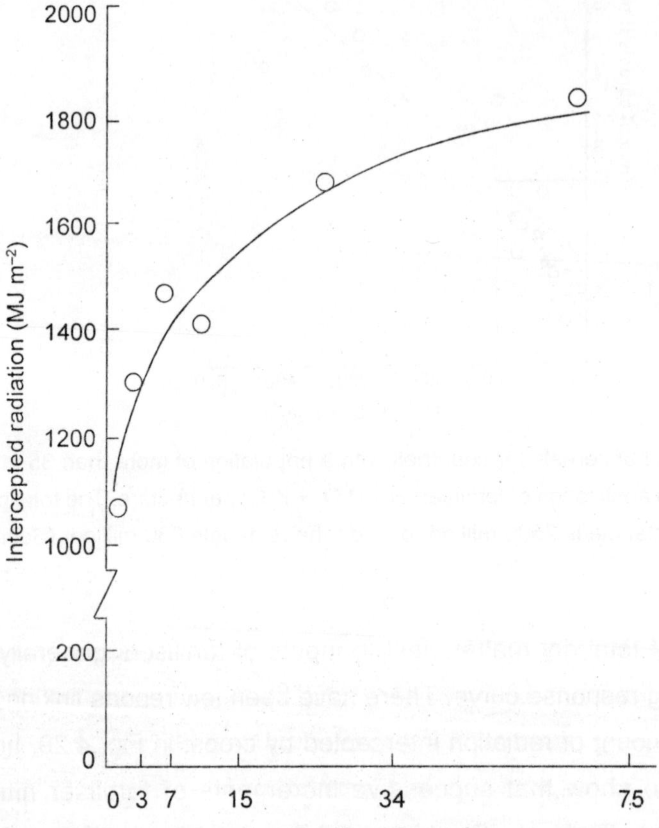

Fig. 4.29: Relationship between applied N fertiliser and the total amount of radiation intercepted during the course of growth for the maincrop potato variety Cara. (*Source*: Harris).

Water

The role of the green plant in trapping radiation is facilitated by the structure and the arrangement of leaves which permits the crop to intercept as much radiation as possible. An inevitable consequence of this arrangement is that a complete canopy of leaves attached to a root system well supplied with water will lose water at about the same rate as an open water surface.

The rate of water loss is fixed by the amount of energy available to vaporise water together with the rate at which water vapour is carried away from the surface of the leaves. For a short green crop which completely covers the ground with leaves and is liberally supplied with water, the amount of water loss can be calculated from standard weather data recorded at meteorological stations. Some allowance must be made for the fact that the stomata of the plants close during the night, which serves to cut down the rate of water loss to some extent. If these conditions are met, then the potential transpiration of crops can be calculated (Penman).

Potential transpiration is expressed in millimetres (mm) of water per unit of time (day/week) and would be expected to increase as temperature rises and incoming radiation increases with the advance of the growing season in temperate latitudes (see Fig 4.13 and 4.30). In arid condition, the loss of water from irrigated crops may be enhanced through the influx of hot, dry air from adjacent unwatered areas, an effect often referred to as an 'oasis effect'. In the growing seasons, rates of potential transpiration are typically 1-3 mm water/day: at the other extreme in very hot arid regions, the rate may be 10-12 mm/day.

Knowing the rainfall and potential transpiration (Fig. 4.30), it is possible to construct a daily, weekly or monthly budget showing the balance between income in the form of rain or irrigation and output in the form of potential transpiration. If the result is positive, it shows that more water has entered the soil, from rain or irrigation, than has been lost by transpiration. However, this surplus water should be lost by drainage (assuming this to be adequate), and any positive value will be regarded as returning the soil moisture deficit to zero. If, however, more water has been lost than gained, a 'soil water deficit' of that magnitude will have built up.

The soil acts as a reservoir of water, buffering the crop to some extent between short-term fluctuations in supply and demand. The upper limit of water that can be held by a soil as fixed by the 'field capacity' or the water content of the soil at which it is holding the maximum amount of water against the pull of gravity. Plants can utilise the water in the soil profile, although often with increasing difficulty as the water is removed and is held in increasingly smaller pores and consequently at higher tensions. Plants will cease to remove water from the soil profile when the soil water content has been lowered to 'the permanent wilting point'. The tension at which this occurs is about −1.5 MPa. At this moisture content plants will wilt even when their leaves are placed in a water-saturated atmosphere. The amount of water held between field capacity and the permanent wilting point is said to be the 'available water-holding capacity' of the soil. The amount of available water per unit volume of soil is strongly dependent on the soil texture and organic matter content.

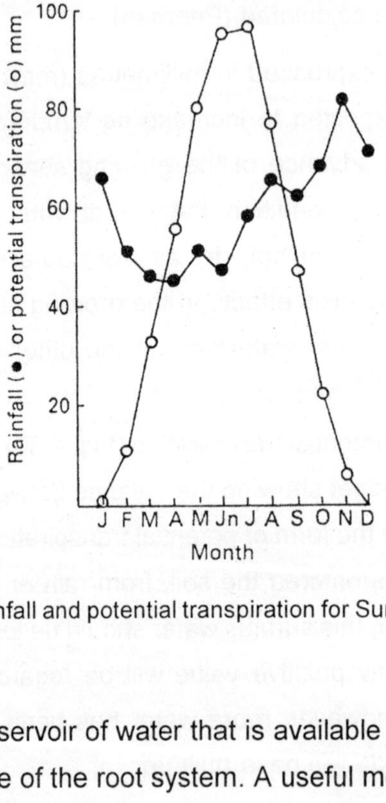

Fig. 4.30: Mean monthly rainfall and potential transpiration for Surrey and east Berkshire, England. (Drawn from data of Smith).

How much of this reservoir of water that is available to plants can be drawn upon will depend upon the size of the root system. A useful measure of roots is root length per cm^3 of soil. This can be very extensive, amounting to total root length of many kilometres beneath each square metre of soil surface area. For any given set of soil conditions, plant species differ considerably in the size of their root systems and the depth to which they penetrate the soil. This is undoubtedly an important factor (but by no means the only one) governing the ability of crop species to withstand water stress. A feature common to all plants is the progressive decrease in the size of the root system with depth (Fig. 4.31).

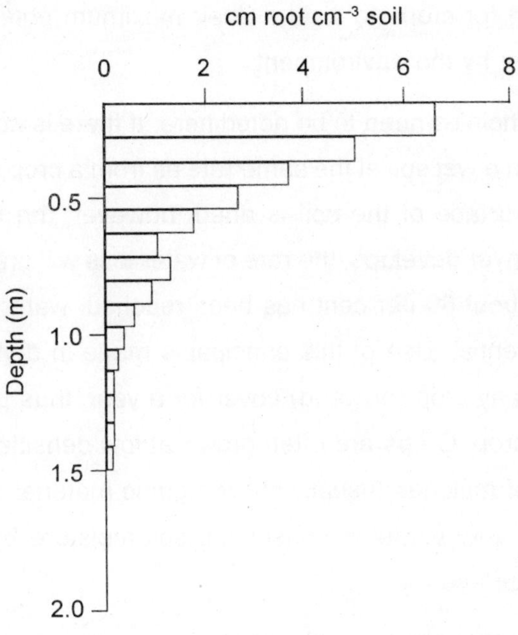

Fig. 4.31: Distribution of root length with depth of wheat at anthesis. (Source: redrawn from Gregory).

Under conditions where the rates of transpiration are extremely high, e.g., on clear days with high temperatures, crops may wilt, even though the soil is close to field capacity. This is because, once the soil has dried within the immediate vicinity of the root, water has to move to the root across a layer of dried soil, which substantially reduces the influx of water to the root.

In the absence of any input of water, the diurnal water stress in the plant increases and its ability to recover during the night decreases. Such stresses on the plant inevitably affect growth. In the first instance, the efficiency of conversion of radiation into dry matter is affected, principally through the closure of stomata which cuts down the inflow of carbon dioxide, but as deficits increase, metabolic activities become increasingly disrupted. The plant also reacts in other ways to reduce stress which also affect growth. The major consequence of these effects is to curtail the expansion and accelerate the senescence of leaves. Thus water stress affects yields by decreasing the amount of radiation intercepted and the efficiency with which the intercepted radiation is utilised. There is therefore a strong positive relationship between transpiration and yield, and for crops to yield at their maximum potential, they must transpire at the potential set by the environment.

Two other important principles need to be noted here. If there is no crop (or weed) cover, water will be lost from a wet soil at the same rate as from a crop or from an open water surface. Once the surface of the soil is dried, however, the rate of loss will decline rapidly. As a crop cover develops, the rate of water loss will gradually increase and once a leaf cover of about 60 per cent has been reached, water is being lost at rates not far below the potential. Use of this principal is made in dryland farming by keeping the soil free from any crop (or weed) cover for a year, thus using two years' water supply to grow one crop. Crops are often grown at low densities in low rainfall situations, where the use of mulches (usually bulky organic material spread over the surface of the soil) is also very useful in conserving soil moisture by preventing its evaporation from the soil surface.

The second principle is that plant growth may be severely curtailed at soil water deficits which still allow crops to transpire at about the potential rate. The reason for this is that soil water is depleted from the superficial, mineral-rich layers of the soil progressively downwards, while roots may still extract sufficient water for growth from the mineral-deficient subsoil. As water is essential for the uptake of minerals, which enter the plant in solution, depletion of water in the upper layers of soil may result in a severe depression of growth due to water-induced nutrient deficiencies.

A major concern of research into crop water stress is to determine when to irrigate and how much water to apply. To achieve this, attempts have been made to define 'limiting deficits', or the soil water deficits at which growth is potentially seriously restricted, for crops grown in Britain. In general, for crops considered economic to irrigate (e.g., potatoes), soil water deficits are maintained at less than 50 mm if possible.

Negative Biological Factors

Most of the negative influences on yield exert their effect through reducing the ability of the crop to intercept radiation, that is, through reducing the leaf area of crops. Traditionally, reasonable levels of pest, weed and disease control were achieved by growing plants in rotation of contrasting crops, but in modern capital-intensive agriculture there has been a shift in some countries, such as Britain, to increased specialisation with a concomitant reduction in the opportunity to rotate different crops. For example, increasing areas of cereals, and particularly winter cereals, provide greater opportunities for the carry-over of pests and diseases from one crop to another, and certain weeds become more difficult to control.

The steady introduction of a wide range of both selective and non-selective herbicides has made it possible to control most weed problems in most crops. In developed countries, where labour is relatively scarce and expensive, herbicide use has become a normal input on most arable crops.

Non-cultural control of pests and diseases has relied on two principal methods, the breeding of disease- and pest-resistant crop varieties and the development of crop-protection chemicals. Both methods of control suffer from the phenomenal rate of reproduction of many disease and pest organisms, which often contain sufficient genetic diversity for the development of new strains resistant to pesticides or capable of attacking hitherto resistant cultivatars. This requires increased rates of chemicals and eventually the discovery of new chemicals and the breeding of new varieties in a race in which ultimate victory is rare or impossible. Chemicals suffer from two additional disadvantages: they may be harmful to organisms which do no harm or are beneficial to crops, and some chemicals resist breakdown and may give rise to insidious long-term harmful effects (Carson). More rigorous testing of new compounds for such effects has had the effect of reducing the flow of new compounds into agriculture as well as increasing their development costs.

An outstanding 'bad' example of these problems is the use of chemical control of the cotton bollworm (*Heliothis zea*) which is practically immune to all available insecticides, following a spiral of more and more frequent applications of chemicals, in extreme cases up to 60 applications in one growing season. The high costs entailed, and the ultimate failure to control the pest has in many instances made it uneconomic to grow the crop, has elevated formerly harmless organisms to pest status, has affected the quality of the environment and posed serious hazards to the health of the agricultural workers (Luckman and Metcalf). Such problems have not reached this acute stage in Britain, mainly because pests are much less of a problem in cool temperate climatic conditions. However, the development of resistant aphids—an important crop pest—to certain insecticides has occurred. More attention is now being paid to the possibility of controlling pests and weeds by biological rather than chemical methods. Biological control has been defined as "the purposeful use of an organism or organisms to reduce a plant or animal population that is inimical to man" (Samways). This principle is illustrated in Fig. 4.32.

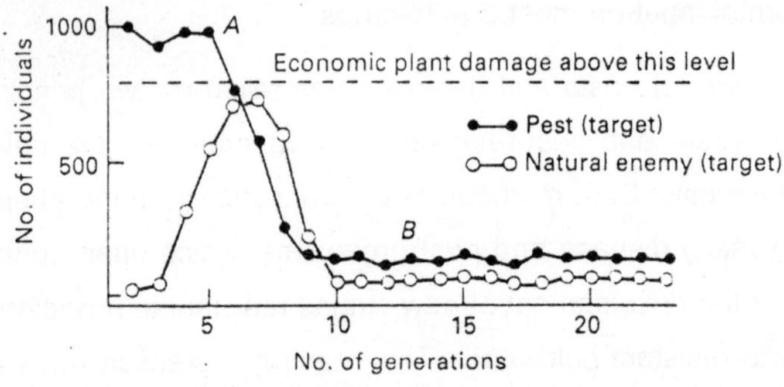

Fig. 4.32: Diagram of successful biological control. The increase in the population of the natural enemy (O) case a decrease in the pest (●) from a damaging level (A) to a non-damaging level (B). With now fewer pests to sustain it, the natural enemy population declines. Both populations then continue at a low equilibrium level. (*Source*: Samways. The figure is modified after Varely G.C., Gradwell G. R. and Hassell M.P., *Insect Population Ecology*).

The biological control of weeds has generally been most successful with perennials and especially with introduced weeds. While most attention has been given to the insect control of weeds, a wider range of agents, including mites, nematodes, plant pathogens and parasitic plants are currently the objectives of research.

There have been several examples of the successful control of insect pests by biological methods, particularly, for example, on relatively sedentary insects such as scale insect pests and mealy bugs in citrus fruit crops. In Britain, perhaps the most widespread application is the control of the glasshouse white fly (*Trialevrodes vaporatiorum* Westw) by the introduction of a parasitic chalcid wasp (*Encarsia formosa*

Gahan). The small scale of such operations and the close environmental control available has obviously facilitated implementation in this particular case. However, the lack of commercial incentives to develop such methods has restricted the research done on these problems, and a greater input of resource into such research may be expected to extend the range of applications of biological control.

The philosophy of pest control which appears to be emerging is neither chemical nor biological control but 'integrated control', in which appropriate chemicals are used alongside biological control. Much, too, is expected from the new technology of genetic engineering with respect to the containment of pests and diseases without the unwelcome side effects noted in connection with the use of conventional pesticides.

Little attention has been given in this discussion to the application of ecological thinking to the solution of pest and disease problems. In general, the more diverse the cropping system in a given environment, the more stable it is likely to be and the less prone to suffer from major pest and disease epidemics. Modern cropping systems arc often vcry far removed from such stable ecological systems and therefore need constant inputs of 'artificial' rosaries or maintain reasonably stable production. The understanding and harnessing of 'natural' methods of maintaining stability from pests and diseases is a challenging problcm for which solution become increasingly pressing.

Completion of the Production Cycle, Harvesting and Storage

The end of the production cycle may come when the crop has naturally senesced (see hypothetical example in Fig. 4.16), and has no green plant parts with the capacity to intercept radiation and photosynthesise, as exemplified by the cereal crop. Other crops may senesce naturally by the time of harvest, but in some instances may still have a functioning crop canopy, depending on variety, season and the market outlet for the crop. This situation is exemplified by the potato crop: early varieties tend to senesce within the normal growing season if allowed to develop to maturity, but they are usually harvested as soon as an economic yield has been obtained, and the crop canopy is destroyed chemically or mechanically to allow the crop to be harvested. The canopy of a late-maturing variety may persist until killed by frost, or artificially. In the biennial sugar beet crop, the leaves persist all though the harvesting period

(September to December in Britain), and are capable of photosynthesising, although the crop growth rate in the late autumn is negligible owing to low temperatures and low incident radiation. Some yield will be scarified by lifting in September and October. At the other extreme, the whole crop canopy may form the harvested product, as for example, in grass or maize cut for silage. The particular condition in which a crop is harvested and whether some yield has been sacrificed is therefore dependent upon the crop species and the purpose for which it is grown.

Method and ease of harvesting are important considerations which have to be taken into account when deciding certain details in the method of production. For example, the adoption of very wide row spacing in the potato crop is due largely to the need to reduce the production of clods which interfere with the size grading of the tubers and lead to damaged tubers. The deeper ridges made possible by growing the crop on wide rows, also make it easier to obtain a good soil cover over the tubers, thus preventing their exposure to light and becoming green. Ease of harvesting and reduction of damage is also associated in this crop with the production of a deep, stone-free tilth.

The need to spread the date of harvest will be affected by the choice of crop, e.g., early potatoes, winter barley, winter wheat, spring beans, maincrop potatoes and sugar beet could give a sequence of crop harvests which span a period from June to December in Britain. Within any one crop species, the choice of variety can influence the date of harvest, although this does not constitute the only way in which the plant breeder has influenced the harvesting operation. For example, the production of short, stiff-strawed varieties of cereals has considerably reduced the tendency of these crops to 'lodge' and reduced the difficulty associated with harvesting such crops and the attendant loss of yield due to failure to collect all the grain, while long before plant breeding was consciously organised, selection of cereals for a non-brittle rachis was a vital factor in improving their harvestability. Breeders can also affect the degree of determinacy exhibited by naturally indeterminate crop species, such as peas and beans, which can condense the age range of the economic yield component and improve the harvestability of the crop. In the potato crop, the breeder may improve the ease of harvesting by concentrating the tubers in the ridge, by reducing the length of stolons of which the tubers are borne, by improving the shape of the tubers and by increasing their resistance to mechanical damage.

A weed-free crop at the time of harvest is also a vital prerequisite to the smooth running of this operation. Clearly, examples of the interaction between methods of growing the crop, its genetic make-up and the success of the harvesting operation can be multiplied but enough has been said to indicate the importance of such interactions, and the need to be aware of them selecting the crop species, variety and method of production (Fig. 4.16).

Once harvested the crop is frequently, although not invariably (e.g., zero-grazed grass or forage), stored or conserved before use or sale to some later date, which can be a matter of days, months or even years.

Biologically, the major causes of deterioration in store are natural senescence, microbial and non-vertebrate degradation. Natural senescence is associated with the hormonal balance in plant organs. Microbial degradation is associated with saprophytic organisms which accumulate on crop tissues during growth, but which fail to colonise them: during storage, however, given appropriate conditions, they will multiply and decompose the product. In store, it is therefore necessary to eliminate the decomposers, or retard their activity. The particular strategy adopted will depend upon the nature of the stored product, its end use and cost.

For seed crops, water content and temperature are the two most important variables. Bacteria require very high relative humidities (approaching 100%) for optimal growth, while storage fungi operate best at somewhat lower relative humidities (about 85%).

The storage moisture content of most crop seeds in equilibrium, with relative humidities between 65 per cent and 70 per cent (which should successfully eliminate microbial growth), is between 13 per cent and 15 per cent but is 8 per cent for crops such as oil-seed rape and linseed, crops with a high oil content. Storage temperature has an important bearing on whether moulding takes place, a marked rise in temperature leading to an increase in equilibrium relative humidity, to the advantage of the fungi. Fungal growth leads to a rapid rise in temperature, as the respiration rate of fungi is much higher than that of stored grain. While most seed crops are stored under cool dry conditions, there may be some advantages in storing grain intended for animal feed at high moisture contents (e.g. 18-20%): this may involve storage at very low temperatures, in airtight conditions or with the addition of a chemical preservative such as propionic acid.

To store vegetative material in its natural state, without processing for human consumption, poses additional problems, because the material usually has a high water content (e.g., in potato tubers about 80%). The main product of this type stored on farms in the UK is the potato and here the main causes of weight loss are respiration, evaporation, sprout growth and disease. The overriding requirement is to store under cool conditions (i.e., between 4-8°C) and to maintain a relative humidity of between 95-98 per cent. Again, the end use may affect the storage requirements; low temperatures, for example, give rise to reducing sugars which bring about unacceptable darkening in crisps and chips, storage temperatures in the range 5-10°C are required here with the use of chemical sprout suppressants.

Materials with a high water content, such as potato tubers, are much more prone to damage than are low moisture content grains. At the start of storage, use is made of the principle that wound healing is favoured by high temperatures and in the first ten days of storage, potatoes are allowed to heat up to temperature of about 18°C, from their own respiration, by withholding ventilation.

The main invertebrate pests of storage are the grain mites and storage insects. In cool, damp climates, all grain is potentially at risk from mites which can survive at the normal moisture contents at which grain is stored. The storage insects (saw-toothed beetle, *Oryzaephilus surinbamensis* L and the grain weevil *Sitophilus granarius* L) are of tropical or warm temperate origin and need temperatures of 12°C or more in order to breed. Control is by cooling, drying and store hygiene.

Vegetative material, such as grass or lucerne, is conserved mainly as hay or silage; the former involving the principle of drying and storing at low moisture contents, and the latter encouraging the production of organic acids which effectively pickle the product.

Most farmers are concerned with storage of products for up to about one year. There is, however, major concern by plant breeders that, due to the widespread use of improved varieties and the resultant disappearance of local varieties, important sources of genetic variability may be irrevocably lost. This has led to the setting up of so-called 'gene banks' for the long-term storage of seeds with widely differing genotypes.

Conclusions

In this brief account of the underlying principles on which crop production is based, the central theme has been the crop as a collection of individual units which together form a green crop canopy. It is suggested that the maximum rate at which dry matter can be 'manufactured' by the crop canopy is achieved when the canopy completely covers the surface of the soil with sufficient layers of leaves to allow negligible amounts of radiation to filter down to the surface of the soil, under conditions where the water supply is non-limiting. Given a knowledge of the climatic variable such as temperature, and the length of the cropping seasons, it has been shown that it is possible to estimate the maximum potential biological yield of a crop for defined environments. It would appear unlikely that this potential would be raised significantly in the foreseeable future. Increases in crop yields, which have been spectacular over the last forty years, may be attributed to two main principles: firstly, diverting a greater proportion of the total biological yield to the economic fraction, and secondly, to increasing the size and duration of crop canopies. The first objective is largely attributable to the work of the plant breeder whose major contribution to increasing the yield of cereals, the worlds most important crop type, has been to increase the harvest index, through the introduction of short, stiff-strawed varieties. The second is largely attributable to the chemist and chemical engineer, who have, by proving relatively cheap fertilisers, made it possible to remove mineral nutrients as a limiting factor for growth, while the provision of crop protection chemicals has, as their name implies, undoubtedly reduced the negative effects of weeds, pests and diseases from restricting the potential size of crop canopies.

The principles underlying the production of crops which maximise their yield potential are, as have been illustrated, relatively straightforward, the major difficulty is the means whereby limiting factors are removed, and their implications for the long-term sustainability of production and the quality of the product in terms of its nutritional value and risks to health.

Thus in capital-intensive agriculture, typified by countries in temperature northern latitudes, with relatively stable populations, the emphasis has been shifted from maximising food production per unit area, since many items are in surplus, to lowering the cost of production. There is some considerable pressure for agriculture to reduce the use of crop production chemical (e.g., nitrogen) and crop protection chemicals, not only as a means of reducing crop surpluses, but to reduce the amounts of naturally-occurring (e.g., nitrates) and exotic chemical (e.g., pesticides) entering the food chain or the water supplies, since their presence is considered to be hazardous to the environment and to health. In the tropics and sub-tropics there is a similar concern with inputs, partly because they are unavailable or too expensive (see Fig. 4.15), and partly because the use of pesticides has not provided sustainable solutions to crop protection problems and has given rise to attendant health risks, as already noted. Here the problems are particularly acute since the population pressure is intense, and the ability to develop new land resources limited and potentially non-sustainable because of erosion problems where, as is most likely the case, trees have been removed to make way for crops.

The next 50 years should see increasing emphasis on deriving principles through which crops can be produced in sufficient quantities to sustain a world population which may almost double over that period, by methods of production which are sustainable and environmentally benign. Problems will be particularly acute if population changes are accompanied by marked climate changes which are fairly confidently predicted. This is an immense challenge for all engaged in crop production, the most fundamental of man's activities, and it is vitally important that the resources and enthusiasm with which to meet it will not be lacking.

❑❑❑

5

The Principles of Animal Production

Introduction

The primary purpose of animal production systems is to convert plant products into animal products usable by man. Animals provide foods, such as milk, meat and eggs; clothing, from wool, hair, fur and hides; and power, from horse, cattle and other species used for draught purposes.

Over the centuries since animals were first domesticated, their role has extended from that of ensuring man's survival to that of enhancing his enjoyment of life. The Indian farmer still relies on the power of his cattle to till has cropland, on their milk to keep his children alive, and on their dung for fuel. Developed countries such as Britain, have alternative sources of power and fuel, the technology to provide with a nutritionally adequate diet entirely from plant products, and the skills to manufacture synthetic alternatives to wool and leather, yet use of (if not dependence on) farm animals is much greater than that of the people of India. Table 5.1 shows the contribution of animal products to the national diet in the UK. Table 5.2 calculates the plant input to these animal products. These tables show first that the principle used to open this chapter, that animals are used to convert plant products into animal products usable by man, does not have a sound biological basis. But the tables also show that the principle is economically correct, for the British farmer annually converters £3,875 m of crop products (some of which are scarcely saleable anyway) into £7,471 m worth of animal products. These animal products account for 62 per cent of the monetary value of UK agricultural output.

Table 5.1: Daily nutrient consumption per person in households in the UK (and the percentage provided by animal products)

Energy (MJ)	8.69 (37)	Calcium (mg)	864 (63)
		Iron (mg)	11.1 (30)
Protein (g)	67.6 (62)	Thiamine (mg)	1.26 (27)
Fat (g)	97.2 (59)	Riboflavin (mg)	1.77 (63)
Carbohydrate (g)	246.0 (10)	Vitamin A (mg)	1377 (65)
		Vitamin D (mg)	2.92 (39)

Source: MAFF, *Household Food Consumption and Expenditure.*

Table 5.2: Annual feed input and animal output in UK agriculture

	Feed input		Animal output	
Source	Quantity (Mt of dry matter)	Value (£m, 1988)	Category	Value (£m, 1988)
Pasture herbage			*Livestock*	
Grazed	20	800	Cattle	1919
Conserved	12.5	700	Sheep	714
			Pigs	871
			Poultry	815
			Others	122
			Sub-total	**4441**
Other fodder crops			*Products*	
Roots[a]	1.7	150	Milk and milk products	2483
Kale, etc.	0.6	40	Eggs	465
Straw	1.5	35	Wool	47
			Others	35
			Sub-total	**3030**
Cereals				
Grains	9.0	1100		
By-products	3.0	300		
Protein concentrates				
Oilseed residues	2.8	600		
Fish and meat meals	0.5	150		
Total		**3875**		**7471**

[a] Includes sugar beet by-products.

To understand the principles of animal production we need some basic knowledge of the animal body and its functions. The *growth and development* of the body is particularly important because it determines the nature of many of the products used by man (e.g., meat and wool). Growth is achieved by *nutrition*. For animal production to be self-perpetuating, *reproduction* is essential; in birds, this also gives a product (eggs), and in mammals it is followed by lactation (i.e., milk production). Reproduction also offers opportunities for changing animals by the processes of genetic *improvement*. Finally, the animal has to be protected from the elements of its *environment*, and particularly from diseases. The italicised words above provide the conceptual framework that will be used in this chapter to explain the principles of animal production.

Growth, Development and Function

All animals begin life as a single cell, the fertilised ovum. This cell divides; the new cells enlarge and then divide again. As the process is repeated the animal increases in size and weight and is said to grow. For the first few divisions the cells remain identical to one another but quite early in life (e.g., by day 14 in cattle) they begain to differentiate to form an embryo. The different types of cells form distinct tissues, such as nerve, bone, muscle and fat. After 45 days the cattle embryo has become attached to the mother's uterus by the placenta, and becomes more recognisable as an animal. The tissues are arranged to form the body parts, such as head, trunk and limbs, and also the internal organs, such as heart, lugs, stomach, liver and kidneys. During the last one-third of gestation, the foetus grows rapidly (Fig. 5.1, solid line).

By the time it is born, the animal will be complete in the sense that it will have all the tissues, parts and organs needed for post-natal life. Some of its components will have begun to function before birth; for example, the foetal heart begins to beat as early as day 21 of the 285-day gestation period of cattle. Other components will be brought into use at birth; thus the foetus receives the oxygen it needs via its mother's blood supply, but at birth the newborn animal's lungs must immediately take over this function. A third group of the components present at birth will not function until later in life, the obvious example being the reproductive organs.

After birth the animal continues to grow rapidly, the growth of mammals being sustained by their mother's milk. If a good supply of food is maintained after weaning, rapid growth will continue until the animal achieves its mature weight (Fig. 5.1 solid line). If the food supply fluctuates, perhaps as a result of seasonal changes in grass growth, the animal's growth will not follow the idealised curve, but will also fluctuate (Fig. 5.1, broken lines).

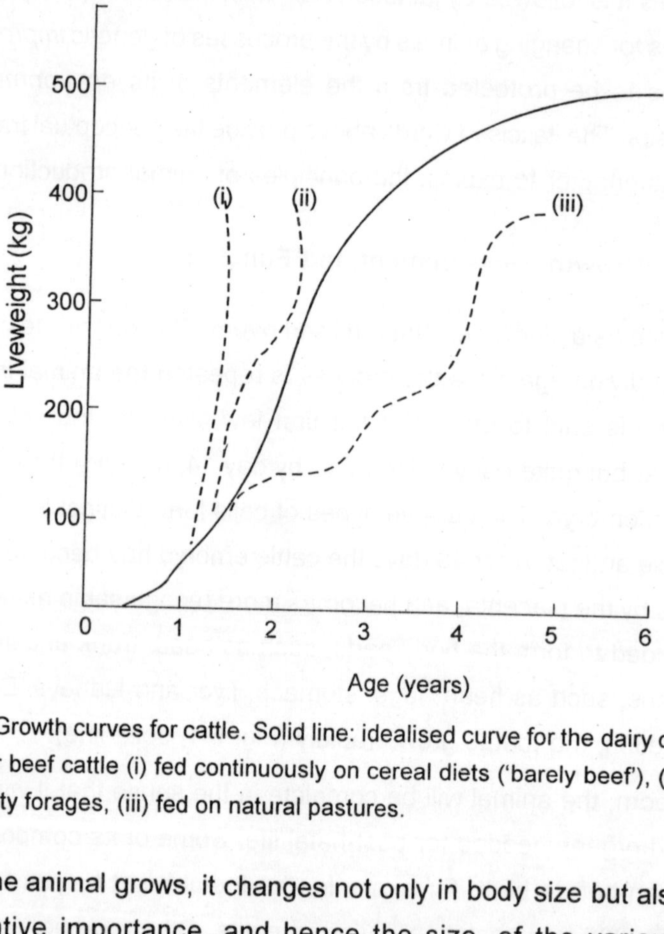

Fig. 5.1: Growth curves for cattle. Solid line: idealised curve for the dairy cow. Broken lines: typical curves for beef cattle (i) fed continuously on cereal diets ('barely beef'), (ii) fed on diets based on high-quality forages, (iii) fed on natural pastures.

As the animal grows, it changes not only in body size but also in body properties. The relative importance, and hence the size, of the various body tissues and components changes as the animal grows from embryo to adult. For example, an animal cannot use a skeletal muscle until it has bones to which to attach it and nerves to control it. Thus in the foetus the order of priority for growth of tissues is nerve-bone-muscle-fat. As a result, the calf of lamb at birth has a large head (brain and bone), long spindly limbs (bone but little muscle), and very little fat. After birth, bone growth continues until puberty (i.e., sexual maturity) is reached. There is also a relative increase in muscle growth. Fat is the last tissue to be laid down, initially as internal deposits within the abdomen, later under the skin and around the muscles, and last of all, within the muscles.

The processes by which animals change their body proportions are known collectively as development. For each of the domestic species it is possible to discern a typical pattern of growth and development in which the weight of a particular organ or tissue of the animal comprises a predictable proportion of the total weight. Nevertheless, the pattern for each species is likely to vary according to breed type, sex and nutrition. In cattle, for example, a small, early-maturing breed such as the Aberdeen Angus will have a mature weight of about 400 kg, whereas a large, late-maturing breed such as the Charolais will have a mature weight of 600 kg. This means that if the two breeds are compared at 400 kg, the Aberdeen Angus, being mature, will be fatter than the Charolais. Within these breeds, females will be fatter than steers of the same weight, and the latter will be fatter than bulls. Males also differ from females in other growth characteristics; for example, males have greater development of the head and shoulders. But the major variations in growth patterns are due to differences in fat deposition; in other respects, for example, in bone-to-muscle ratio, differences between breeds and sexes are relatively small.

The same is true of the influence of nutrition on growth. In general, animals on a high place of nutrition, hence growing rapidly, will be fatter at a given weight than animals that have grown slowly to that weight. This effect is seen most clearly in the pig, and with bacon pigs it is usually necessary to restrict food intake and growth rate to prevent over-fatness of the carcass.

Growth and Meat Production

The shape and form of today's meat animals is a compromise between the needs of the animals themselves—to eat, move about, reproduce, etc.—and the requirements of man for meat. In species that reproduce rapidly, and hence can be changed in form more readily, modifications to meet man's requirements may sometimes seem excessive; for example, the selection of turkeys for massive breast muscle development has produced birds that have difficulty in copulating. Extreme as this example may be, it illustrates the importance of consumer demand for muscle (i.e., lean meat). Earlier generations of consumers also demand for fat in meat (to provide energy for physical labour) but the demand for fat has fallen to the pint where today's consumer wants little or no visible fat (e.g., subcutaneous fat) on his meat but still values the flavour and juiciness imparted by intramuscular fat. As the latter is the last fat to be laid down it cannot be provided without fat being deposited elsewhere. There are also other contradictory requirements. The consumer's primary criterion of meat quality is tenderness; this is best achieved by slaughter of young animals, but meat from young animals lacks flavour (compare veal with beef, or lamb with mutton). A further complication is introduced by the requirement of the butcher for carcasses that can be cut up with minimal waste (e.g., as bone or trimmed-off subcutaneous fat).

The strategy adopted by the farmer to meet the requirements of the consumer will vary according to the animal species. With poultry, he uses specially selected hybrids (breeds) that can be grown rapidly to a slaughter weight that yields the carcass desired. With pigs the breed type and slaughter weight are again important, but nutrition (as indicated earlier) is a critical determinant of carcass fatness. For the bacon market in particular, a few extra millimetres of backfat can make the difference between top-grade and poorer (hence lower-priced) carcasses. With cattle and sheep the farmer's problems are complicated by numerous breed types (each suited to its particular environment) and by the influence of fluctuations in feed supply on optimum time of slaughter. For each individual, the farmer will have in mind some target degree of development (as judged mainly from subcutaneous fat cover) that is likely to suit market requirements.

Animal Nutrition

The essential processes of nutrition are depicted in Fig. 5.2. The animal ingests foods, which are mainly of plant origin, digests them into simple substances that can be absorbed, and then *metabolises* them either to form body constituents and products or to provide energy. Waste products are excreted.

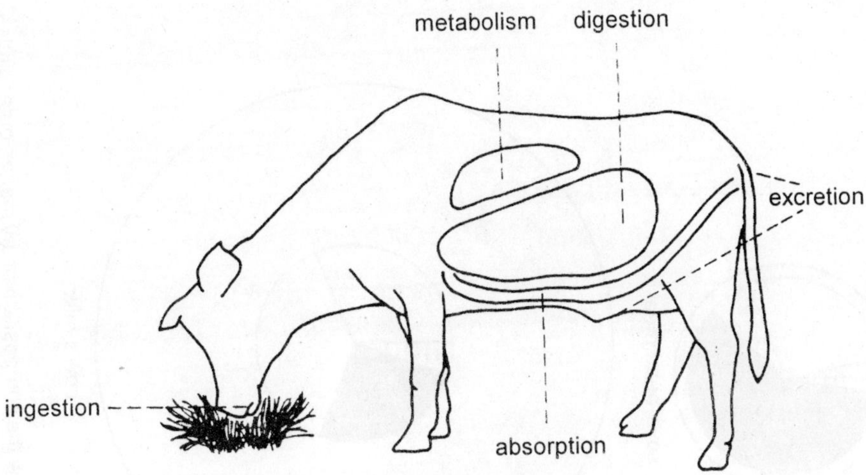

Fig. 5.2 Elements of nutrition.

As explained earlier (Chapter 2 and 4), plants consist of three major classes of nutrients; proteins, carbohydrates and fats. The carbohydrates can be sub-divided into the fibre of the plant cell walls, which is difficult to digest, and more soluble carbohydrates (like starch and sugars) that are easily digested. Plants also contain a wide variety of other constituents, some of which are needed by animals (the vitamins) and others of which may be harmful (toxic constituents). Finally, plants contain a variety of mineral elements such as calcium, iron and copper, many of which are needed by animals. In both plants and animals, all the nutrient materials listed above are associated with water.

The relative concentration of nutrients in foods derived from plants, and in animal products, are shown in Fig.5.3.

Digestion and Absorption

Foods are digested in the stomach and intestines of the animal, known collectively as the alimentary tract or gut. The main purpose of digestion is to break down food constituents present as large, complex and insoluble molecules into simple, soluble substances that can be absorbed through the wall of the gut into the animal's tissues. The main agents of digestion are a special group of proteins known as enzymes.

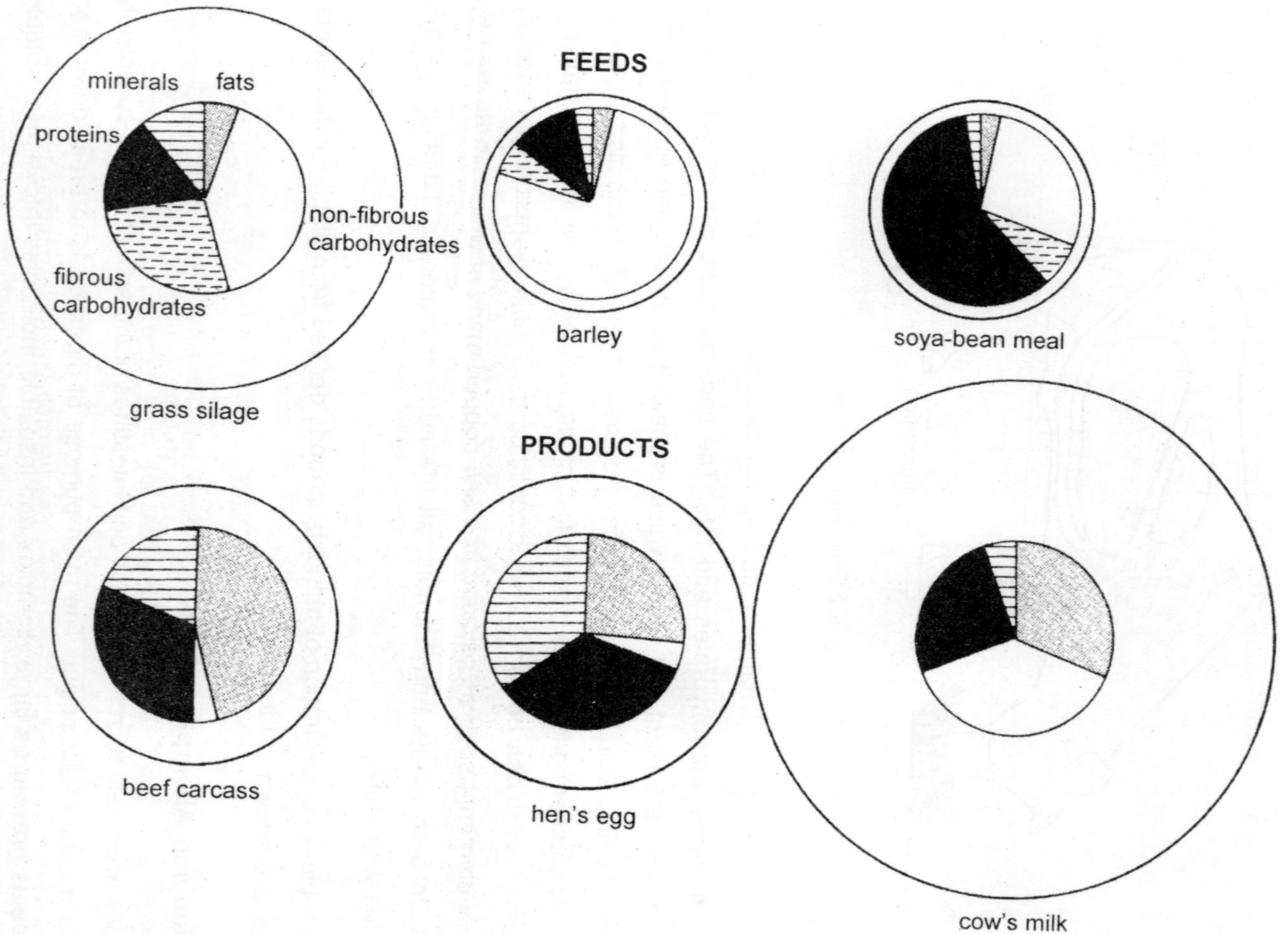

Fig. 5.3: Chemical composition of typical animal feeds and animal products (inner circles show the composition of dry matter; outer circles show the water associated with the dry matter).

These are secreted into the gut, either by the animal itself of by micro-organism such as bacteria, which are present in parts of the gut. In an animal with simple stomach, like the pig (or like man), the greater part of digestion is carried out by enzymes produced by the animal itself, and these enzymes are summarised in Table 5.3.

Table 5.3: The major digestive enzymes

Enzymes	Origin and site	Substance digested	Substance produced
(a) Acting on proteins			
Rennin	Stomach (calf)		
Pepsin	Stomach	Proteins and	Peptides and
Trypsin	Pancreas and small intestine	peptides	amino acids
Peptidases	Small intestine		
(b) Acting on carbohydrates			
Amylase	Pancreas and small intestine	Starch	Maltose
Maltase	Small intestine	Maltose	Glucose
Surcase	Small intestine	Sucrose	Glucose and fructose
Lactase	Small intestine	Lactose	Glucose and galactose
(c) Acting on fats			
Lipase	Pancreas and small intestine	Triglycerides	Monoglycerides and fatty acids

Proteins are large molecules composed of many hundreds of thousands of smaller compounds known as amino acids. The digestion of proteins, which starts in the stomach and is continued in the small intestine, involves the production of intermediate-sized compounds (known as peptides), which are then further broken down to amino acids.

Like the proteins, many carbohydrates are large molecules built up from the simpler substances we know as sugars. The starch of cereal grains, for example, is formed by the linkage of several thousand molecules of the sugar glucose. The enzyme amylase breaks starch down to a substance called maltose, which contains two glucose molecules and is known as a disaccharide sugar. The enzyme maltase splits maltose into its constituent glucose units. Cane or beet sugar is also a disaccharide, and in the small intestine is split by the enzyme surcase into two simpler sugars, glucose and fructose.

It was explained earlier that carbohydrates are of two major types, those that are soluble and easily digested (like starch) and those that are insoluble and less-easily digested. The second group form the fibre of plants. The units from which they are constructed are again the monosaccharide sugars; for example, the fibre carbohydrate cellulose is made up from many glucose molecules. In the fibre carbohydrates, however, the linkages are formed in such a way that they cannot be broken by the digestive enzymes of animals. The fibre carbohydrates also contain some substances not found in soluble carbohydrates. The most important of these is lignin, a dark-coloured encrusting substance which imparts the characteristic of 'woodiness' to plant fibres. Fibre carbohydrates therefore tend to pass through the gut of animals without being digested, and are excreted in the faeces. We shall see shortly that this pattern can be modified by the enzymes of micro-organisms present in the gut.

The fats (or lipids) of feeds consist of substances called triglycerides. A triglyceride molecule may be envisaged as a three-pronged fork with three sausages attached to it. The 'fork' is glycerol, an alcohol related to the carbohydrates, and the 'sausage' are substances called fatty acids. These fatty acids give the fat its physical characteristics; if their molecules are small or have gaps in their structure (i.e., are unsaturated: see Chapter 2), the fat will have a low melting point and be liquid at normal temperatures (i.e., it will be a vegetable oil). Fats are insoluble in water (a matter of common observation). To digest fats the animal secretes an enzyme (lipase) that splits them into glycerol and fatty acids. The acids may then be neutralised with sodium or potassium to form soluble salts which can be absorbed.

The animal also has to absorb the vitamins and minerals in its food. Many of the minerals have first to be released from combinations with other feed constituents. Those that readily form soluble compounds, like sodium, are easily absorbed, but others, like calcium, magnesium and iron, are liable to be incompletely absorbed by the animal.

Animals that live largely on fibrous plants, like the ruminants (cattle, sheep and goats), have developed methods for digesting fibrous carbohydrates with the aid of micro-organisms (and their enzymes). The anterior part of the stomach of ruminants is modified to form three additional compartments, the rumen, reticulum and omasum. The rumen is a particularly large organ; that of a 500-kg cow holding about 80 litres of liquid digesta. The rumen contents consist of food, saliva swallowed with it, and a large population of micro-organisms (mainly bacteria, but also some fungi and the minute animals known as protozoa). The ruminant animal begins the breakdown of food by chewing it as it is ingested and chewing it again when it is regurgitated from the rumen (the process known as rumination). The micro-organisms become attached to the food particles and secrete enzymes that break down the food in a manner analogous to the host animal's enzymes, but with two important difference. The microbial enzymes attack a wider range of food constituents, including cellulose and some related fibrous carbohydrates (although not lignin). Secondly, the breakdown process goes further modified (by a process called fermentation) to yield a mixture of acids (acetic, propionic and butyric, called collectively volatile fatty acids). These are absorbed by the host animal. Other important products of fermentation are the gases carbon dioxide and methane. These must be expelled from the rumen by belching; if they become trapped in the rumen they cause the potentially fatal condition known as 'bloat'.

Protein digestion by rumen microbes gives rise initially to amino acids, but these are fermented to yield volatile fatty acids and the gas, ammonia. The ammonia is then used by the microbes to re-synthesise amino acids, and then to construct their own microbial proteins. When the rumen contents flow through the remaining stomach compartments, thence to the small intestine, the microbes are attacked by the host animal's enzymes and broken down again to amino acids. This rather complicated sequence of events has two important consequences for ruminant animals. First, it ensures that the host animal is supplied with a 'standard brand' of microbial protein that is not much influenced by the type of food consumed. Second, it gives the ruminant an opportunity to augment its protein supply from other sources of ammonia in the rumen. Some simple chemicals, like urea and ammonium sulphate, form ammonia in the rumen, and thus may be added to the diet of ruminants to increase microbial protein synthesis.

A few additional features of microbial digestion may be noted briefly. Fats are hydrolysed in the rumen and the glycerol is fermented, but the long-chain fatty acids escape fermentation (although they may be converted from the unsaturated to the saturated form). Water-soluble vitamins are synthesised in the rumen.

Animals other than ruminants can benefit to some extent from microbial fermentation, but this takes place *after* normal digestion in the final section of the gut known as the large intestine. .

Metabolism and Excretion

The term metabolism embraces all the chemical reactions that take place in the animal's body. These reactions are carried out by enzymes produced in the cells of the animal; thus nearly all cells are involved in metabolism, although concentrations of cells in organs such as the liver and mammary gland may have special roles to play. The reactions of the body are numerous and complex; here we can only outline the main functions of metabolism.

Some reactions are designed to break down chemical compounds derived from food or body tissues with the specific purpose of providing the body with energy; these processes, by which chemicals become fuels, are called katabolic. Others produce new tissues or repair existing tissues; these are synthetic in character and are called anabolic. Katabolic and anabolic reactions are linked by energy-carrying intermediary compounds. These usually contain phosphorus, a typical example being adenosine triphosphate (ATP).

A typical cycle of katabolic reactions is that used to oxidise glucose to carbon dioxide and water (i.e., the reverse of photosynthesis; see Chapter 4). About 70 per cent of the energy of glucose is transferred to ATP and other energy carriers, and 30 per cent is released as heat. A part of the same cycle is used to oxidise fatty acids, again with the transfer of energy.

A typical anabolic cycle is that used to synthesise protein from amino acids. Energy is required to forge the links between amino acid molecules, and this is provided by ATP. For protein syntheses to take place smoothly and efficiently, the amino acid molecules must be delivered to the cells in the correct forms and numbers, but there is no guarantee that amino acids absorbed by the animal from its gut will meet this requirement. The animal therefore possesses mechanisms for creating some of its own amino acids by transferring the distinguishing amino (NH_2) group for one compound to another (a metabolic process known as transamination). But about ten amino acids cannot be formed in this way by the animal and must therefore be present in the mixture absorbed from the gut. These ten, of which important examples are lysine, methionine and threconine, are known as the essential or indispensable amino acids.

Protein synthesis is a complicated process because the amino acids involved must be linked together in a predetermined sequence. Thus a protein made up of 500 amino-acid molecules might start the chain with lysine, followed by threconine and methionine, and so on. Each body protein is unique in its amino-acid make-up. Faced with the problem of linking together a total of 500 components, comprising 20 different types, an engineer would use a plan or blueprint. The animal (or plant) uses as a guide, or template, substances called ribose nucleic acids (RNA). The RNA are stringlike molecules which may be envisaged as having various spaces cut into them, each type of space being of a size and shape to accommodate a specific amino-acid molecule. In protein synthesis, amino-acid molecules locate themselves along a strand of RNA in a precise order, and are joined together in the same order to produce a protein molecule.

Other important synthetic reactions in the body are those involved in the formation of the body's reserve carbohydrate (glycogen), of body fats, and of the sugar of milk, lactose; this last synthesis takes place in the mammary gland. These syntheses, and that of protein, differ in one important respect, the stability or permanence of the product. Glycogen, as a short-term energy store, has a brief life and within hours of its synthesis may be required for catabolism as a source of energy. Fats are longer-term energy reserves. Lactose is in a different category, for after being synthesised it is soon removed from the body in the milk. Proteins, as the 'working parts' of the body, have lives that vary in length according to their activity. The protein of an active tissue such as the gut, wears out rapidly and must be replaced at frequent intervals. At the other extreme, the protein of wool is virtually without activity and is never replaced or 'turned over'. Proteins withdrawn from tissues are katabolised to amino acids, many of which are re-used to synthesise new proteins. The overall picture of protein metabolism in the body is one in which proteins are synthesised from a 'pool' of amino acids to which they eventually return these components. Re-utilisation is incomplete and there is a 'leakage' from the pool which must be balanced by an inflow from the food; in a growing animal, inflow of amino acids exceeds leakage, and protein synthesis exceeds protein katabolism.

The katabolic reactions of the body are dependent ultimately on a supply of oxygen, the carbon and hydrogen of the compound katabolised being oxidised to carbon dioxide (CO_2) and water. The respiratory system of the body, centred on the lungs and using the blood as a distribution network, provides oxygen to the tissues and removes most of the CO_2 and some of the water. The excretory system of the body, centred on the kidneys, removes some CO_2 and much of the water, and is also responsible for removal of other waste products discussed later.

Blood consists of a fluid medium, plasma, containing several types of cells, together with some extracellular proteins. It also carries nutrients such as glucose and lipids. The red blood cells (erthythrocytes) carry the iron-containing protein haemoglobin which can bind with oxygen. Normal blood contains 100-150 g haemoglobin per littre, which can carry 130-200 ml of oxygen. The blood becomes oxygenated when it passes through the lungs, then delivers the oxygen to cells throughout the body. Of the remaining constituents of the blood, the white cells (leukocytes) and the proteins known as globulins protect animal against infective agents, and the blood platelets (thrombocytes) and the protein fibrinogen are involved in the clotting of blood.

The kidneys provide a filtering system for the blood. They pass into the urine the waste products of metabolism, such as urea derived from the breakdown of amino acids, and they also remove nutrients in excess of requirements, such as potassium.

More generally they regulate the composition of blood by, for example, removing acidic substances when the blood threatens to become too acidic. To function effectively the kidneys must have a sufficient throughput of water.

Vitamins in Nutrition

The vitamins are a group of 15-20 substances that animals need but cannot synthesise; they must therefore be present in the diet, or alternatively, the diet must contain substances known as vitamin precursors which the animal can transform into the vitamins. Vitamins are required in small quantities; for example, the requirement of the pig for the vitamin cobalamin is about 20 micrograms (i.e., 0.02 mg) per day, or 1 part in 1 million parts of its diet. Most of the vitamins are required in the animal in some regulatory role, for example, as components of the enzymes referred to earlier. If an enzyme lacks its vitamin component, it will be unable to play its role in digestion or metabolism, thus depriving the animal of a supply of nutrients and causing it to become ill.

It is not possible here to describe in detail the individual vitamins and the diseases induced by their deficiency, but Table 5.4 provides a summary. Note that vitamin C (ascorbic acid) does not appear in the table, because farm animals, unlike man, monkeys and the guinea pig, can synthesise this substance. Vitamins A, D and E are known as fat-soluble vitamins.

The vitamin requirements of ruminant livestock are simpler than those of pigs and poultry, because all the so-called B group vitamins are synthesised by the bacteria of the rumen. The B vitamin, cobalamin, however, contains the element cobalt, and if the ruminant's diet is deficient in this element, its rumen bacteria cannot manufacture cobalamin. Furthermore, all reminants are susceptible to B deficiencies in early life, before their rumen flora has become established. Thus deficiencies of the water-soluble vitamins (B group) occur in ruminants only under rather special circumstances. Of the fat-soluble vitamins the precursor of vitamin A, carotene, and also the tocopherols known collectively as vitamin E, occur in relatively high concentrations in green plants. Ruminants are therefore unlikely to be deficient in A or E during the summer grazing season, but winter feeds such as hay, straw, roots and cereal concentrates are commonly deficient.

Table 5.4: Characteristics of some vitamins of importance in animal nutrition

Common name (and chemical name)	Precursor	Role in the animal	Sings of deficiency	Food sources	
				Good	Poor
A (retinol)	Carotene	Protects mucous membranes; vision	Night blindness; roughened skin	Green plants; fish liver oils	Cereals and other concentrates
D (calciferol)	Animal and plant sterols	Absorption of calcium; bone mineralisation	Rickets and other bone deformities	Fish liver oil sun-dried forages	Cereals
E (tocopherols)	None	Anti-oxidant	Muscular dystrophy	Green plants; grains	Animal products; root crops
B_1 (thiamin)	None	Enzyme co-factor	Nerve degeneration	Most feeds	Refined cereals
B_2	None	Enzyme co-factor	Nerve degeneration; skin lesions	Green plants; yeast	Cereals
Nicotinamide	Tryptophan	Co-enzymes	Dermatitis	Most feeds	Maze
B_6 (pantothenic acid)	None	Co-enzymes	Loss of hair; dermatitis	Most feeds	None
B_{12} (cobalamin)	None	Co-enzymes	Growth check; dermatitis	Animal products; bacteria	Plants

The contrast between the high carotene content of fresh grass and the much lower carotene content of hay made from the fresh grass demonstrates an important principle of vitamin nutrition, which is that the vitamins or their precursors can easily be destroyed or modified during the processing of feeds. In this example carotene is destroyed when the grass is bleached by sunlight. Sunlight, however, improves the vitamin D content of both plant and animal tissues, because irradiation converts the sterol precursors into the active forms of the vitamin. Thus sunlight on grass produces vitamin D and sunlight on the skin of animals has the same effect. Animals are therefore more vulnerable to vitamin D deficiency in winter or when kept indoor (although animals have the capacity to build up reserves of fat-soluble vitamin in their fat depots). Vitamin D also illustrates another principle of vitamin nutrition, which is that an animal's requirement for a vitamin D assists the absorption and metabolism of the bone-forming minerals, calcium and phosphorus. If diets are deficient or unbalanced in these minerals, animals need more vitamin D.

Pigs and poultry are more vulnerable to vitamin deficiencies than are ruminants. They are kept indoors and not fed on forages, and thus need supplements of the fat-soluble vitamins. They have no rumen, with microbes synthesising B vitamins, and therefore require supplements of these, especially riboflavin. In practice, vitamin deficiencies in all farm animals are avoided by careful choice of natural feeds and judicious supplementation. At one time, supplements came from natural sources like fish oils (rich in A and D) and yeast (rich in B vitamins), but many vitamins can now be produced by industrial synthesis.

Mineral Nutrition

The mineral elements present in animals are divided into two categories, the major elements, that are present at relatively high concentrations (usually expressed as g/kg) and the minor (or trace) elements, that are present at much lower concertinos (usually expressed as mg/kg). The members of each group, and their approximate concentration in the animal body, are listed in Table 5.5. Some elements, like calcium and phosphorus in bone, have a structural role. Others, like sodium and potassium in blood, have a regulatory role (e.g., regulating the acid/alkaline balance of the body).

Many elements, particularly trace elements, occur as constituents of proteins (e.g., sulphur in several amino acids), enzymes (e.g., selenium, see Table 5.5) and pigments (e.g., iron in haemoglobin).

The bodies of animals may contain other elements; some of these are harmless contaminants such as silicon from sand, but others, like radioactive elements from atomic fall-out, are potentially toxic. Even essential elements may be toxic if present in excessive concentration. For example, copper is needed in the diets of sheep at a concentration of about 5 mg/kg, but a dietary concentration of 20-30 mg/kg will eventually kill sheep from excessive deposition of the element in the liver.

Mineral deficiencies bring about characteristic diseases in animals, which are summarised in Table 5.5. When animals obtain most of their food from natural vegetation, as do hill sheep in Britain, their mineral supply is very dependent on the composition of the soil. Soils deficient in an element are likely to give plants that are also deficient in that element. In some instances, as with phosphorus, the deficiency will affect plant growth before it effects the animal; in others, as with iodine, the plant has no requirement for the element, and the animal is the first to suffer. Some minerals, such as selenium, are accumulated in plants and make them toxic to livestock.

Animals that eat feeds from a variety of sources, such as dairy cows given home-grown forages and purchased concentrates, are less likely to suffer from soil-induced mineral deficiencies. Nevertheless there are some categories of feeds that are consistently food or poor sources of particular elements, regardless of their origins. Thus the seeds of plants (e.g., cereal grains) are generally good sources of phosphorus but poor sources of calcium, whereas forages show the reverse tendency (see Table 5.5). Some natural feeds, such as fishmeal, are recognised as being good sources of a wide range of elements. Deficient diets are fortified with specially formulated mineral supplements. Alternative methods of supplying minerals to animals are to apply them to the soil (as fertilisers), to provide concentrated mineral blocks or licks, or to administer them directly into the animal's gut or tissues. Some of the trace elements can be reincorporated in a soluble glass 'bolus' which, when placed in the rumen of cattle or sheep, is slowly dissolved to release the elements over a period of months.

Table 5.5: Characteristics of mineral elements of importance in animal nutrition

Element	Concentration in animal body	Role	Signs of deficiency	Feed sources: Good	Feed sources: Poor
Major elements					
Calcium	15g/kg	Bone, nerve function	Rickets; paralysis ('milk fever'); soft-shelled eggs	Green plants; milk; fish and meat meals	Cereals; roots
Phosphorus	10g/kg	Bone; co-enzymes	Rickets; depraved appetite (pica)	Animal products; cereals	Hay; straw
Magnesium	0.4g/kg	Bone; co-enzymes	Tetany (staggers')	Protein concentrates; legumes	Spring grass
Sodium	1.6g/kg	Osmotic pressure; acid-base balance	Poor growth; depraved appetite	Fish meal	Plants
Trace elements					
Iron	50 mg/kg	Blood haemoglobin	Anaemia; poor growth	Green plants	Milk; cereals
Copper	3 mg/kg	Blood; enzymes	Anaemia; hair depigmentation; nerve degeneration ('swayback')	(Soil dependent)	Milk
Cobalt	0.1 mg/kg	Cobalamin (vitamin)	(As for cobalamin deficiency)	(Soil dependent)	
Zinc	30 mg/kg	Enzymes	Growth check; skin lesions	Most feeds; yeast	None
Iodine	0.5 mg/kg	Thyroid hormone (thyroxine)	Enlarged thyroid (goitre)	Marine feeds	(Soil dependent)
Manganese	0.3 mg/kg	Bone enzymes	Bone deformities	Green plants	Animal products
Selenium	17 mg/kg	Anti-oxidant; enzyme (glutathione peroxidase)	Muscular dystrophy	(Soil dependent)	

Essential elements not listed, because they are rarely deficient, are potassium, chlorine, sulphur, molybdenum and chromium.

To be useful to the animal, minerals, like other nutrients, must be absorbed from the gut. Many minerals are brought into solution by the acid conditions prevailing in the gut, but others may be so tightly bound to other feed constituents as to be unavailable for absorption. Thus phosphorus in cereal grains may be present as insoluble phytic acid; in ruminants, phytic acid is broken down by micro-organisms and the phosphorus made available, but this is not the case in pigs and poultry. Another problem of availability arises when one element interferes with the absorption of a second. The classic example is that of copper, which in the rumen may form an insoluble compound with sulphur and molybdenum and thus fail to be absorbed.

Practical Nutrition: Principles of Rationing

In this chapter so far, nutrition has been discussed mainly in qualitative terms. Animals have been shown to need organic nutrients (carbohydrates, fats, proteins and vitamins) and also mineral elements. But nutritionists or farmers have to know the quantities of nutrients needed by their livestock, and the quantities of feeds that will supply these nutrients. The main principles of rationing are that the animal's nutrient needs can be calculated from its productive functions (i.e., growth, etc.), or alternately, that an animal's productivity can be calculated from its known intake of nutrients. To start with a simple example, 1 kg (= 0.9685 litre) of cow's milk contains the quantities of nutrients shown in Table 5.6. From these figures one might assume that to produce 1 kg of milk a cow would need to consume 34 g of protein and 40 g fat; for example, if hay contained 170 g protein per kg, the cow would need (34/170) = 0.2 kg hay to produce a litre of milk. In fact, the calculation is more complicated than this, for three reasons.

(a) *Nutrient losses in digestion and metabolism:* It was explained earlier that feeds are generally not completely digested by animals, and that some nutrients pass through the animal and are excreted in the faeces. Also, nutrients undergo losses when they are metabolised by the animal. For example, when the amino acids of food protein are 're-modelled' to form microbial protein, and the amino acids of the latter are re-modelled to form, say, milk of muscle protein, some amino acids will be discarded and their components excreted in urine. Allowance must be mad for these losses when calculating nutrients or food requirements.

(b) *Nutrients and energy:* To synthesise milk fat the cow can use not only dietary fat but also dietary carbohydrate (and even dietary protein). As precursors of milk fat, the nutrients interchange with one another approximately in proportion to their energy value. Table 5.6 shows that fat contains more than twice as much energy per unit weight as carbohydrate. It is generally more convenient to consider requirements for nutrients in terms of energy. Thus energy provides a 'common currency' for expressing nutrient requirements and the nutrient content of feeds.

Table 5.6: Nutrients and energy in 1 kg of cow's milk

	(g)	×	(kJ/g)	=	(kJ)
Fat	40		38.1		1524
Lactose	47		16.5		776
Protein	34		24.5		833
					3133

The links between food energy and the energy content of animal products are illustrated in Fig. 5.4. The total or gross energy content of feed can be measured by burning a known amount and recording the quantity of heat produced. Some of that energy will be present in feed constituents not utilisable by the animal and thus excreted in solid, liquid or gaseous excreta. When these amounts are deducted, the remainder is known as the *metabolisable energy* (ME) of the feed (i.e., it represents the nutrients metabolised by the animal). A further form in which energy is lost from the animal is as heat; in the metabolism of absorbed nutrients, the transfer of energy from one substance to another is not completely efficient and some energy is lost as heat. When the heat losses are deducted the remainder is known as net energy.

The energy value of feeds (and also of human foods, see Chapter 2) is commonly stated as metabolisable energy. Thus the grass in Fig. 5.4 contains 11 MJ metabolisable energy per kg dry matter (11 MJ ME/kg DM). To express energy requirements of animals in the same terms, the energy value of their products (e.g., milk) has to be 'scaled up' to allow for the heat losses. Table 5.6 shows the energy content of milk to be 3,133 kJ/kg (or 3.13 MJ /kg), but the metabolisable energy required by the cow to produce 1 kg milk is 5 MJ.

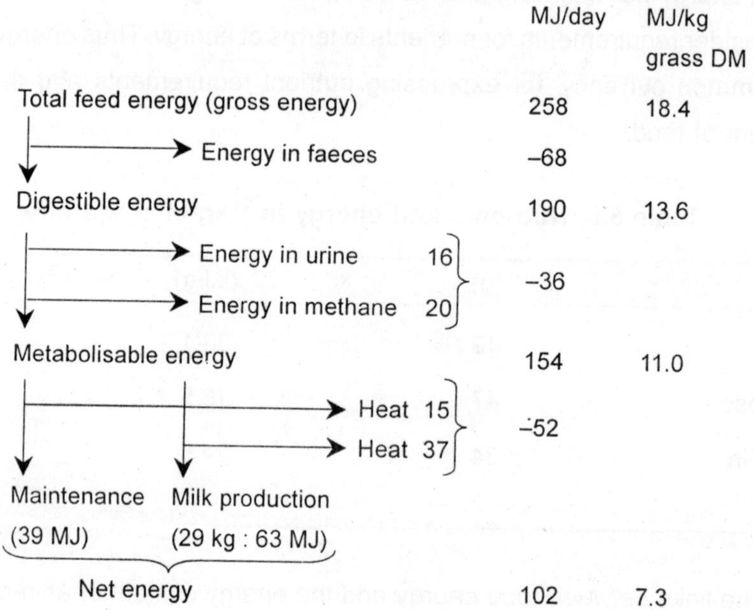

Fig. 5.4 An example of the partition of energy in a dairy cow fed on grass (14 kg dry matter/day).

(c) *The concept of maintenance:* Animals need nutrients and energy to manufacture their recognisable products (e.g., milk and meat), but they also have requirements for body maintenance. An adult man uses virtually all of his nutrient intake for this purpose. The metabolisable energy of his food is used to fuel essential body functions, such as the circulation of blood, the renewal of worn-out body components, and some muscular activity. Farm animals are less often in this unproductive state, but even when they are producing, the first call on their nutrient intake will be for maintenance. A 600-kg dairy cow needs about 55 MJ ME/day for maintenance; if it is producing 20 kg milk, it will need a further 5 × 20 = 100 MJ ME/day (and will thus be using the energy of about one-third of its nutrient intake for maintenance and the other two-thirds for milk production).

Rationing in Practice

To ration animals scientifically we need to draw up a list of their nutrient and energy requirements, and then devise a ration to meet requirements from feeds of known nutritional value. Table 5.7 summarises the requirements of a 600 kg cow producing 20 kg milk per day. The energy requirements have already been explained. Protein requirements of ruminates may be divided into protein that is digested in the rumen and reformed into microbial protein (known as rumen-degradable protein) and that which escapes rumen fermentation but is digested in the small intestine (undegradable protein). To simplify the example, only the major mineral elements are shown in Table 5.7, but the animal would of course need trace elements and vitamins.

Table 5.7: Daily nutrient requirements (g) of a 600 kg cow yielding 20 kg milk

	Maintenance	Production	Total
Energy (MJ ME)	55	100	155
Rumen-degradable protein	430	780	1210
Undegradable protein	–	320	320
Calcium	15	33	48
Phosphorus	12	32	44
Magnesium	11	14	25
Sodium	4	13	17

Let us suppose that the feeds available for the cow are hay and a simple, farm-mixed concentrate based on cereal and soyabean meal. The composition of these feeds in shown in Table 5.8.

Table 5.8: Composition of feeds for a dairy cow (g/kg dry matter)

	Hay	Concentrate
Energy (MJ/kg DM)	9	12
Rumen-degradable protein	70	95
Undegradable protein	20	40
Calcium	4.5	1.0
Phosphorus	2.0	4.0
Magnesium	0.9	1.5
Sodium	3.0	1.2

The calculation of the quantities of the two feeds required might adopt the following sequence:

(a) Assume that sufficient hay is given to meet the maintenance requirement for energy.

(b) Deduct from the total requirements of the cow the quantities of nutriments supplied by the hay.

(c) Calculate the quantity of concentrate needed to meat the remaining requirement for energy.

(d) Determine the quantities of nutrients supplied by the concentrate, and compare them with requirements.

These four are illustrated in Table 5.9.

The final ration (6 kg hay dry matter and 8.4 kg concentrate dry matter) has a rather wasteful excess of undegradable protein, and a relatively large (but harmless) execs of sodium. Calcium and magnesium are in deficit, and the ration could be supplemented with 33 g calcium carbonate (containing 400 g Ca/kg) and 21 magnesium oxide (containing 520 g Mg/kg).

In this example, feeding standards (Table 5.7) and data (Table 5.8) were used to derive a ration for an animal with a predetermined level of production; this procedure is known as *ration formulation*. The same tables can also be used in a different way, to predict the performance of animals given a predetermined ration. For example, from Table 5.8 we calculate that a 500-kg cow given daily 8 kg hay dry matter and 10 kg concentrate dry matter would have a total energy intake of 172 MJ. If its maintenance requirement remains at 55 MJ/d, the energy available for milk production is 117MJ. As mentioned earlier, the cow needs 5 MJ for each 1 kg of milk produced, so 117 MJ should be sufficient for (117/5) = 23.4 kg/d (although the new ration would have to be re-balanced for other nutrients). This procedure is called *performance prediction.*

Table 5.9: Calculating a ration for a dairy cow

	Requirement	Supplied by hay	Remaining requirement	Supplied by concentrate	Excess or deficit
Dry matter (kg)	–	6	–	8.4	–
Energy (MJ)	155	54	101	101	0
Rumen-degradable protein (g)	1210	420	790	790	0
Undegradable protein (g)	320	120	200	340	+140
Calcium (g)	48	27	21	8	–13
Phosphorus (g)	44	12	32	33	+1
Magnesium (g)	25	5	20	12	–8
Sodium (g)	17	18	–1	10	+11

In practice, both types of ration calculation may be more complex than illustrated here, further details are given in standard textbooks on animal nutrition.

Reproduction

The male and female sex organs do not become fully functional until animals are well on their way to achieving their mature size. The attainment of sexual maturity occurs at what is known as puberty. In cattle, puberty occurs at a fixed body size regardless of age; for British Friesians it occurs at a live weight of about 270 kg. In sheep, puberty is again determined by body size, but its occurrence is complicated by the fact that sexual activity in the female of this species is restricted by day length (to a period from September to February in the UK for many breeds of sheep). Thus a lamb born in April may, if well grown, achieve puberty in the following October, but a delay of two months may cause it to be held over until the following September. Pigs achieve puberty at a fixed age (about 200 days) rather than fixed weight, although females (gilts) may be jolted into sexual activity by sudden contact with males. The fowl lays its first egg at about 160 days of age, but this event is controlled by day length; a long day (14 h of light) or an increasing daylength will advance puberty, while short days or decreasing daylength will delay it.

Hormones

Reproduction, like many of the functions of the animal body, is controlled by the 'chemical messengers' known as hormones. Typically, a hormone is a substance that is synthesised in one part of the body and transported in the blood stream to another part of the body, where it initiates a reaction or process. Many hormones are produced by special organs known as endocrine glands that are dedicated to the production of one or more hormones. Examples are the thyroid gland in the throat, the pituitary gland at the base of the brain, and the paired adrenal glands located close to the kidneys. However, hormones may also be produced by organs that have other major functions. Thus the pancreas produces digestive enzymes as well as the hormone insulin (which control carbohydrate metabolism), and the reproductive organs (the ovaries and testes) synthesise hormones as well as ova and spermatozoa.

Some hormones are relatively simple substances that can be synthesised in the laboratory. Other are larger and more complex molecules, such as peptides and proteins; although these cannot easily be synthesised by man, many can now be produced by bacteria that have been subjected by man to genetic engineering. Hormones may be thought of as having a key-like structure that allows them to attach to lock-like structures on their target organs known as receptor cells. In this way their action is restricted to organs carrying the correct receptors.

What starts or stops the secretion of a hormone? In some cases the starting signal may come from another hormone. For example, the secretion of many hormones from the pituitary gland is initiated by so-called 'releasing factors' produced by a nearby organ called the hypothalamus. The breeding season of the ewe is initiated by declining daylength, which stimulates the light-sensitive pineal gland at the back of the head to produce the hormone melatonin, which in its turn stimulates the production of reproductive hormones by the pituitary gland.

The actions of the reproductive hormones are illustrated in Fig. 5.5, and described in the section that follow.

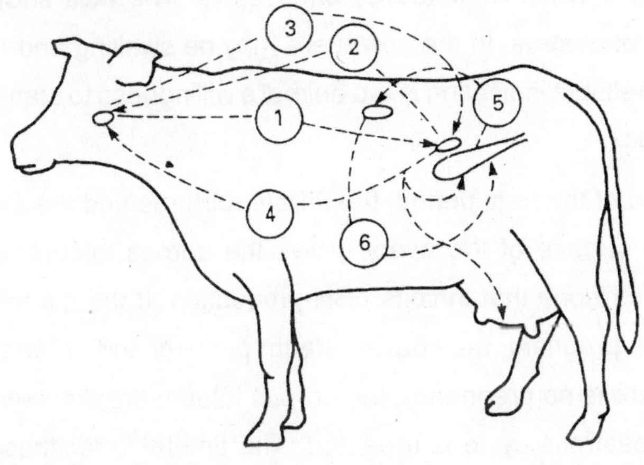

Fig. 5.5: Hormonal control of oestrus, conception and pregnancy in the cow. (1) Pituitary gland produces follicle stimulating hormone (FSH) that promotes growth of follicle in the ovary. (2) Oestrogen produced by the follicle stimulates production of luteinising hormone (LH) by the pituitary, and prepares the uterus. (3) LH causes ovulation. (4) Corpus luteum on the ovary produces progesterone which inhibits FSH, maintains pregnancy and eventually (with oestrogens) prepares the mammary gland for lactation. (5) Foetal pituitary produces adrenocorticotrophic hormone (ACTH), which causes maternal adrenal gland to produce glucocorticoids. (6) Glucocorticoids cause parturition.

The Female

In the female, puberty is marked by the start of oestrous cycles. The immature ovary contains large numbers of special cells known as ova, each ovum being surrounded by a layer of follicular cells. Under the influence of the follicle-stimulating hormone (FSH) produced by the pituitary gland, one or more follicles will enlarge, each forming a cavity filled with fluid between the ovum and the follicular cells. In the cow, for example, a mature follicle may be 16-18 mm in diameter, and project from the surface of the ovary sufficiently to be felt by a hand passed into the rectum. As the follicle grows it secretes oestrogens, which act on the anterior pituitary gland to stimulate the release of luteinising hormone (LH). The oestrogens also increase the vascularity of the uterine wall (in preparation for anticipated pregnancy) and provoke the signs of 'heat' or sexual receptivity.

Heat periods in the cow are marked by a discharge of watery mucus from the vulva, by excitability which includes mounting other cows, and – most particularly – by a willingness to stand when mounted by other cows. The ewe shows no generally recognisable signs of oestrus. In the sow there may be swelling and reddening of the vulva, but the most reliable indication is the animal's willingness to stand when pressure is applied to her back.

Towards the end of the heat period, the follicle ruptures and the ovum is released. In its place on the surface of the ovary grows the corpus luteum, which produces progesterone, the hormone that inhibits FSH production. If the ovum is fertilised and the animal becomes pregnant, the corpus luteum persists and prevents the growth of more follicles. If there is no pregnancy, the corpus luteum breaks down, FSH is again produced and the oestrous cycle is repeated. The timetable for these events varies between species, as shown in Table 5.10.

Table 5.10: Reproductive cycles for female domestic animals

	Cattle	Sheep	Pigs
Length of oestrous cycle (d)	.21	16.5	21
Length of heat period (h)	16	36	48
Ovulation (h after start of heat)	30	36	36
Typical number of ova fertilised	1	2	12
Gestation period (d)	282	147	113
Interval from parturition to first heat (d)	40-70	-	4-7

In cattle, sheep and pigs it is now possible to influence the occurrence of oestrus (and the numbers of ova produced) by the use of exogenous hormones (i.e., hormones not product by the animals itself). In sheep, the insertion into the vagina of pessaries impregnated with progestagens (hormones that help to prolong the life of the corpus luteum) blocks the oestrous cycle. If a flock of ewes so treated has its pessaries withdrawn simultaneously, the ewes will come into oestrus simultaneously and can then be mated together (two days after removal of pessaries) in order to lamb together on a prescribed date. This procedure, known as *oestrus synchronisation*, allows the shepherd to plan his lambing programme more precisely. A similar technique has been used in pigs, but with the progestagen included in the feed. In cattle, oestrus synchronisation can be achieved by an injection of a different kind of hormone, known as a prostaglandin, which causes an active corpus luteum to regress.

Another exogenous hormone preparation, known (from its origin) as pregnant mare's serum gonadotrophin (PMSG), possesses FSH and LH activity. When given to sheep or cattle it increases the number of ova shed at a single oestrus. Carefully controlled use of PMSG can increase litter size in sheep to three or four, and induce twins in cattle.

The Male

The gonadotrophins, FSH and LH, have roles in the male which parallel those in the female. The production of spermatozoa in the tubules of the testis is stimulated by a combination of gonadotrophins (FSH and LH) and androgens; the androgens are produced by the so-called interstitial cells (lying between the tubules) in response to stimulation by LH. Androgens, such as testosterone, also promote the growth of (and maintain) the penis, scrotum and secondary sexual characteristics, and regulate the secretion of the gland associated with the testes, namely, the prostate and the seminal vesicles.

Each spermatozoon consists of a head, containing a cell nucleus, a midpiece, containing the mitochondria, their enzyme systems, and the tail. The mitochondria provide the energy which moves the tail and gives the spermatozoon its motility. When spermatozoa leave the testis they are stored in the epididymis, and they are subsequently diluted with, and activated by, the secretions of the accessory glands, to form semen. The semen is expelled from the penis by the muscular contractions constituting ejaculation. In the bull, each ejaculate is 4-8 ml in volume, and contains about 1 million spermatozoa per mm^3 (i.e., 6×10^9 per ejaculate). In the ram the volume is less (0.5 - 2.0 ml) but contains 3 million spermatozoa per mm^3. In contrast the ejaculate of the boar is large in volume (150 – 300 ml) but contains only 100,000 spermatozoa per mm^3.

Insemination

In natural mating, the ejaculate is delivered by the penis into the vagina or uterus, and the spermatozoa move up the fallopian tubes to meet the descending ova. In artificial insemination, this process is interrupted; the semen is collected from the male and examined, diluted (and possibly stored) before being introduced into the female. The advantages of artificial insemination are that a single ejaculate can be used to inseminate 20–200 females (thus extending the influence of superior sires on the next generation), and that the transfer of diseases is minimised. Cattle semen can be frozen and stored (at –196° C in liquid nitrogen) for many years, and it is therefore possible to delay the meat or milk production of a small number of his offspring. The use of frozen semen also allows a bull to be mated with cows located in any part of the world. At present, these advantages are not fully obtainable with sheep and pigs, because their semen cannot be so extensively diluted, and their spermatozoa are damaged by the freezing techniques currently available.

Ovum, Embryo and Foetus

The fusion of an ovum with a spermatozoon is known as fertilisation. The single cell resulting immediately begins to divide, to form an embryo. In the cow the cells of the dividing ovum remain undifferentiated for the first 12 days, but during the periods 10–45 days, primitive forms of all the major tissues and organs are laid down. The heart, for example, begins to function as early as day 21. Also formed are the membranes which surround and nourish the embryo. The amnion holds the embryo suspended in fluid, the allantois collects the excretions of the embryo and also carries blood vessels towards the lining of the uterus, and the serosa (or trophoderm) envelopes the embryo and the other membranes. The allantois and serosa eventually fuse to form the chorion, and the latter becomes attached to the lining of the uterus (at 30–45 days) to form the placenta. The function of the placenta is to bring the blood vessels of the embryo and the mother in close proximity, so that nutrients and other substances may pass from one to the other. In the cow and sheep the conjunction of embryonic and material tissues occurs at a number of isolated spots, called cotyledons; in the pig, the placenta provides continuous contact.

At 45 days of age the bovine embryo weighs only 2.5 g and is about 3 cm in length. From then until birth it is known as a foetus. The foetus' tissues grow rapidly, but at different rates, to produce the characteristic proportions of a newborn calf weighing c. 40 kg.

Pregnancy Diagnosis

In the cow, a veterinarian passing his hand into the rectum can palpate the uterus and detect a developing foetus about 60 days after conception. In the ewe and sow, the reflection of ultrasonic waves can be used to detect pregnancy from about days 70 and 50 respectively. In the cow, and also in other species, the corpus luteum remains active during pregnancy and continues to secrete progesterone, which circulates in the blood, and also reaches the milk, at a relatively high concentration. If a milk sample taken from a cow 20–30 days after mating is analysed for progesterone, a low concentration will indicate the recurrence of oestrus to be expected at that time, whereas a high concentration is indicative (although not conclusively so) of pregnancy.

Parturition

When the foetus reaches maturity, its pituitary gland excretes adrenocorticotrophic hormone (ACTH) (see Fig. 5.5) which stimulates the adrenal cortex to produce a glucocorticoid hormone. The latter acts on the placenta to alter the balance of hormones produced there, and this in turn initiates the separation of the placental tissues and the muscular contractions of the uterus. Parturition can be induced artificially, by injecting the mother with a synthetic glucocorticoid.

Embryo Transfer

In all domestic mammals it is possible to transfer ova or embryos (fertilised or not) from one individual to another. Embryo transfer is now used routinely in cattle; initially surgical methods were used for both collection and implantation of embryos, but non-surgical methods (analogous to those used in artificial insemination) have now been developed.

The techniques is used by breeders to multiply the offspring of outstanding females, such animals being 'superovulated' to produce up to 20 ova, which after fertilisation, are transferred to nondescript cows. Embryo transfer may also be used to give twin pregnancies in cows.

Lactation

During pregnancy, oestrogens and progesterone, together with other hormones, promote the growth of the mammary glands, and a few days before parturition the glands fill with the secretion known as colostrum. The main constituents of colostrum are special proteins known as immunoglobulins, which, when consumed by the young animal, provide protection against bacterial infection.

After 3–4 days, colostrum is replaced by normal milk having the composition described in Table 5.6.

Egg Production in Poultry

The hen has only one ovary (the left) and at ovulation a whole follicle (i.e., the chromosomal disc plus the 'yolk') passes into the oviduct. This tract, which is 80 cm long, consists of five sections, each with a specific function. The funnel-shaped infundibulum collects the follicle, and fertilisation may take place there. The next section, the magnum, is responsible for covering the follicle with albumen (i.e., egg 'white'), and this is followed in turn by the isthmus, which adds a double membrane cover. The fourth section, the uterus or shell gland, adds the shell, and the final section, the vagina, passes the egg to the exterior via the cloaca.

Each follicle takes about 70 days to grow to the size (30 – 40 mm across) at which it undergoes ovulation. From ovulation to oviposition (i.e., laying) takes about 26 hours, of which about 20 hours is required for shell formation. The completed hen egg weighs about 57 g and contains (g/kg): proteins, 120; carbohydrates, 10; lipids, 105; mineral material, 110.

Principles of Genetic Improvement

Comparisons of today's farm animals in the UK with those of former times, 50, 100 or 200 years ago, disclose considerable differences in outward appearance and in productivity. For example, today's dairy cattle tend to be black and white in colour, to have no horns and to give an average milk yield of about 5,000 kg per lactation. Their predecessors of 50 years ago were generally red and white in colour, carried horns, and yielded about 2,000 kg per lactation.

These changes have been due to a variety of factors. The change in coat colour is due to the replacement of one breed, the Dairy Shorthorn, by another, the British Friesian. The loss of horns is due to the routine use of chemical agents to destroy the horn bud in calfhood. The increase in milk yield is due partly to the breed substitution, partly to improved feeding and management and partly to the selection of superior breeding stock in each successive generation.

These factors which have operated in the past are available today to the farmer who wishes to improve his livestock in some way. Basically he may impose external changes in the environment of the animals (e.g., improve their nutrition), or modify the nature of the animals themselves. To achieve the latter type of change he may introduce animals of the desired quality from elsewhere. Such a change need not be as drastic as the replacement of one breed by another, however, and this section is concerned with the alternative methods of improvement, namely, genetic improvement of livestock by selection of superior individual for breeding. The advantage of this method is that once achieved, improvements are maintained without further effort, whereas an improvement obtained by changing the external environment of animals is likely to require continued implementation. Today's British Friesian calf, even if its ancestors have been chemically dehorned for the past ten generations, will still grow horns if left untreated.

The Mechanisms of Inheritance

The calf grows horns because it possesses in its body cells what is known as a 'gene' for horns. A gene is a strand of the string-like molecules known as nucleic acids. We saw earlier that ribose nucleic acid (RNA), which occur in the fluid content of cells, act as templates for protein synthesis. Genes, however, are composed of deoxyribonucleic acid (DNA) and are found in the nuclei of cells. They are joined together to form structures known as chromosomes which are arranged in pairs. In cattle there are 30 pairs of chromosomes, carrying many thousands of genes.

The DNA acts as the fundamental template for protein synthesis, because it provides the temperate for construction of RNA. Thus we may envisage the development of horns in cattle as depending on the presence of one or more specific proteins (perhaps as enzymes), which can be synthesised with the aid of RNA, provided the cell nuclei contain the appropriate gene, or strand of DNA. One of the fundamental properties of DNA is that it can reproduce itself, by acting as a template for another molecule to be produced; this ensures that all the cells of the calf contain identical DNA, because they all originate from one cell, the fertilised ovum. The only exception to the rule of every cell in an individual carrying the same genes occurs in the formation of the gametes (ova and spermatozoa). Fertilisation depends on the fusion of the nuclei of an ovum and sperm, and if each of these two nuclei has its full complement of DNA, the nucleus of the fertilised ovum would contain twice the normal complement. This is avoided by a halving of the DNA in the gametes which is achieved by the segregation of the pairs of chromosomes; thus the gametes of cattle contain 30 individual chromosomes.

Each pair of chromosomes is identical in the sense that each strand of DNA that we call a gene has a corresponding strand in the other member of the pair. The two genes may be identical, in which case the gametes produced by an individual must always carry the same gene; the individual is then said to be homozygous for that gene. Alternatively, the genes may differ in some way. For example, one chromosome might carry the gene for horns while its pair mate carries a gene for 'no horns' (i.e., a DNA strand which fails to promote the synthesis of the protein required for horn development). Such an individual is said to be heterozygous for the horns gene and its gametes could contain either the horns or the no-horns genes.

In a population having an equal distribution of horns and no-horns genes an individual has a one-in-four chance of having two horns genes, a one-in-four chance of having two no-horns genes and two chances in four of having one horns and one no-horns gene. It seems likely that the homozygous animals will have horns or have none (i.e., be naturally polled). But what will the heterozygous animals have: will they be intermediate to the extent of having small horns or even one horn? In fact the no-horns (or polled) gene is dominant to the horns gene (which is, conversely, said to be recessive) and the heterozygotes are therefore polled. For other characters, however, neither gene may be dominant, and the heterozygotes will be intermediate in character to the homozygotes. In cattle an example is provided by the Shorthorn breed, in which a red-coated animal crossed with a white-coated animal (i.e., the homozygotes) produces heterozygotes with a mixture of red and white hairs that is described as a roan colour.

The segregation of chromosomes in the formation of gametes is largely a random process, and the phenomenon of heterozygosity ensures a great many possible assortments of chromosomes and hence of genes. Thus two brothers will each inherit half of their genes from each parent, but the random segregation will ensure that they do not inherit the same assortment of genes, and therefore that they will not be genetically identical. Only twins derived from the same fertilised ovum (called monozygous twins) are genetically identical. Nevertheless, the likeness between brothers, or between parent and offspring, is strengthened by the fact that many genes are linked on the same chromosome. The effect of this linkage is that characters are inherited in 'packages' rather than in a completely random fashion.

One of the most obvious effects of linkage is the association between the sex of an individual and its other characters. One pair of chromosomes does not obey the rule described above, of one member of the pair matching the other. In male mammals the chromosome called the X chromosome is paired with a distinctly different Y chromosome; a male therefore produces sperm half of which have an X chromosome and half a Y chromosome and is said to be heterogametic. Female mammals, on the other hand, have two X chromosomes; their ova all have X chromosomes and they are said to be homogametic. An ovum that is fertilised by sperm carrying an X chromosome becomes a female, and an ovum fertilised with a Y-containing sperm becomes a male. This is illustrated in Fig. 5.6.

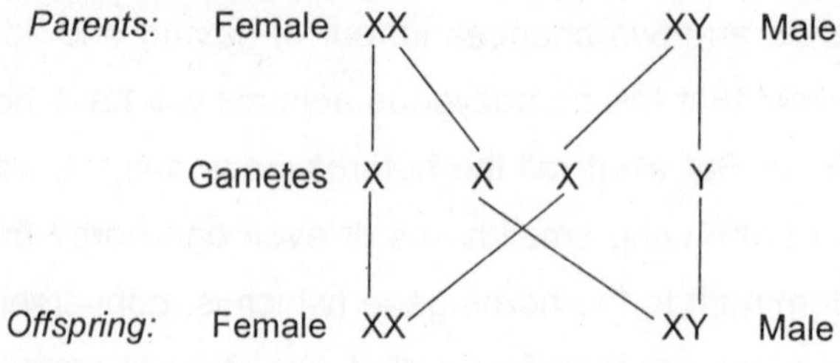

Fig. 5.6: The determination of sex by chromosome combination.

The Y chromosome carries very little genetic information, but the characters it does carry can be expressed only in the male. In addition, recessive genes carried on the X chromosome have a greater chance of being expressed because they are not paired with a dominant partner.

Although characters such as horns and coat colour provide some striking examples of single-gene inheritance, the important production traits of farm animals, such as growth rate, milk yield or litter size, are determined by the combined action of many genes. Another important distinction between, say, coat colour and milk yield is that whereas the former is unaffected by the environment, the latter is susceptible to many environmental influences. A cow with a genetic potential for high milk yield may fail to achieve her potential because she is not given enough food or because she is affected by some disease such as mastitis. While a farmer may have no difficulty in identifying cows carrying the gene for a particular coat colour, it is much more difficult for him to select cows carrying the gene combination required for high milk yield. Furthermore, with the multigenic inheritance of milk yields, the chances of obtaining the same combinations of genes in successive generations are extremely small.

We can now distinguish between the genotype and phenotype of a character. The genotype refers to the genes present in the individual and the phenotype to the way the character controlled by the genes is actually expressed. In the coat-colour example referred to earlier, genotype and phenotype correspond exactly, the heterozygotic roan being recognisable as an intermediate to the homozygotic red or white. In the horned/polled example, however, we cannot distinguish by eye the homozygote polled animal from the heterozygote which is also polled, and phenotype does not correspond fully with genotype. In the case of the production traits, the correspondence between genotypes and phenotype is disturbed also by environmental influences. This last correspondence is often expressed in terms of what is called heritability. The heritability of a character is measured on a scale from 0 (least correspondence) to 1 (perfect correspondence). Heritability is a measure of the proportion of the difference between the animal's own performance and the population average that will be inherited by its offspring. Some examples of heritability for a selection of production traits are given in Table 5.11.

Table 5.11: Approximate values for the heritability of important production characters in farm animals

Heritability range	Species			
	Cattle	Sheep	Pigs	Poultry
0-0.2	Calving interval	Litter size	Litter size	Egg production
0.2-0.4	Milk yield Live-weight gain	Fleece weight	Live-weight gain	Yolk colour Live-weight gain
0.4-0.6	Milk fat percentage	Wool fibre diameter	Bone length	Egg shell quality
0.6-0.8	Milk protein percentage	Birthocat quality	Backfat depth	–

Production rates, such as milk yield and live-weight gain, have low to moderate heritability, while reproductive traits have very low heritability. Product quality characteristics, such as milk and carcass composition tend to be more highly heritable.

Principles of Selection

In any population of farm animal, a herd of dairy cows, for example, there is a considerable range in individual productivity. The range may be defined by a very small group of exceptionally good producers, an equally small group of exceptionally poor producers, and a large mass of intermediate producers. A frequency distribution for individual milk yield in a herd of 100 dairy cows appear as in Fig. 5.7.

Yields range from 2,500 to 7,500 kg per lactation, with a mean of 5,000 kg. The aim of genetic improvement is to identify the superior animals, to use them in preference to the rest as the parents of the next generation, and by doing so to raise the average level of production of the next generation.

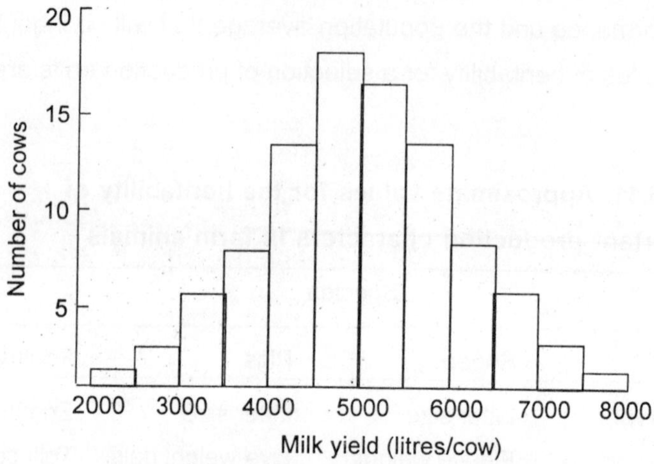

Fig. 5.7 Frequency distribution of individual milk yield in a herd of 100 dairy cows.

This simple concept of genetic improvement is in practice complicated by several factors. The first has been mentioned already, namely, that phenotypic performance is an unsure guide to genotype. For milk yield, heritability is only 0.25–0.40. A second problem illustrated by the example above is that the scope for selecting females may be quite small. For example, the average herd life of a dairy cow is about four lactations, so one in four of all calves born must be retained as herd replacements. As only half the calves born are females the proportion selected is reduced to 1 in 2 (= 50%), and may be further reduced by loss of herd replacements by death or illness. Genetic improvement in milk yield will also depend on the selection of superior males, since these will provide half the genes of the new generation, yet males have no phenotype for milk yield.

A further factor to be borne in mind is that selection of cattle on milk yield alone may lead to determination in other important traits. In Britain, dairy cows provide calves for beef production, so their offspring must be able to grow rapidly and yield a satisfactory carcass. Another important characteristic of dairy cows is that they should conceive at regular intervals of a year.

Genetic improvement in milk yield of dairy cows will therefore depend primarily on the heritability of milk yield and the difference in yield between the population average and the average of the individuals used for breeding. The latter difference, known as the selection differential, will in turn depend on the efficiency with which superior animals (of both sexes) can be identified, on the extent to which breeding can be concentrated on a small proportion of the population, and on the extent to which characteristics other than milk yield must be taken into account in selection. Finally, the rate of genetic environment will depend on the generation interval, which in cattle is about five years (in poultry, in contrast, the generation interval is less than one year, and rate of improvement is correspondingly faster).

Selection Methods

The simplest method of selecting animals for breeding is to base selection on individual performance. Such performance testing is feasible only if the character(s) used for selection are expressed in the appropriate sex (e.g., males may be performance-tested for meat production but not for milk production). Furthermore, it is effective only if the heritability of the character(s) is high (>0.25). If heritability is low and/or the character is not expressed in the sex to be selected from, then selection must be based on the performance of close relatives of the individuals concerned. Such family selection may involve preceding generations (pedigree selection), contemporaries (sib testing; sib being a corruption of sibling, a sister or brother), or the offspring of the individuals concerned (progeny testing).

The selection differential for males is generally greater than that for females because only a small proportion of all available males need to be used for breeding. Even with natural mating, one bull may serve 100 cows a year, and with artificial insemination he will be mated with several thousand females. Most programmes of genetic improvement therefore place emphasis on the selection of males.

Performance Testing

A typical test of this type is carried out at a special test centre to which males are sent by individual breeders. The males are fed on a standard ration and their growth rate over a predetermined age- or weight-range is recorded. The growth rate of each individual is then compared with the group average. If large differences between breeds may be expected, as in cattle for example, the test is confined to one breed. An example of a test conduced on bulls by the UK Meat and Livestock Commission is shown in Table 5.12. The four bulls tested varied quite considerably around the breed average, especially in fat depth.

Table 5.12: The results of a performance test of Hereford bulls, carried out by the UK Meat and Livestock Commission

Bull (number)	Withers height at 365 days old (cm)	Live weight at 400 days old (kg)	Daily gain (kg)	Feed conversion efficiency (kg feed/kg gain)	Fat depth (cm)
1	116.5	514	1.3	6.7	7.8
2	110.5	513	1.3	6.9	5.2
3	116.0	499	1.2	7.0	3.5
4	110.5	480	1.1	6.5	5.4
Breed average	114.4	507	–	–	5.7

By standardising some major features of the environment (housing and feeding), the performance test gives the breeder a better chance of identifying genetically superior individuals than he would get testing his animals on his own farm. Nevertheless, the degree of standardisation is usually insufficient to justify the comparing of a male's performance with that of individuals tested at another time of place. Another drawback of the performance tests is that it does not allow the quality of the animal product (i.e., carcass quality) to be assessed directly. This really requires a progeny test but there are some simpler alternatives. When boars are performance tested they may be accompanied by some of their littermates, who are slaughtered for carcass appraisal at the end of the test (i.e., a sib test). It is also possible to make an assessment of carcass quality of males on test by the use of non-destructive methods such as ultrasonic probing of fat and muscle.

Progeny Testing

This type of testing can be illustrated by the tests currently carried out on dairy bulls by the Milk Marketing Boards. Each year, about 130 bulls selected on the milk production of their mothers are each mated by artificial insemination with a randomly selected group of 300 cows in seven or more commercial herds. The daughters of these matings eventually calve and complete a location. Their milk yield and composition is then compared with those of unrelated heifers in the same herds to give what is known as an improved contemporary comparison (ICC). The ICC of a bull might be expressed as + 250 kg milk, + 2 kg fat, + 8 kg protein, for example, and this would indicate the extent to which he would be likely to improve the lactation milk yield, milk fat yield and milk protein yield of his daughters when mated to unselected cows. In practice about 18 bulls are selected, from the 130 tested, for general use by artificial insemination.

The attraction of the progeny test is that it provides a direct estimate of the bull's genotype and hence of his breeding value. The disadvantages of progeny tests are the numbers of animals required (which are inflated by some inevitable losses) and the time required. A bull's ICC may not become available until six years after he is first mated, by which time he may be dead. In practice, this problem is overcome by collecting and storing semen by deep-freezing while his progeny test is in progress, and by either releasing it for use or discarding it once the result of the test becomes available.

Pedigree Selection

As mentioned earlier, the initial selection of young bulls for progeny testing is likely to be based on the performance of the bull's mothers. This is a form of pedigree selection, in which the records of ancestors are used to identify both promising families and individuals. The individual breeder of large farm animals generally relies on pedigree selection, because he has insufficient animals to conduct his own progeny tests. Pedigree selection was used by the eighteenth-century breeders who established many of our well-known breeds of livestock. It has the advantage of allowing rapid selection from record accumulated previously, and such records may be more detailed than those provided by a progeny test (e.g., milk records for several successive lactations of the same cow).

Selection Indices

When selection is based on more than one criterion it is often desirable to combine the measures of each character into a single or index value for each animal. An example is provided by a performance-testing scheme for young boars. Two boars—plus a castrate and a gilt from the same litter—are grown from 27 kg to 91 kg liveweight at a central testing station. The boars as assessed for growth rate, food conversion ratio and backfat depth. The castrate and gilt are slaughtered and their carcasses are dissected. All measures are weighted for heritability, economic value and correlation with other measures, and the total score is expressed as a deviation from the current average score for boars of the same breed, which is maintained at 100 points. Below-average boars (<90 points) are slaughtered. Boars well above average (>200 points) are likely to be used for artificial insemination. In between these extremes, the better boars are likely to enter specialised breeding herds, while boars nearer the average will be used in commercial herds.

Breeding Plans

Although the selection of breeding animals is often the most difficult part of genetic improvement it must, to be fully effective, be accompanied by a breeding plan.

Inbreeding

The interbreeding of close relatives may sometimes be used to give a concentration of desirable genes. For example, a bull may be mated to his own daughters and granddaughters, or brother may be mated with sisters. Inbreeding was used by some of the breed pioneers, who wished to establish uniform flocks or herds carrying distinctive traits. The problem with inbreeding is that it can concentrate undesirable, as well as desirable, genes, which means that improvements in the desired production characters may be accompanied by a decline in other traits, such as resistance to disease; in addition, inbreeding is likely to increase the frequency of inherited physical defects.

Breeding plans based on inbreeding are rare today but may be used to produce contrasting families which are then interbred (i.e., used for crossbreeding).

Linebreeding

Like inbreeding this involves the mating of close relatives but the relationships are less close. The purpose of linebreeding is to try to preserve in successive generations the combination of genes responsible for an outstanding individual.

Crossbreeding

In its most extreme form this type of breeding can involve the mating of different species, as, for example, when European-type cattle (*Bos taurus*) are crossed with the native cattle of tropical countries (*Bos indicus*). More commonly, it involves the crossing of contrasting breeds; an example is the crossing of Border Leicester rams with hill breeds of sheep to produce 'half-breds' that are suitable for use on improved pastures. Crossbreeding within a breed may, as suggested above, be used to combine the merits of inbred lines.

Much crossbreeding is carried out for reasons other than genetic improvement. Thus Hereford (beef) bulls may be crossed with Friesian (dairy) cows to produce offspring better suited to meat production than is the purebred Friesian. The crossbred males are not used for breeding and so have no influence on subsequent generations. The crossbred (Hereford × Friesian) females might be used as breeding cows in beef herds, where they would be mated with another beef bull to produce offspring that were three-quarters beef type.

The choice of a suitable breed of bull to cross with crossbred cows illustrates one of the main problems of crossbreeding. Crossbred animals are normally not bred together because their offspring are likely to vary in type between the extremes of the grandparent breeds. It is sometimes appropriate to 'backcross' the crossbreeds to one of the parent (e.g., Hereford × Friesian heifers crossed with a Hereford bull).

Alternatively a bull of a third breed, such as a Charolais, might be used. In sheep breeding the Border Leicester × hill breed 'half-breeds' are commonly crossed with a ram of breed selected for its carcass quality, such as the Suffolk. A ram used to produce animals for slaughter is known as the 'terminal' sire. For replacement half-bred ewes, the flock master must return to the original Border Leicester cross.

An example of crossbreeding intended to give genetic improvement is provided by what is called 'grading-up'. Nondescript cows are mated to a bull of recognised breed, their offspring are mated to another bull of this same breed, and later generations are similarly crossed. Eventually the crossbreeds are genetically so close to the breed selected that they are indistinguishable from members of that breed, and may be registered in the breed's herd book.

On of the main purposes of crossbreeding is to gain the advantages of what is called 'heterosis' or 'hybrid vigour'. If two separate breeds or lines within a breed are homozygous for many genes (i.e., each line has genes on its chromosome pair mates which are the same, but differ from the genes carried in the same positions by the other line), the interbreeding results in a high degree of heterozygosity. Such heterozygous animals often perform better (by as much as 20%) than the average of their parents. Thus the hybrid females that the pig farmer may buy from a specialist breeding company can be expected to produce litters at weaning containing more piglets, each individually heavier, than would be expected from their parent lines.

Genetic Engineering

Until recently both animal and plant breeders were handicapped in their ability to transfer or concentrate desirable genes by their inability to identify the genes responsible for particular traits. In the past few years, however, it has become possible to recognise the gene (or section of DNA) that, via RNA, is responsible for the synthesis of a particular protein. Moreover, it has proved possible to remove that section of DNA from a chromosome and to transfer it to another organism. Protein hormones provide some examples of this process of gene transfer. The pituitary hormone called somatotrophin (or growth hormone) controls the growth rate of young animals. The gene for somatotrohin has been identified and can be transferred from one species to another. For example, the gene for the growth hormone of the rat was transferred to the fertilised ova of mice, where it was incorporated in the mouse chromosome. Mice carrying the rat growth hormone gene grew up to be 'giant' mice. A less extreme example of gene transfer is the introduction into the chromosomes of pigs of extra copies of their own output of growth hormone ad hence causing them to grow faster. A contrasting example is the transfer of completely 'foreign' genes from one species to another. Thus sheep can be given the genes for the production of some of the special proteins of man that are responsible for the clotting of blood (and in which haemophiliac individuals are deficient). If these genes are expressed (i.e., function) in the mammary gland of the sheep, it will secrete the special protein in its milk, and these can be extracted and given to haemophiliacs.

These examples illustrate the potential for the new science of genetic engineering to improve animal production. At present, the realisation of this potential is hampered by the problem of identifying the genes responsible of controlling important production trails, such as milk synthesis. Another main problem is that of ensuring that transferred genes operate correctly in the recipient animal. Thus the growth hormone gene in its natural state is expressed only in the pituitary gland (i.e., only this gland secretes growth hormone although all the cells in the animal's body carry the gene). However, foreign genes for growth hormone tend to be expressed in tissues additional to the pituitary gland, thus producing the hormone in uncontrolled amounts.

As genetic engineering is a powerful tool, it must be used with great caution and control.

Conclusion

It is hoped that this very brief account of the principles of animal breeding will have given the reader some insight into breeding practices on farms. No great emphasis has been placed on individual breeds, although the development and maintenance of distinct breeds has played a most significant role in genetic improvement. Individual breeders, often relying on somewhat subjective methods of selection, still make an important contribution to animal improvement, but the future lies more in the hands of breeders who collaborate—in national schemes or small groupings—by submitting their breeding stock to performance and progeny tests. A relatively new development is the group-breeding scheme, in which several breeders contribute their best females to a nucleus herd which is used to breed sires. Both sexes can be tested under standardised conditions, and tested sires can be returned to members of the group for use in their own herds. For poultry and to an increasing extent for pigs also, genetic improvement is now carried out by commercial companies, which supply stock to farmers.

As mentioned earlier the rate of genetic improvement in poultry is rapid. Growth rate in broilers, for example, is estimated to have been increased through breeding by about 2.5 per cent per year over the past 20 years. In dairy cattle breeding, detailed progeny tresting and intensive selection of bulls could theoretically increase milk yield by about 2 per cent per year, but in practice the improvement has been < 1 per cent per year. For beef cattle and sheep, genetic improvement is probably even slower. A recent development that is likely to increase the rate of improvement in ruminants is the use of techniques for multiple ovulation and embryo transfer. By allowing genetically superior dairy cows, for example, to produce as many as 20 daughters a year, the selection of females can be considerably intensified.

Principles of Disease Control

The first principle of maintaining the health of farm livestock is that prevention is better than cure. To maintain health, however, farmers and stockmen must be familiar with the more common diseases of livestock, the circumstances in which they occur, and the agents responsible for them. Farmers and stockmen must be able to recognise ill health when it does occur, to make a preliminary diagnosis, and to apply remedial measures, either on their own initiative or on the advice of a veterinary surgeon. Although the veterinary surgeon must be called in to deal with severe or prolonged illness, he or she does not expect to treat minor conditions of ill health; it is generally the farmer who must distinguish between the major and minor cases.

Signs of Ill Health

The first maifestation of ill health in an animal may be a change in its behaviour. It stands apart from the herd, adopts an atypical posture, is reluctant to move, or shows no interest in food. Ruminants may stop chewing the cud; there may be a sudden reduction in prouctivity (which can be detected quickly if it is a fall in milk yield but not so quickly if it is a reduction in growth rate).

Close examination of the animal will show changes in appearance. Head, wings or tail may droop, eyes may be dulled, the coat (of cattle) may be dull and rough rather than shining and smooth. Swellings or sores may be seen. The nature of excretions and secretions may change. Thus the faeces may be liquid (diarrhoea) or contain blood, urine may contain blood or become cloudy, milk contain clots or blood. Abnormalities of movement, ranging from shivering to uncontrollable convulsions, may be seen.

As in man, some simple tests may help to establish the nature of the disease. A high respiration rate may indicate fever and/or lung infection, and elevated pulse rate and body temperatures are also indicative of fever. Normal values for these measures are shown in Table. 5.13.

Table 5.13: Normal values for respiration rate, pulse rate and rectal temperature in farm animals

	Respiration rate (per min)	Pulse rate (per min)	Body temperature (°C)
Horse			
adult	8 – 16	28 – 42	37.5–38.5
foal	10 – 15	40 – 58	37.5–39.0
Cow			
adult	26 – 30	60 – 90	37.5–39.5
calf	30	100	39.0–40.0
Sheep or goat	10 – 20	68 – 90	39.0–40.5
Pig	10 – 20	60 – 90	38.0–40.0
Fowl	15 – 48	120 – 160	41.5–42.5

Ill Health at Birth

An animal may be abnormal at birth, because either it has a genetic defect which has affected its early development, or it has been damaged *in utero* by some maternal inadequacy or through competition from its littermates. An abnormality acquired *in utero* is described as congenital. Some examples of genetic defects are the misshapen head of the 'bulldog' calf, or the imperforate anus of piglets; they cannot be treated, but recording their occurrence and selecting against them will reduce their future incidence. Some congenital disorders such as 'swayback' in lambs (muscular incoordination resulting from maternal copper deficiency –see Table 5.5) are also irreversible.

Genetic and congenital defects, however, account for only a minor proportion of the animals that die at or shortly after birth. In pigs, for example, of all such neonatal deaths, these defects are responsible for only about 12 per cent (i.e., they account for losses of about 0.2 pigs per litter). Attending animals during parturition will save many young from death. Early recognition of those that are poorly equipped, because of size or lack of reserves, to adapt to extra-uterine life is also very effective in preventing losses. Animals disadvantaged at birth may be temporarily removed from their dams (e.g., lambs to a warmer place) and given special treatment.

Injuries

Poorly designed or constructed housing can be a major cause of injuries to livestock. Slippery or broken floors, projecting metal-work, and unsafe feeding yokes or neck chains must be avoided. Animals may often be injured by their companions. A badly designed furrowing crate allows the sow to crush piglets when she lies down. The close confinement imposed in some modern systems of production may be responsible for injurious vices in livestock; some examples are tail-biting in pigs and feather-picking (perhaps leading to cannibalism) in poultry. Injuries may be prevented by modifications of mutilations of animals, such as de-horning in cattle and partial beak removal in poultry. A common cause of injuries due to fighting is the mixing of animals that are strangers to each other. A final point to remember is that physical injuries may predispose animals to diseases due to other causes. For example, a crushed teat (sometimes a self-inflicted injury) may allow the invasion of the mammary gland by bacteria which cause mastitis (see below).

Diseases due to Micro-organisms

The principal pathogens in this category are the bacteria (i.e., minute unicellular plant organism) and the viruses. The latter are extremely minute particles, consisting largely of DNA, which can multiply only inside the cells of other organisms. Two further types of pathogen in the same category are protozoa (i.e., minute unicellular animal organism) and fungi.

These pathogenic organisms commonly gain access to animals through their respiratory, alimentary and urinogenital tracts, and may be transmitted to other animals by direct contact (e.g., venereal diseases), or via food, water or air (e.g., the virus causing foot and mouth disease may be carried by wind for long distances). Some organisms may always be present in animals but become harmful only if their numbers are excessive and/or they move to a more sensitive part of the animal (e.g., invasion of intestinal tissues by bacteria living in intestinal contents). Other pathogens may be capable of surviving for long periods outside animals; an example is the bacterium causing anthrax, which forms spores capable of surviving in soil for many years.

Some diseases may be transmitted between farm animals and man and are known as zoonoses. In Britain the most important examples are tuberculosis, brucellosis and a group of intestinal and other disorders caused by Salmonella bacteria.

When pathogens invade animal tissues they commonly stimulate the production of what are known as antibodies. These are proteins which react with pathogens and suppress or kill them. Pathogens are also contained, and thus inactivated, by the white cells of the blood.

The production of antibodies offers several methods for control of diseases. After an animal has been infected by a particular pathogen, and has combated the infection by production of antibodies, the latter may remain in its blood and confer immunity to further infection. Antibodies may also be passed from one animal to another. This transfer occurs naturally from mother to offspring, either via the placenta or via the initial secretion of the mammary gland known as colostrum. During the first day or so of life, the young animal is able to absorb intact proteins (the immunoglobulins of colostrum). In adult animals, antibodies may be transferred by collecting blood serum from an individual that has experienced infection in the past and injecting it into an individual recently exposed to infection. Animals which acquire immunity passively, at birth or by transfer of antiserum, receive protection that is immediate but which is also short-lived, because no production of antibodies has been stimulated.

Animals may be actively immunised by injecting them with pathogens that have been killed or modified but are still capable of stimulating the production of antibodies.

Control of Disease Caused by Micro-organisms

Having established the background to diseases caused by micro-organisms, we can now consider the strategy used to control them. If pathogens fail to make contact with animals there will be no infection, so the first means of control is isolation. A policy of isolation may be exercised at several levels. An island country, such as Britain, can prevent the importation of livestock (or require them to spend time in quarantine to ensure they are free of disease); this policy has (so far) proved particularly successful in keeping out many diseases caused by viruses, such as rabies and cattle plague. In the case of diseases spread mainly by close contact between animals, an individual farmer may be able to isolate his flock or herd if it is known to be free of the disease. This strategy is used to control virus pneumonia in so-called 'minimal disease' pig herds. Again, on the individual farm, a single animal that contracts a disease may be isolated from the herd until it recovers. Many farms have a special building for this purpose, and it may be used also to quarantine purchased stock.

A second type of strategy used to control disease requires the slaughter and hygienic disposal of infected animals. This strategy may be used to control diseases which occur as sporadic outbreaks but which can spread rapidly if given the opportunity. In Britain, foot and mouth disease is controlled by immediate slaughter of all susceptible stock on a farm once cases have been diagnosed on that farm. A slaughter policy may also be used to eliminate diseases that occur in a relatively small number of individuals. In Britain, bovine tuberculosis and brucellosis have been reduced to a very low incidence by a policy of routine testing of animals for these diseases and slaughtering those found to have them.

In the case of diseases caused by organisms widely distributed in nature, and which can survive apart from animals (for example, in faeces), a strategy of routine hygiene is required. The routine used to control infection of calves by organisms such as *Salmonella* bacteria, which cause intestinal upsets and scouring (diarrhoea) may be used as an example. Young calves are housed separately from older animals (to prevent infection from elsewhere) and are kept in individual pens with individual feed buckets (to prevent one calf infecting another). Their buckets are cleaned frequently (to prevent bacterial growth in feed residues) and the calves are inspected frequently (to allow detection and isolation of infected animals). A further precaution, which is not always practicable, is to rear calves in batches; in between each batch, the calf house is clear of animals and can be thoroughly cleaned and disinfected.

Intestinal disorders in calves are caused by a wide variety of organisms, and it is at present difficult to protect the animals by vaccination. This technique is used, however, to control diseases caused by specific organisms. In lambs, for example, there is a small group of diseases (pulpy kidney, lamb dysentery and tetanus) caused by bacteria in the *Clostridium* group, which can be prevented by vaccination at birth. In poultry, vaccines are used to control Marek's disease (a viral infection which affects the nervous system and causes paralysis), infectious bronchitis and several other diseases.

If a vaccine is used routinely for a long period, the incidence of the disease concerned may be reduced sufficiently to justify a change to an alternative control strategy. An example of such a disease is brucellosis in cattle, which was at one time prevented by vaccination but is now controlled by the slaughterer of infected animals.

The last line of defence against disease is the use of therapeutic agents or drugs. For some livestock, principally poultry, drugs are given routinely, as a constituent of the diet, an example being the use of dimetridazole to control the protozoal disease, blackhead, in turkeys. Drugs may also be used routinely (i.e., to prevent infection) but intermittently; thus dairy cows may have their teats injected with antibiotics at the end of lactation, in order to treat existing infection and prevent infection during the dry period. Finally, drugs may be used selectively to treat infections.

The use of antibiotics to treat animals has been questioned because they may encourage the production of resistant strains of bacteria that may subsequently infect man. Antibiotics used in human medicine are not now used routinely for farm animals in Britain but may still be used for selective treatment. The treatment of mastitis in dairy cattle, for example, is based on the use of antibiotics.

Diseases due to Parasites

The major internal parasites of livestock in Britain are helminth worms, such as the roundworms that inhabit the stomach and intestine (and also the lungs), tapeworm of the small intestine, and flatworms exemplified by the liver fluke. The major external parasites are small invertebrates such as ticks, mites, lice and flies.

Worms

The nematode roundworms found in the stomach and small intestine of sheep and cattle damage the lining of the gut, thus interfering with absorption and also causing loss of blood. The adults lay eggs which are excreted on the pasture and may hatch to produce larvae within a few days or lie dormant over winter. The infective larvae are ingested during grazing and develop into adults to complete their life cycles. Lungworms have a similar cycle but the adults live in the respiratory tract, where they cause bronchitis and predispose their host to pneumonia.

The life of tapeworms and the liver fluke is complicated by the need for an alternate host. The adult fluke, for example, lives in the duct system of the liver, its effect ranging from chronic 'un-thriftiness' to severe liver damage and death of the host. Its eggs pass through the bile duct and thence to the faeces. The larvae derived from them invade a type of small snail found commonly on ill-drained pastures, and subsequently develop into cysts which are ingested by the grazing animal.

Ideally, pastures would be kept completely free of infective larvae, but in practice this is difficult to achieve because a newly sown pasture is soon re-infected by the stock brought on to it. Nevertheless, helminth infestation of livestock can be kept to an acceptably low (i.e., non-injurious) level by the use of husbandry practices that break the life-cycle of the worms. Life-cycles can also be broken by the destruction of alternate hosts. For example, the draining of pastures discourages invasion by snails and hence reduces fluke infestation.

Like bacteria and viruses, some helminths stimulate the production in the host of antibodies. Older animals are therefore less affected by worms than are their offspring, and rotational grazing strategies are often designed to give young animals the 'cleanest' area. Young cattle, when turned out to graze in the spring, are particularly susceptible to a lungworm infestation which causes the coughing condition known as 'husk'. For this parasitic disease, however, it has proved possible to produce a vaccine which is now used routinely in cattle.

During the past thirty years there has been a remarkable development of anthelmintic drugs, most of which are administered by oral 'drenching' but the most recent of which can be given by injection. Although the first lines of defence against worms are the husbandry practices suggested above, anthelmintics provide valuable reinforcement, especially when stocking rates are so high as to present special problems of control.

External Parasites

As a group, these parasites cause ill health in farm animals through three major effects. First, they may irritate the animal to such an extent that its feeding behaviour is affected and its productivity reduced (e.g., biting flies). Secondly, the parasites may have a direct damaging effect on the animal's tissues (e.g., ticks may suck blood and cause anaemia, warble fly larvae burrow through the skin of the animal). Thirdly, external parasites may act as vectors (i.e., carries) of disease caused by micro-organisms. The most serious example of insect transmission of disease is trypanosomiasis (sleeping sickness) of ruminants (and also man), which is caused by protozoa carried by tsetse flies. While this does not occur in Britain, there are several diseases here (e.g., tick-borne fever and louping-ill in sheep) which are transmitted by external parasites.

Parasites that can survive only on selected host species can be eliminated by routine treatment of their hosts with insecticides. The disease known as sheep 'scab', which is caused by mites, was eliminated from Britain (and other countries) by compulsory 'dipping' of all sheep in insecticide solution, but has recently recurred. The warble fly of cattle, which spends part of its life-cycle in animal tissues, is currently the subject of an eradication campaign.

Nutritional Diseases

Three types of disease may be considered under the heading of nutritional diseases. The first group are the deficiency diseases discussed earlier in this chapter (e.g., Tables 5.4 and 5.5). The second group are the diseases caused by ingestion of poisons. Thirdly, there are the so-called 'metabolic' disorders which are characterised by abnormalities in the absorption, metabolism and excretion of nutrients.

Poisons

The range of substances toxic to livestock is extremely wide, and only the major classes of poisons can be indicated here. Grazing animals may consume weeds, growing in or around their pastures, that contain toxins (e.g., bracken and ragwort). Normal crop plants may contain substances that become toxic if they accumulate in plants, or one otherwise consumed in excessive quantity.

Many members of the Brassicae (i.e., cabbage family) contain an unusual amino acid derivative that causes anaemia in ruminants. Grasses may accumulate toxic quantities of nitrates, and pasture legumes (like red clover) can contain phyto-oestrogens, which interfere with the reproduction of grazing animals. The toxicity of a normally safe plant or feed may sometimes be due to fungal infestation. The classical example of this is ergotism, the nervous disorder of man and animals that is caused by consumption of grains infested with the ergot fungus, *Claviceps purpurea*. A more serious threat today are the carcinogenic aflatoxins produced by fungi found on groundnuts.

Poisoning may also be caused by contamination of feed. In the field, airborne pollutants such as fluorine from aluminium smelters, may make crops toxic to livestock. In the feed-mixing plant, pig feeds fortified with 175 mg/kg of copper as a growth promoter may contaminate feed intended for sheep, which are sensitive to diets containing >20 mg Cu/kg. Animals may also ingest toxic substances from sources other than feed; for example lead ingested by licking painted surfaces is a common cause of poisoning in livestock. The many chemicals used on farms to control weeds, plant diseases and pests represent a continuous hazard to animals.

Methods for preventing poisoning are largely self-evident. The farmer must be aware of the harzards present on, or likely to arrive at, his farm and must be prepared to protect his livestock accordingly.

Metabolic Disorders

The complexity of the chemical transformations taking place in the animal is evident from earlier parts of this chapter. Nutrients absorbed from the gut are used continuously to construct new or replacement tissues and products, or are shunted into or withdrawn from stores; nutrients derived from tissue breakdown are salvaged for re-use, and unusable materials are excreted. Occasionally something goes wrong. We have seen already that a dietary shortage of a particular nutrient can be responsible for serious ill health. But even when there are no dietary deficiencies, an enzyme deficiency, for example, may mean that some vital product fails to be delivered in sufficient quantity to the site where it is required. The disease which follows such a failure of metabolism is known as a metabolic disorder.

Metabolic disorders generally arise from some overloading of the animal's systems. The high-producing animal, such as the high-yielding dairy cow, is particularly susceptible. The likelihood of metabolic upsets is also very dependent on the speed at which the increased load is imposed. At calving, the dairy cow moves rapidly from a state of relatively low nutrient intake and nutrient demand, to the more intense metabolism imposed by the high milk yield or early lactation. She is therefore particularly susceptible to metabolic disorders in early lactation. It would be wrong to assume, however, that the low producer is unlikely to suffer from metabolic problems. In an animal fed at a maintenance level, and with few reserves, a slight change in nutrient demand can upset metabolism as disastrously as can a much greater change in a normally well-fed animal. Although the dairy cow is used here to illustrate the main types of metabolic disorder, these diseases are found in all forms of animal production.

When a dairy cow begin to lactate, the nutrient demands of milk secretion rise faster than does food intake and thus nutrient supply. The cow therefore uses her reserves. Body fat is transformed into milk fat and may also be used for maintenance purposes; tissue proteins might also be transferred to milk, but scope for this is limited by the absence of any substantial protein store; minerals, especially calcium, are moved from bone to milk.

To metabolise fat effectively, the cow needs an adequate supply of carbohydrate in the form of glucose or propionic acid. If the balance is disturbed by too much fat, the latter is imperfectly metabolised to substances which may be toxic to the animal; some of these substances belong to the class of chemicals called ketones, and the resulting disorder is called ketosis. A further manifestation of too rapid a removal of body fat from reserves may be an undesirable accumulation of fat in the liver.

If calcium cannot be drawn from bones fast enough to meet the drain in milk, then the calcium concentration of body fluids is reduced. Nerve and muscle functions depend on adequate calcium, and a fall in blood calcium concentration (hypocalcaemia) causes muscular spasms and incoordination; this condition is commonly called milk fever. A related condition is associated with low blood magnesium (hypomagnesaemia); it occurs commonly in grazing cows and is called grass staggers.

Additional metabolic disorders arise from imperfections in the metabolism of rumen micro-organisms. Excessive production of gas, often trapped in a foam, causes bloat. Too rapid fermentation of diets rich in soluble carbohydrates to lactic acid and the subsequent absorption of this acid by the animal may cause acidosis.

As metabolic disorders are commonly associated with high levels of production, isolated cases may be accepted by the farmer as part of the cost of increased productivity, and are treated as they occur. Thus, cows with milk fever may be given an intravenous infusion of calcium in the form of calcium borogluconate, which relieves the tetany and may maintain blood calcium concentration at a normal level for sufficient time for calcium supply and demand to return to balance. First-aid measures, however, are inadequate if the incidence of milk fever or other metabolic disorders is severe, and some means of prevention must be sought.

Although metabolic disorders are not generally due to dietary deficiency, they can often be prevented by modifying the diet. For example, the dairy cow can be prepared for the calcium requirement of lactation by feeding her on a *low* calcium diet before calving. This has the effect of mobilising bone calcium, and ensures that the mobilising machinery is operating before a large demand is suddenly placed on it. A rather different example is provided by the ewe in late pregnancy that may suffer from ketosis (in the form known as pregnancy toxaemia) because she is drawing heavily on her fat reserves to provide for growth of foetuses. A small supplement of a starchy concentrate given at this time will help to restore the balance in metabolism between fat and carbohydrate.

As production levels rise, and as other diseases are eradicated, metabolic disorders (sometimes called production diseases) assume greater importance in animal production. A device that is sometimes used to diagnose them when they are still sub-clinical (i.e., manifested in a form too mild to be recognised by the farmer) is to carry out what is called a metabolic profile test. In a dairy herd subjected to such a test, blood samples would be taken from several animals having different levels of milk production, and the samples would be analysed for a series of ten or so metabolites, including calcium, glucose, etc. The extent to which the overall profile of these metabolites differ from normal is used to indicate where preventive measures are needed. Although interesting in principle, this test has not proved to be as effective in practice as it was hoped.

Preventive Medicine

The first principle of animal health control, that prevention is better than cure, is now enshrined in the important branch of veterinary practice called preventive medicine. Instead of merely treating sick animals, the veterinarian advises the farmer on the procedures he himself can use to prevent infections and disorders, and visits the farm at regular intervals to inspect the stock and apply any specialised tests or treatments that are required (e.g., pregnancy diagnosis). The veterinarian may also discuss with the farmer the wider aspects of animal husbandry such as nutrition, ventilation of buildings and pasture management.

❑❑❑

6

The Principles of Grazing Management

Introduction

Grazing systems can be thought of essentially as combination of populations of plants and animals occupying a particular climatic and edaphic environment. The major plants and animals are the obvious inhabitants of these systems and attention will be concentrated on them in this chapter, but the coherence of any system is heavily dependent upon the activities of associated large populations of macro- and micro-flora and fauna.

Such systems are found across a very wide range of condition, ranging from the semi-arid grasslands of North Africa, the Middle East and northwest China to the intensive, largely artificial grasslands of northern Europe and New Zealand. In all cases, grazing management is concerned with the manipulation of plants and animals to some and often, but not always, enhanced output of animal product for use by Man. We will concentrate attention largely upon the management of sown temperate grasslands, which are usually characterised by very simple populations of both plants and animals, but some principles will also be illustrated by reference to natural communities which offer more variety, at least in plant characteristics.

In agriculture, the success of a grazing system may be judged in terms of the output of animal product – principally meat, milk, wool or hides. This output is the result of a chain of processes involving the production of herbage by populations of plants, its consumption by populations of animals, and its conversion into animal product. These proceses interact directly with one another so that management decisions which improve efficiency at one stage of the production chain may result in a reduction in efficiency at another. Thus an understanding is required of the factors influencing efficiency at each stage, the degree to which they are amenable to manipulation, and the form and extent of their associations with one another.

The Sward

Temperate grasslands are populated primarily by low-growing, perennial, sward-forming plants, most of which can perpetuate themselves by vegetative propagation as well as by seed production. Vegetative propagation depends upon the development of secondary buds, either in leaf axils on the man stem of the parent plant (for example, tillering activity in many grasses and sedges), or on specially-adapted branching organs (e.g., stolons in some grasses and white clover, or underground rhizomes in other grasses and legumes) which also aid the active spread of individual plants. The sequential development of successive generations of buds gives rise to the typical branched form of most trees and shrubs. However, a particular advantage of the better-adapted temperate forage species is the ability of new tillers of stolons to form their own root systems, thus quickly becoming independent of the parent plant and providing ground anchorage. This characteristic of the sward-forming species provides an important basis for flexibility of response to defoliation by grazing animals. A further important characteristic is that the vegetative buds are carried very close to ground level, or even below it, thus protecting the buds from risks of defoliation or treading damage.

Perenniality is a consequence of continued tiller or stolon development, because few individual growth sites have an effective productive life of more than 12 months. In perennial plants the supply of new buds is virtually inexhaustible, and their development into tillers is limited primarily by competition with neighbours for light, water or nutrients. The initiation of new tillers is itself triggered by defoliation of the parent and exposure of the bud site to light, a further example of a successful adaptation to grazing conditions. With these adaptive mechanisms in mind, it is helpful to consider the grazed sward principally in terms of populations of individual tillers in the case of grasses, and their nearest analogues, rooted nodes, in the case of the stoloniferous legumes.

Leaf production is virtually continuous over the lifetime of a vegetative tiller, individual leaves being produced in sequence from the stem apex which is borne close to the base of the tiller and encased in the sheaths of older leaves. Following its appearance from within the sheath of 7-21 days during the main growing seasons, depending upon climatic conditions, but a much longer period of time in the winter. Once it is fully extended an individual leaf has a relatively short period of existence before irreversible degenerative changes begin which lead rapidly to senescence and death. A vegetative tiller of perennial ryegrass characteristically carries only three live leaves, one of which is in an active phase of expansion, one fully expanded or approaching full expansion, and one mature. A similar pattern of leaf appearance and senescence is apparent in tillers of other grass species, though the number of live leaves per tiller may vary from species to species, and on individual stolon branches of white clover. This pattern of development and death proceeds irrespective of whether individual leaves are defoliated or not. Dead leaves eventually become detached from the tiller and work down to the litter layer at the base of the sward, or they may be taken down by earthworms into the soil.

Under appropriate conditions individual tillers may develop from a vegetative to a reproductive state in which the stem apex elongates to form a flowering head. In many species of grass the stimulus to flowering is increasing daylength in the spring, and in some species, such as *Lolium perenne*, an initial cold stimulus is required. In other species, such as *Poa annua*, tillers may develop rapidly to reproductive state at almost any time of the year. Reproductive changes involve progressive elongation and signification of the main stem, resulting in a rapid increase in the height of the sward canopy. Once the seed head borne at the top of the flowering stem has ripened, leaf production has effectively ceased and the tiller produced through the final states of senescence and death. After a phase of development of reproductive tillers, plant growth can be sustained only from any vegetative tillers which remain, by the development of new shoots from the bases of the old reproductive tillers, or eventually by the development of new plants from shed seed.

Plant growth is dependent upon the supply of assimilates resulting from the process of photosynthesis in active leaves exposed to the light. Since leaves develop from growing points situated close to ground-level they must penetrate though the overlying sward canopy in order to reach the light, and successive leaves tend to over-top those earlier in chronological sequence. Characteristically the proportion of live leaf increases from the bottom to the top of the sward canopy and the proportions of stem, sheath and dead material declines. The bulk density (mass per unit volume) of foliage also decline progressively from the base of the sward upwards. As a consequence of stem elongation, reproductive swards are usually taller than vegetative swards and have a lower bulk density. Individual tillers tend to adopt a more erect habit of growth, and there is a more heterogeneous admixture of stem and leaf material. All of these factors may influence both herbage production and the herbage intake of grazing animals.

Young actively-growing leaf tissue contains relatively limited amounts of fibre, and is therefore easily digested by ruminants. The ratio of structural to non-structural material increases with increasing leaf maturity, particularly once the degenerative changes of senescence begin and the more soluble components are progressively lost by respiration and remobilisation. Thus digestibility of leaf tissue may be as high as 0.9 during elongation, declining to 0.7 in senescence and 0.5 in dead leaves in the litter layer. The digestibility of reproductive stem declines rapidly with advancing maturity as a consequence of its high content of structural tissue and degree of lignification, so that the digestibility of herbage in reproductive swards falls progressively with increasing maturity and increasing ratio of stem to leaf tissue (Fig. 6.1).

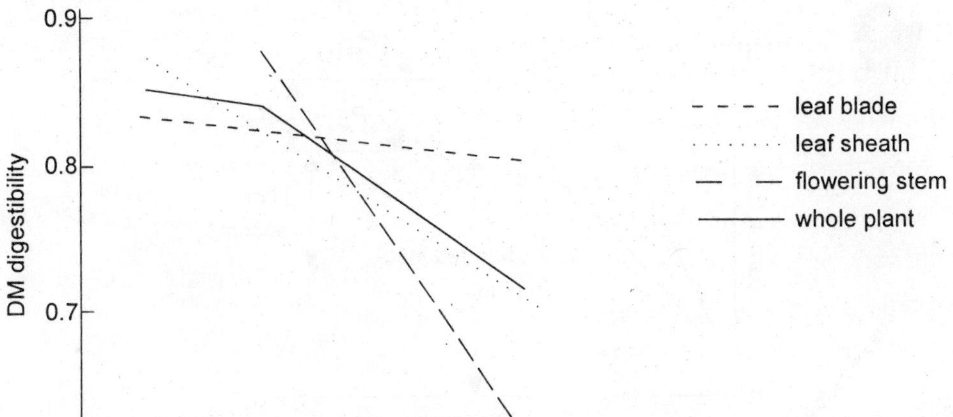

Fig. 6.1: Changes with time in the digestibility (DM) of the whole plant and its components in S_{24} ryegrass during primary spring growth.

Herbage Growth and Utilisation

The processes of growth and senescence occur simultaneously and virtually continuously in temperate swards. Conventional measurements of herbage 'production' – strictly net production or accumulation over a finite period of time – depend upon the balance between the rates of growth and senescence. Successful management involves optimising this balance, so it is important to know how the rates can be manipulated, but neither is particularly easy to measure in the field.

One approach is to measure the rates of carbon uptake in photosynthesis and release in respiration, and the balance between them, the net carbon assimilation. These rates are not strictly analogous to the rates of herbage growth, senescence and net accumulation, but cumulative net carbon assimilation must set an upper limit to production because carbon forms such an important component of plant tissue and plant energy supply. The net rate of carbon assimilation per unit of ground area (P_{nc}) increases progressively to a maximum with increasing leaf area index in a developing sward as the efficiency of light interception increases, but the accumulation of herbage in the sward will eventually start to decline as mutual shading reduces the rate of carbon uptake per unit of leaf area and as the losses attributable to respiration and decomposition increase (Fig. 6.2).

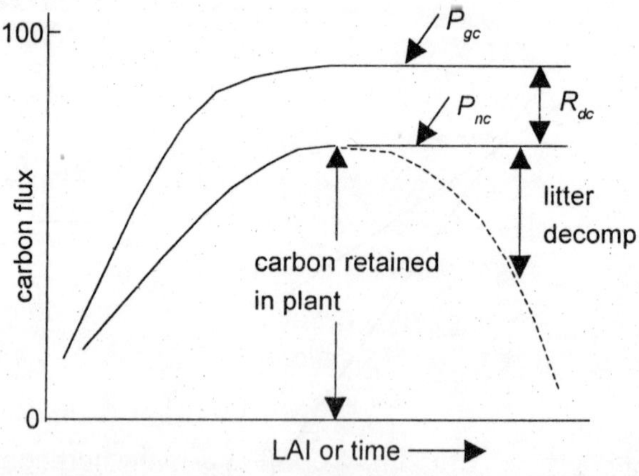

Fig. 6.2: Diagrammatic representation of changes in the rates of gross photosynthesis (P_{gc}), net photosynthesis (P_{nc}) and respiration (R_{dc}) in a grass sward over a period of growth following defoliation. Values shown are rates of carbon flux (g^{-1} m^{-1} h^{-1} CO_2) relative to maximum P_{gc} = 100.

Herbage production can also be viewed as the product of the number of growth sites (i.e., of grass tillers and clover nodes) per unit ground area and the rate of production of plant tissue per site, both of which can be measured. Continuous heavy stocking of a perennial ryegrass/white clover sward results in reduction in the size of individual tillers but an increase in tiller population density, an increasingly prostrate habit of growth and an increase in the ratio of young to old leaf tissue. These adaptive changes help to maintain both the efficiency of light interception and the photosynthetic efficiency of leaf tissue, and hence the rate of herbage growth per unit area (Fig. 6.3).

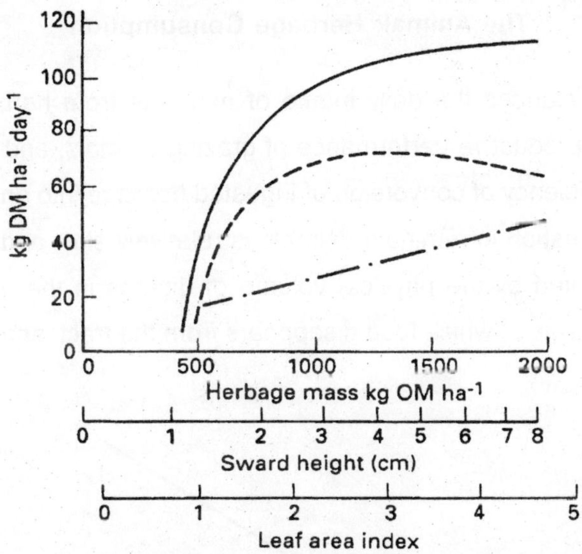

Fig. 6.3: Relationship between herbage mass (kg OM ha^{-1}) and rates of herbage growth (——), senescence (– · – · –) and net production (-----) (all as kg DM ha^{-1} day^{-1}) in a mixed grass/clover sward continuously stocked with ewes and lambs. Values for sward surface height (cm) and leaf area index are also shown. (Source: adapted from Bircham and Hodgson).

However, if herbage mass is maintained below 1000 kg OM ha^{-1}, the tiller population may decline and the rate of herbage growth then fall away rapidly. The rate of senescence of mature herbage is directly proportional to the herbage mass maintained. Thus, when the balance between the rates of herbage growth and senescence is taken into account, net herbage production per hectare is usually maximised at a herbage mass of 1200-1500 kg OM ha^{-1} (LAI 3-4) under continuous stocking, though the variation in net production is in fact relatively small over the range 1,000-2,000 kg OM ha^{-1} (LAI 2-5) (Fig. 6.3). These results relate to swards of perennial ryegrass and white clover, species which show substantial flexibility in morphology and population dynamics in response to variations in management. Information of the kind shown in Fig. 6.3 is not available for other species.

Intermittently cut or grazed swards are subject to continuing change of state and therefore have less opportunity for adaptation. Nevertheless, the effect of management upon rates of tissue production and loss can be described in much the same terms.

The Animal: Herbage Consumption

In most circumstances the daily intake of nutrients from herbage is the major determinant of the productive performance of grazing animals, and has a substantial influence on the efficiency of conversion of ingested herbage into animal product. The process of food digestion in ruminant animals is relatively slow and the consumption of food may be limited by the physical volume of digesta in the alimentary tract or some part of it. The rate at which food disappears from the tract, and thus allows more

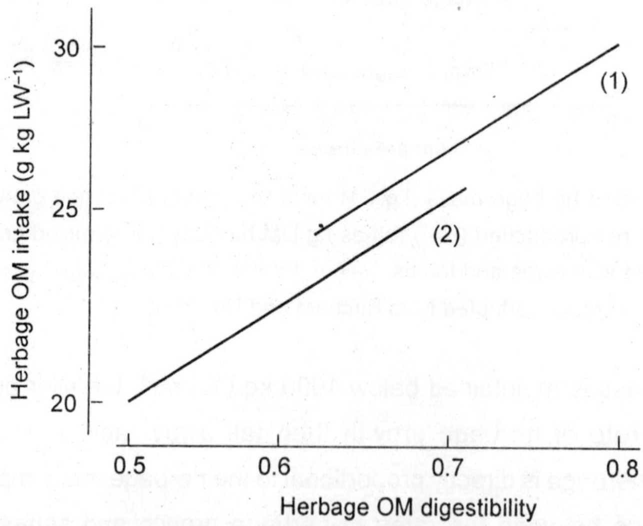

Fig. 6.4: Relationship between the digestibility of the diet ingested (OMD) and the daily herbage intake (g OM kg LW-1) of weaned calves strip-grazing swards of perennial ryegrass. (1) Primary growth; (2) regrowths.

to be consumed, reflects its digestibility, the rate of digestion and absorption of the digestible components, and the rate of passage of indigestible components down the tract. These factors in turn are influenced by the structural strength of the plant material, and its resistance to comminution by chewing. All of these characteristics tend to be correlated in long forages, and collectively they exert an important influence upon forage intake (Fig. 6.4). In ruminants fed on mixed diets of forages and concentrates, intake may also be limited at high nutrient concentration by the level of circulating metabolites in the blood, but this limitation is seldom if ever important in animals eating all-forage diets, particularly under grazing conditions.

The bulk density of the herbage within a sward canopy is relatively low, particularly in the surface horizons of the sward, and grazing ruminations may take up to 30,000 or more individual bites of herbage and spend up to 12 hours or more grazing per day in order to satisfy their appetite. Factors which limit the weight of herbage in individual bites are therefore likely to limit daily herbage intake unless the animal can compensate by increasing the rate of biting or the time spent grazing. Intake per bite is influenced by variations in both sward height and the bulk density of herbage in the sward (Fig. 6.5), principally through the effects of these award variables upon the dimension of

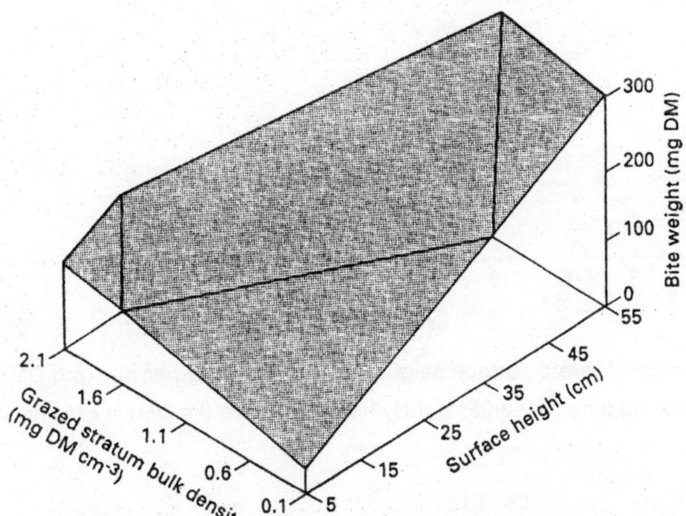

Fig. 6.5: The influence of variations in sward surface height (cm) and herbage bulk density (mg DM cm^{-3}) on intake per bite (mg DM) in grazing sheep. (*Source:* adapted from Burlison *et al.*).

individual bites and upon the weight of herbage per unit volume encompassed at a bite. Compensating changes in biting rate and grazing time are possible, but they are seldom of sufficient magnitude to adjust fully for limitations in intake per bite, which usually appears to be the major determinant of daily herbage intake (Fig. 6.6).

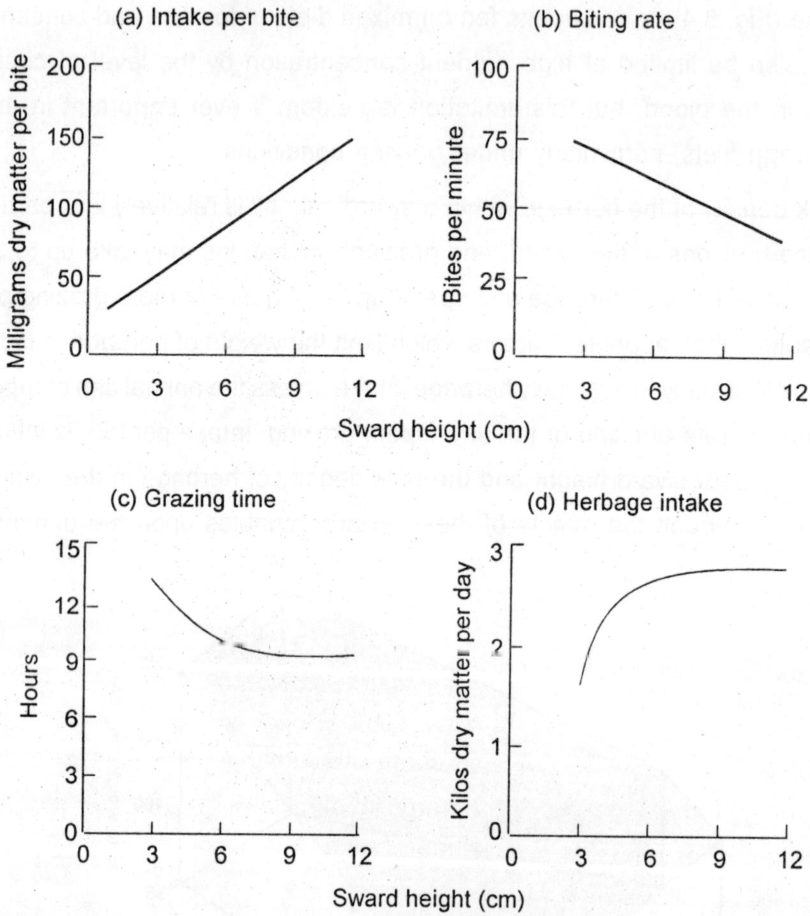

Fig. 6.6 The influence of sward surface height (cm) on (a) intake per bite (mg DM), (b) biting rate (bites min^{-1}), (c) grazing time (h^{-1}) and (d) daily herbage intake (kg DM) in grazing sheep. (Source: Penning).

The responses shown in Figs. 6.4 and 6.6 relate principally to homogeneous vegetative swards where animals appear to exhibit little deliberate discrimination between the alternative botanical or morphological components. In more heterogeneous swards grazing is likely to become more selective, selection being based upon the animal's assessment of the surface characteristic of individual plants, the presence of specific volatile compounds affecting olfactory or gustatory assessment, and the ease of prehension of particular components. The position of individual components within the sward canopy will also affect the change of defoliation. Selective grazing will work to the animal's advantage by increasing the nutrient concentration of the diet but this advantage is likely to be offset to some extent (and, in extreme cases, completely) by the reduction in intake per bite and/ or in biting rate which is likely to accompany selection.

Herbage intake per animal is therefore likely to be influenced primarily by the stage of maturity of the sward, the vertical distribution of herbage, and the disposition of the component species and morphological parts within the canopy. In simple swards these effects may be simplified to the relationship shown in Fig. 6.6. It will be seen that the factors which characterise a well-grazed sward—a dense population of tillers bearing relatively young leaves—are likely to be as beneficial to herbage intake as to herbage production. However, the close grazing necessary to maintain a sward in this condition is unlikely to maximise herbage intake per animal because it will result in a relatively shallow sward canopy.

Efficiency in Grazing Systems

Examples of the efficiency of each stage of the production process are given in Table 6.1, taking ratio of (energy out)/(energy in) as a common basis for management. The examples refer to temperate conditions. Management is concerned with the interrelationships between the efficiencies of the three stages and the factors influencing overall efficiency.

The efficiency of conversion of the sun's energy into plant tissue is sensitive to variations in temperature and in the concentration of carbon dioxide in the atmosphere. It is always low but, because of the enormous quantities of light energy striking the earth and the consequent benefits which should accrue to even a limited improvement in efficiency, there has always been great interest in the possibility of manipulating the

Table 6.1: Estimates of the energetic efficiency of (a) herbage growth; (b) herbage utilisation; and (c) conversion of ingested herbage into animal product: temperate grassland

Stage of production	Ratio $\frac{\text{energy output}}{\text{energy input}}$
(a) Herbage growth[a]	
$= \frac{\text{energy in plant tissue}}{\text{photosyntheically} - \text{active radiation}}$	0.2-0.04
(b) Herbage utilisation[b]	
$= \frac{\text{energy in herbage consumed}}{\text{energy in herbage grown}}$	0.4-0.8
(c) Conversion to animal product[c]	
$= \frac{\text{energy in product}}{\text{energy in herbage consumed}}$	0.02-0.05

[a] Mean value for year; during active growth efficiency may be 0.04 - 0.08

[b] Grazing animals

[c] Sheep. Estimate bases on annual food input to ewe and lamb (s); during grazing season, efficiency on a daily basis may be 0.05 – 0.15.

sward to increase herbage production in a predictable fashion. Under a cutting management the amount of herbage harvested is almost invariably higher at low harvesting frequencies than at high harvesting frequencies (Fig. 6.7), but there is less unanimity about the effects of severity of defoliation on the amount of herbage harvested. The relatively high average herbage mass and LAI on infrequently cut swards should result in efficient light interception and a high rate of carbon uptake (Fig. 6.2), but some of the advantages of infrequent cutting may also be attributable to improved harvesting efficiency.

There appears to be much less scope for improving the efficiency of herbage production by modification of grazing management, though herbage growth rates may still be depressed by prolonged, severe defoliation (Fig. 6.3).

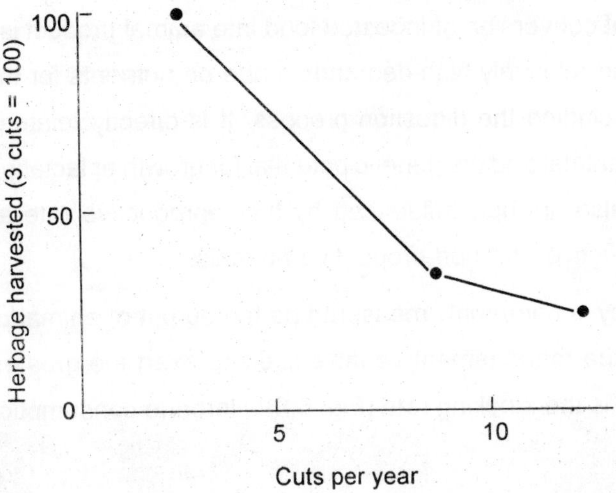

Fig. 6.7 Relationship between the number of cuts per year and the herbage harvested (value at 3 cuts per year = 100).

The efficiency of utilisation of the herbage grown is in general substantially higher than the efficiencies of the other two stages in the production process, but this is probably the stage at which most unobserved loss of efficiency occurs. In continuously stocked swards the rate at which herbage is harvested by grazing animals (kg/ha/day) appears to be maximised at a herbage mass and LAI at which the herbage growth rate is already depressed, but at which the ratio of consumption to growth is high. When utilisation is expressed as the ratio of the rates of herbage consumption and growth, as in Table 6.1, the maximum efficiency appears to be of the order of 0.8. Maximising herbage consumption per hectare therefore depends upon avoiding managements which are severe enough to depress herbage growth seriously at one end of the range and lax enough to increase herbage losses (or depress the consumption/growth ratio) at the other. Fortunately, it seems that the effective zone of approximate balance between these extremes under grazing conditions gives considerable scope for manoeuvre (Fig. 6.3). The relationship outlined in Figs. 6.3 and 6.6 are now used to provide objective management decisions based on defined sward conditions appropriate to particular animal production targets. It is worth noting in passing that most farmers still tend to err too close to the upper limits of the ranges of sward height or herbage mass shown in Fig. 6.3.

The efficiency of conversion of ingested food into animal product is also inherently low, due partly to the relatively high demands made on nutrients for the maintenance of body function, including the digestion process. It is directly related to the level of herbage intake per animal and the genetic potential for growth or lactation. In a breeding herd or flock it is also strongly influenced by the reproductive rate and the relative proportions of productive and non-productive animals.

Overall efficiency is commonly measured as the output of animal product per unit area of grassland, the management variable likely to exert the greatest impact upon output per unit area is the stocking rate (Fig. 6.8). Herbage consumption per unit area

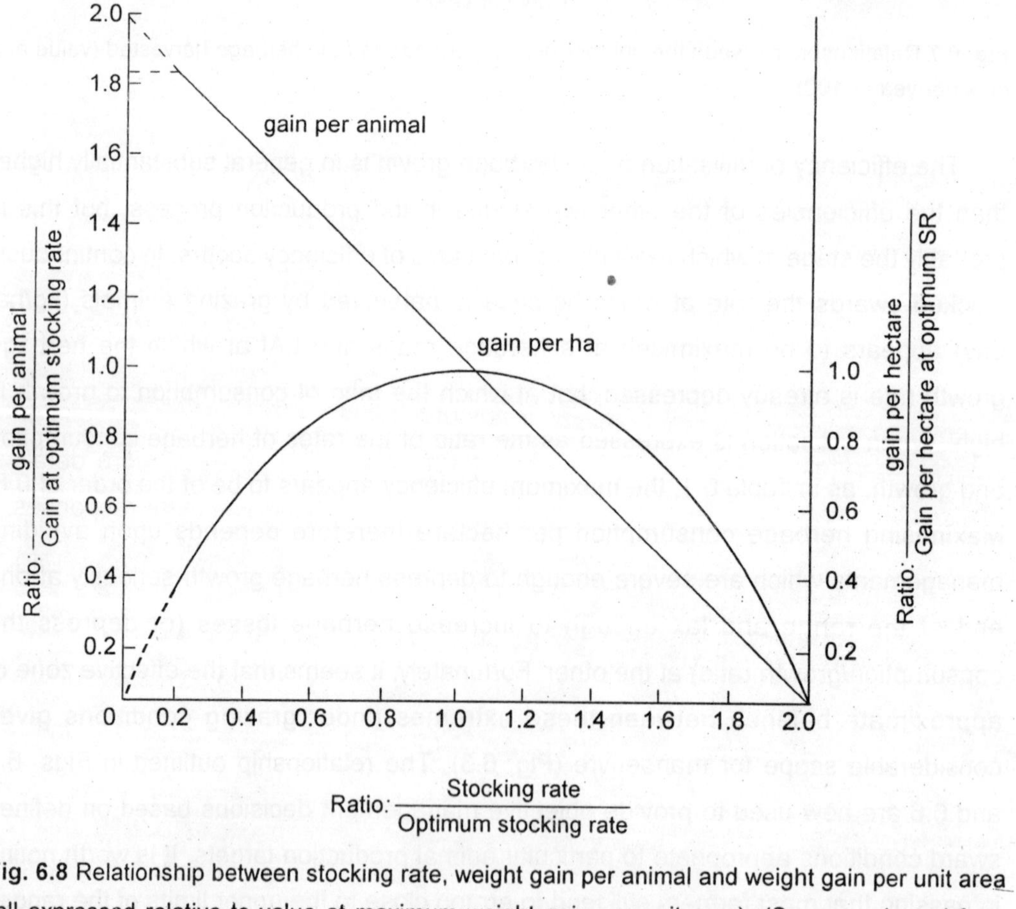

Fig. 6.8 Relationship between stocking rate, weight gain per animal and weight gain per unit area (all expressed relative to value at maximum weight gain per unit area). (Source: adapted from Jones and Sandland).

is the product of intake per animal and stocking rate (SR). It almost invariably increases with increasing SR, despite some reduction in intake per animal as a consequence of the associated decline in herbage mass. In meat animals, performance (LWG) per animal declines steadily with increasing SR over virtually the full range of studies and there is very little indication of 'plateaux' of animal performance at low SR, except perhaps at the very lowest levels. This reflects the progressive reduction in intake per animal with increasing SR. The relationship between output (LWG) per ha and SR takes the form of a rectangular hyperbola, being zero (predictably) at zero SR and at the SR giving zero LWG, and a maximum at the SR midway between the two. This defines the point of maximum biological efficiency for the system; it represents a compromise, because none of the three main stages in the production process are likely to be operating at maximum efficiency at this point. Wool production per animal is less sensitive than LWG or meat production to variation in SR. The effects on reproductive performance or milk production are more complex than those shown in Fig. 6.8, depending on the opportunity for the animals concerned to call on body reserves to maintain foetal viability or lactation.

The stocking rates at which financial measures of gross output, gross margin and profit per hectare are maximised will lie progressively to the left of that maximising biological efficiency and, although this generalised relationship forms a very useful basis for making comparisons between levels of management inputs, the assessment of stocking rates appropriate to particular farm enterprises requires more detailed information on input costs and output values, and on anticipated output responses.

❑❑❑

7

The Inputs to Agriculture

Introduction

The inputs to agriculture are of various kinds. The economist recognises three major inputs: land, labour and capital. A sustainable and economically efficient agricultural system is one which combines these three major inputs in the most cost-effective manner.

The biologist recognises different set of three major inputs, all of which are necessary to sustain life in both the plant and animal kingdom. These major inputs are water, nutrients and energy. A sound and sustainable system of husbandry is one which combines these three biological inputs in the most cost-effective way.

This chapter will deal with the latter group of inputs although there are obvious interactions between the two sets. For instance, the substitution of labour (manpower) by machinery (whether animal-drawn or tractor driven) has implications for two of the economic inputs (labour and the capital costs of machinery) and also involves one of the major biological inputs, energy (whether in terms of food or feed for man or animal or fossil energy for machinery operation).

Water

About four-fifths of the world's surface is covered by sea. The surface of this large area of salt water, where radiant energy is absorbed from the sun and fresh water evaporated into the atmosphere, is the source of the water supply essential for any agricultural system. The term 'hydrological cycle' describes the circulation of water as it evaporates from the sea, and is then transported by the wind in the form of clouds until it condenses as water droplets and is precipitated as rain, snow or hail.

The water which falls on the land may run off into streams to find its way back to the sea via rivers and lakes, but the most important agricultural aspect of the hydrological cycle is the infiltration of water into the soil. A proportion of this water is taken up by crops and other vegetation while the surplus finds its way underground to rise once more in the form of springs which add to the water flowing down the rivers and back to the sea. Crop plants do not merely absorb water from the soil but in turn transpire it, the rate of transpiration being mainly determined by the heat energy available for evaporation, the ability of the air currents to move water vapour and the capability of the soil to supply water to plant roots.

Water is one of the dominant factors in determing climate and the amount of water precipitation is mainly (but perhaps not completely) beyond the control of man. Different agricultural practices can ameliorate the local effects of water deficiency by irrigation, shade and shelter and the drilling of bore holes to supply drinking water for use by both animals and man.

About 97 per cent of the total water in the world is found in the sea and, being saline (with about 8.5% of salts in solution, primarily sodium chloride), sea-water as such is unsuitable for drinking by man and most farm animals and also unsuitable for all but a very small range of salt-tolerant crop plants.

By contrast, the water in lakes, rivers and underground storage areas only amounts to about 0.05 per cent of the world's total water supply. This small percentage needs putting into perspective, since it amounts to the surprisingly large total of 500,000 cubic kilometres, which, if evenly applied over the land surface, could supply all the water needs of the world's human population (currently about six billion) and all the world's agricultural requirements for food and fibre. Because of the poor distribution of water supplies over the world's surface, certain areas have too much rain water to support normal mixed farming systems (e.g., area currently under tropical rain forest) while other areas suffer almost permanent drought conditions (e.g., the deserts of Africa and the near East). Ancient civilisations were past masters of the art of building dams, digging wells and constructing irrigation ditches -- the Incas in South America and the ancient Egyptians are two outstanding examples.

There is a great deal of difference between true water requirements (both for man and agriculture) and water usage. Where water is scarce it tends to be carefully used but where it is freely available 'on tap', water is used for a variety of non-essential purposes. Thus, in western Europe, water usage varies between about 100 and 250 litres per persons per day, while in certain parts of the USA, usage varies between 400 and 1,000 litres per person per day. At these higher levels of usage, 14 per cent was for municipal and domestic requirements, 38 per cent for agricultural purposes and 48 per cent to meet the needs of industry.

The annual rainfall, and the dynamics of the water balance at different seasons, vary widely across the world. Thus, in the UK, annual precipitation varies from below 600 mm in parts of East Anglia to over 2,500 mm in the Western Highlands of Scotland and the mountains of Cumbria and North Wales. Even in a small country like the UK, certain parts of the country can be suffering severe water shortage, reduction in crop yields and a restriction of potable water supplies while other parts of the country, at the same time, can be suffering from excessive precipitation, flooding and hence an inability to carry out timely arable operations.

For plant growth the water available at crucial growth periods is of critical importance, with major needs corresponding with time of germination and periods of maximum growth, normally in the spring. Conversely, at other times of the year, such as during harvest and subsequent cultivations in preparation for the following crop, even moderate levels of precipitation can be detrimental. However, it is usual practice to consider water requirements of crops in terms of mean annual precipitation rather than in terms of critical seasonal requirements, and for most crops annual rainfalls of between 250 mm and 500 mm are the minimum levels required, but levels of over 750 mm give greater room for manoeuvre by the farmer.

In terms of animal requirements for water, minimum daily requirements are essential. Animals can go for some time without feed but for no longer than a day or so without water, the exception being tropically-adapted animals such as camels and dromedaries. It is usual to relative water requirements of livestock to their dry matter intake—the more dry matter eaten the more water required. On this basis, animals which are non-lactating require between four and eight litres of water per kilogram of dry matter ingested. Lactating animals require an extra litre of water for each litre of milk produced. Since the two requirements are additive, a high-yielding cow giving 50 litres of milk per day would require about 70 litres of water to meet her essential needs.

Chapter 10 deals with the subject of climatological change. Clearly if current predictions are right and most of the world is becoming warmer and perhaps drier, the water requirements for both man and agriculture will become more critical. In the past, many agricultural practices, such as the traditional method of ploughing up and down the slope, were designed to remove excessive surface water during the spring and autumn periods. If warmer and drier seasons result from climatic change, the problem will be to minimise water loss. There would then be need to adopt new agricultural practices, such as ploughing on the contour instead of up and down the slope.

Again, if water becomes more scarce then much greater care must be taken in terms of the production of toxic agricultural substances, such as slurry and silage effluent, since large quantities of water will not be available for dilution. At the present time, parts of the Thames Valley in south-east England face major problems in dry weather where the input of over 500 million litres per day of treated sewage effluent only receives a dilution with 'pure water' of 3 in 1. In future, in addition to recovering more water and polluting less of the environment with wastes, the increasing water demand will have to be met by the careful development of year-long water storage systems and the reduction of the annual loss of fresh water, during the spring period, to the sea.

To summarise, most of the water in the world is too saline for use by man or agriculture. Only about 2 per cent of the world's total store of water is available for man to exploit. Little can be done to control the major components of the hydrological cycle and man can only exert his influence on the reception and distribution of the precipitation which reaches the land surface. By better agricultural systems of land use and by taking measures to control pollution we can ensure the absorption of this rainfall as clean water into the soil and we can control its flow towards the oceans by developing systems of storage and distribution.

In the very long term there may be prospects of desalinating sea water, such as by the use of nuclear power, but the cost of this would be immense. Although such measures may be a partial solution in certain low-rainfall tropical countries (e.g., Israel), there is very little chance that these methods will ever become economically viable for use on a world scale.

Feed

Very few animals are fed on a single feed – even cattle grazing a sown grass pasture will vary their diet with hedgerow plants and weed species. If they are cows in milk, they are likely to be fed on a concentrated diet, in addition to fresh or conserved grass, for about half the year.

It is therefore important to appreciate the complementary nature of the various types of feed. In the case of ruminants the diet will normally consist of a combination of a forage and a concentrate in different proportions according to the system of management employed. In the case of pigs and poultry kept intensively, the diet would usually consist of a concentrated mixture, but this itself will comprise many different feeds. If mixed on the farm, the diet may comprise five or six feeds roughly mixed together, but in a purchased compound the number of feeds employed by the miller will probably number 20 or more. The object is to produce a 'balanced diet' with the correct ratio of protein to energy and the required amounts of major minerals, trace elements and vitamins.

Feed is traditionally broken down by the chemist into its constituent chemical fractions as shown in Table 7.1. However, apart from water, these traditional chemical components are only useful in providing a framework for classification and have no particular significance *per se*. Thus, for example, the ash fraction can be very useful (if it provides the right major and minor elements in the right proportions) or it can be a relatively useless filler if, for instance, it contains a lot of indigestible silica because of soil contamination.

Feeds are usually classified according to their origin, and the individual system chosen is of less importance than the correct understanding of the definitions employed.

One of the commonest errors is to regard a single 'straight' feed (e.g., barley) as a 'concentrate' merely because the nutrients contained in barley are more concentrated than those contained in a forage, such as grass.

Table 7.1: Components of different fractions in the analysis of feed

Fraction	Components
Moisture	Water
Ash	Essential elements and minerals, non-essential elements and minerals
Crude Protein (CP)	Proteins, amino-acids, amines, nitrates, nitrogenous glycosides, B vitamins, nucleic acids
Ether Extractive (EE)	Fast, oil, waxes, organic acids, pigments, sterol, vitamins A, D, E, K
Crude Fibre (CF)	Cellulose, hemicellulose, lignin
Nitrogen-Free Extractives (NFE)	Cellulose, hemicelluse, lignin, sugar, starch, pectins, organic acids, resins, tannins, pigments, water-soluble vitamins.

The three main classes of animal feed are:

(a) grass and forage crops;

(b) straights;

(c) concentrates and supplements.

These will each be considered briefly in turn.

Grass and Forage Crops

These crops have been dealt with in Chapter 4-6, inclusive, but it is important to realise that as nitrogen usage decreases on grasslands, mixed swards of grass and legumes will become more common. In addition, if climatic change brings about warmer and drier summer periods in western Europe, they are likely to see more lucerne grown (an extremely drought-resistant forage crop) and more forage maize. Also, as vegetable production increases and quality standards are enhanced, more arable by-products are likely to become available as 'forage feeds' for animas, although not planted as forage crops.

The advent of big round bales has made the collection and utilisation of cereal straws much easier and as a consequence much of the 9 Mt of straw produced each year is likely to be used as an animal feed. It can be fed either long, chopped or chemically treated with a variety of materials which effectively break down the lignified cell walls and release the cell contents as a source of nutrients. In its normal long form, straw is of low energy value (less than 6 MJ kg of dry matter) and very deficient in protein. If treated, particularly with nitrogenous chemicals rather than with organic acids, the energy level can be elevated to 8 or 9 MJ/kg and in addition a useful source of non-protein nitrogen may be made available. Put another way, straw can be upgraded to the value of a low/medium quality hay.

In general terms numerous waste arable products are capable of being utilised by ruminants (and sometimes by pigs) but the economics of so doing are often suspect, except where the crop can be provided on farm at extremely low cost. Thus animal enterprises situated close to intensive horticultural areas can occasionally use such arable wastes profitably, one particular waste material following another as the season progresses. Transport of such wastes over large distances is, however, normally uneconomic.

Straights

These are single feedstuffs of either animal or vegetable origin which may have undergone some processing. Single straight rarely provide the complete nutritional requirements for farm livestock, although it is suggested that the production of new and genetically improved varieties, such as high-lysine maize, may change this situation. Straights may in turn be subdivided into cereals, cereal by-products, pulses or legumes, oil-seed residues, root crops, fibrous materials, animal by-products, recycled animal waste, synthetic nutrients, liquid feed, and oils and fats.

Cereals (e.g., barley)

These are the most important constituents in animal feeds for all forms of livestock and, although often regarded as 'energy feeds', cereals can provide the preponderance of protein in the diet for non-ruminants simply because the inclusion rates for cereals are often very high. In the UK and western Europe, barley and wheat are the main cereal crops employed as animal feed but in North America maize (corn) predominates. Most cereal crops are used both for human food and animal feed and thus there is a degree of competition for their use between man and animal. The ratio of usage is constantly changing. Thus, before the advent of the tractor, oats were grown primarily as a feed for horse; now most of the declining acreage of oats is used as human food.

Cereal By-products (e.g., wheat offal)

Where cereals are grown primarily for human food they are normally processed to somc dcgree, such as by removing the outer husks or glumes. The resultant by-products are thus available for animal feed but it should be noted that there is a trend towards a greater use of 'whole cereals' for human food because the husks are rich in vitamins and in fibre and therefore regarded by some as useful additions to a balanced human diet.

Other cereal grains are used as the substrate for beer or liquor production and the resultant 'spent grains' may then be available as a useful animal feed. They can be supplied directly on to farms in a wet condition after which they may be ensiled or dried so that they can then be stored and subsequently incorporated into compound feed.

Pulses or Legumes (e.g., field beans)

Although the seed from leguminous crops normally commands an attractive price as human food (peas and beans), nevertheless certain crops, such as field beans, are grown primarily form animal feed use. Additionally, legume seeds which do not come up to the standards required for human consumption may be diverted into the animal market. Such materials are relatively high in protein but they may contain gums, resins and sometimes toxic substances which render them less suitable for animal feed. Whole legume crops, such as clover, lucerne and vetches may be grown as whole crop forage.

Oil-seed Residues (e.g., rapeseed meal)

Traditionally oil-seed crops were used for the production of edible (e.g., olive) or industrial (e.g., linseed) oils. In recent times there has been an escalation in the growing of oil-seed rape and, in the Americas, soya for use as animal feed as its prime purpose. The term used for the entire oil seed crop (without extraction of the oil) is 'full fat' (e.g., full-fat soya).

In the past it was mainly the by-products which were used for animal feeding. Where the by-produced was produced by crushing the seed to expel the oil, the resultant material was known as a 'cake' (e.g., copra cake) and where the oil was extracted by the use of a solvent, the residue was known as a 'meal' (e.g., soyabean meal).

Root Crops (e.g., cassava)

Root crops are capable of producing high yields of energy per unit of land area. The difficulty is that many such crops, particularly cassava, are difficult to harvest because they can be deep-rooted. Over the last quarter of a century larger amounts of cassava have entered world markets, particularly from countries in the Far East such as Thailand. Indeed, cassava has now become a major export crop in several developing countries and is perceived by arable farmers in cereal-growing areas as a major import threat. Most of the cassava imported into EEC countries is supplied to feed compounders but it can also be bought by farmers as a 'straight' via an importing agency.

In recent years the importance of fresh root crops as animal feeds has received renewed attention. Turnips, swedes and mongoloids were traditionally grown as feeds for cattle and horse, hand-harvested and then cut or sliced by hand-operated farm machinery. Now such crops can be mechanically harvested and mechanically cut using the power-take-off (pto) of tractors. Also, such crops are grown on a large scale for harvesting by sheep *in situ.*

Fibrous Materials (e.g., chemically-treated straw)

This class of material covers a large number of crop by-products ranging from processed cereal straws (see above) to tropical products, such as bagasse pith which is a by-product of the sugarcane industry. In Europe sugarbeet pulp forms a useful fibrous by-product of the sugar industry. It can be supplied in wet form, or more usually as dried and pelleted pulp.

Animal By-products (e.g., meat-and-bone meal)

These materials are by-products of the meat and fishing industries, but in certain countries, such as Peru, the national economy is partly dependent upon a fishing industry designed to produce fish meal for export as an animal feed rather than fish for human consumption.

Products of animal origin must be carefully processed if they are to be safely used as animal feed. If the processing is deficit they may become the carrier for pathogenic bacteria (such as *salmonellae* and anthrax) or for viruses (such as the foot-and-mouth virus) or the newly discovered 'slow viruses' or 'prion proteins' which are though to be the cause of diseases such as scrapie (sheep) and bovine spongiform encephalopathy (BSE) (cattle).

In the UK a 'protein processing order' is in operation which ensures correct processing procedures and strict hygiene measures in by-product plants. In addition, it is now forbidden to use certain offals derived from ruminants in animal diets in an attempt to prevent the possible spread of scrapie in sheep to cattle where it is thought to produce the disease BSE.

Re-cycled Animal Waste (e.g., dried poultry manure)

Large quantities of these materials are available on-farm especially on intensive pig and poultry enterprise. Various attempts have been made to develop plants capable of processing these manures to a low-moisture material which can be transported and used as an ingredient in ruminant diets, where it can provide a useful source of non-protein nitrogen together with small amounts of energy and minerals. However, there is a great reluctance on the part of society to accept such practices, on both aesthetic and health grounds, and it is most unlikely that recycled animal wastes will, in the medium- and long-term, be processed into animal feed. It should be stressed, however, that nutritionally there is nothing wrong in using such materials which can provide cheap supplies of essential nutrients, providing the health hazards are completely removed.

Synthetic Nutrients (e.g., urea)

This group of feeds comprises chemicals and fermentation products produced by the chemical and biotechnology industries. Although this class of material currently contains relatively few products, it is likely that the list will enlarge in the future as more micro-nutrients are recognised as essential ingredients in animal feed.

New processes enable the cost of such materials to be progressively reduced. Thus the first synthetic amino acid to be marketed was methionine with a price, in its early days, of several thousand pounds per tonne. However, as world demand has increased and production technology improved, the costs of production have been progressively reduced so that synthetic methionine can now be purchased at approximately one-fifth of the 'real price' originally demanded.

Liquid Feed (e.g., molasses)

The main material in this class is molasses, a by-product of the sugar-processing industry. About 300 kg of molasses are available for every tonne of sugar processes. The two main forms available in the trade are sugarcane and sugarbeet molasses. Where the plants incorporate the molasses back on to the sugarbeet pulp, the combined product is known as 'molassed pulp'.

Another liquid feed is whey, a by-product of the milk-processing industry. Whey can be dried into a powder but can also be supplied fresh, by tankers, on to pig farms where it is fed to fattening pigs.

Oils and Fats (e.g., tallow)

These two terms are interchangeable, since when a fat is in the molten stage it becomes an oil. The chemist would classify both as 'lipids'. The chief material in this class is tallow derived from slaughterhouses. The material could, therefore, also be regarded as an animal by-product; a consideration which also applies to fish oil.

Each livestock class has a limit to the amount of fat or oil which can be added to the diet, usually less than 5 per cent. If large quantities are fed, then appetite can be adversely affected and the excess fat can inhibit the digestion of other components of the diet, particularly fibrous components in the case of ruminants.

Concentrates and Supplements

Concentrates are products specially designed for further mixing before feeding. They are incorporated into diets at an inclusion rate of 5 per cent or more, with planned proportions of cereals and other feeding-stuffs either on the farm or in a compound mill. Protein concentrates contain blended high-protein-content ingredients, such as fish meal, fortified with essential nutrients such as minerals, trace elements and vitamins. Where the rate at which concentrates are used in the mix is as high as 50 per cent, they will contain cereals or cereal by-products. Some protein concentrates are formulated so that, after further mixing with cereals, the resultant product is then suitable for balancing farm forages. They may be added to chopped forage or silage in a mixer-wagon and then fed as a 'complete mix' to cattle. In this system the ratio of concentrate: forage is predetermined and the animals cannot 'pick and choose'.

Supplements are products used at less than 5 per cent of the total ration. They are designed to supply planned proportions of vitamins, trace minerals and non-nutrient pharmaceutical additives. To facilitate adequate mixing into the total ration, the active materials (e.g., drugs) are normally present in supplements in a diluent carrier, such as finely-ground cereal.

Many people misuse the terms 'concentrates' and 'compounds'. For instance, many research workers report experiments in which rolled barely is or is not added to the diet of cattle and describe the barely as a 'concentrate'. Again, many farmers talk of 'concentrates' when they mean 'compounds' which is the form in which a completely balanced diet is supplied by the compounder to the farmer for feeding as such. The fact that compounds may be fed to cattle alongside hay or silage does not make them 'concentrates'.

There is further confusion about the terms 'specification' and 'formulation'. A 'specification' is a description of a given blended feed in nutritional terms. Some 30 or so nutritional parameters may be listed and over 50 different materials may be allowed by the specification either freely without constraint, or restricted to minimum or maximum inclusion levels. A 'formulation ' is the actual list of straight raw materials present in a given blended feed, at fixed levels of inclusion. In other words, a 'specification' is an open formula but a 'formulation' is a fixed recipe. The formulation is usually worked out on a computer, to meet the specification at 'least cost' so as to make the diet as economic as possible.

Agricultural Machinery

Ever since man started to grow his own food, the plough has been his main tillage implement. Other machinery operations assist in cultivating, sowing, harvesting, drainage and irrigation, but ploughing is the fundamental operation. The first crude wooden ploughs were drawn through the soil to loosen it before the seeds were sown. As the tractor has replaced the draft animal as the source of power, so the design and complexity of the plough has increased correspondingly.

It should not be forgotten in this context that many parts of the world still employ crude wooden plough pulled by draught animals and the wider range of cultivating and other implements suitable for draught animal still require improvement in many developing countries. Oxen, mules, donkeys and camel are used as draught animals in addition to horses.

In the first half of last century, the UK was one of the major exporting countries for ploughs and other forms of farm machinery. Many of the inventions in this area were made in the UK. In more recent years, North America and other European countries have overtaken the UK in the manufacture and supply of agricultural machinery. Table 7.2 shows the decline in output in monetary terms of the major sectors of the agricultural tractor and machinery market in the UK.

Although the number of tractors manufactured and registered on UK farms has fallen over the last decade, the average horsepower per tractor has increased from about 70 h.p. in 1978 to 84 h.p. in 1988. This point will be reffered to later as there are clear implications regarding energy inputs and production costs.

Table 7.2: Value of UK agricultural tractor machinery output (non-constant money terms) in £M

Item	1979	1982	1985	1988
Tractors	116	85	83	79
Cultivation equipment	63	52	55	45
Harvesting equipment	87	42	33	28
Milking machines	27	15	8	N/A
Other machinery	65	54	67	96

Source: Agricultural Engineers' Association.

During the same period, the balance of trade (exports minus imports) has remained fairly constant although there have been violent year-to-year fluctuations. However, although British exports have increased by 50 per cent over the period, imports have more than doubled, as is shown by the figures in Table 7.3.

Table 7.3: Value of UK agricultural engineering exports and imports (£M)

Item	1979	1982	1985	1988
Exports	662	594	676	821
Index in constant value	100	64	63	68
Imports	276	384	456	488
Index in constant value	100	102	115	110
Balance of trade	386	250	220	333
Index in constant value	100	52	42	53

Source: Agricultural Engineers' Association.

The figures in Table 7.3 mask major differences in the UK import and export pattern as between tractors and machinery. Thus, about 58 per cent of UK agricultural machinery exports are sent to the EEC while 70 per cent of machinery imports come from same region. On the other hand with tractors, 30 per cent of UK manufactured units are exported to the EEC and an extra 32 per cent to North America. Tractor imports, however, are 82 per cent derived from EEC countries and only 9 per cent from North America. As with other commodities, this pattern of world trade is constantly altering, but the major point to note is that the UK has now lost its pre-eminent position as a major exporter of tractors and machinery on the world market.

In a work of this nature it is not possible to give a detailed description of all the various types of farm machinery currently employed in western European agriculture. This brief review must suffice and for further reading the recent textbook by Witney is recommended.

Power

There is trend towards more self-propelled units, particularly harvesting machinery and also, as already mentioned, towards units of higher horse power. There is a tendency for many farmers to gear up their tractor power requirements to their peak needs, particularly at ploughing and harvesting. The consequence is that for the rest of the year the farm is left with expensive machinery grossly overpowered. Not only is this uneconomic but it leads to a great deal of wasted energy input, a subject which will be reffered to again below. There is a need for more farmers to carry out a 'power audit' of their operations and to gear them more closely to average needs than to peak needs, hiring in machinery, or joining a machinery ring, to enable peak requirements to be met.

There is also a move towards four-wheel-drive power units and, for heavier work, towards track or half-track units and much more attention is given now than previously to the damage done to the soil by the consolidating action of tractor wheels. Wide-area tyres and doubling-up the wheel numbers are now commonplace and in addition 'tramlining' (by which the various operations on the field all take place over the same track) is now general practice.

The price of tractor fuel and oil still dominates the breakdown of tractor operation costs and, as fossil fuels become scarcer, these costs are likely to rise. At present they represent approximately 40 per cent of total operating costs, the other major itmes being deprecation 30 per cent; interest 15 per cent; repairs 10 per cent and tax and insurance 5 per cent. The first four itmes listed are related to horse-power so it follows that the large power units are more costly not only to purchase but also to maintain.

Tillage and Cultivating Implements

The purpose of soil-tillage is to create a soil environment in which plants can grow, to reduce weed competition and to permit crops to be readily harvested. For autumn-sown cereals, direct drilling has shown tillage to be unnecessary on certain soils where weeds are not a problem. Where traditional cultivations are still carried out, the choice include mouldboard ploughs, disc ploughs, rotary cultivators, rotary diggers, disc harrows, chisel ploughs and heavy cultivators.

The plough, in one or other of its forms, still remains the basic item of agricultural machinery. However, the role of the plough is changing. Its earlier prime functions were to invert the soil so as to bury surface vegetation and to provide a 'ridge and furrow' topography in the field in order to assist drainage. Ploughing was traditionally up-and-down the slope, to give the regular undulations or rigs shown on old rural paintings.

Today, the burying of weeds is less important since weeds can be controlled by the use of herbicides, and draining is best carried out on a more permanent basis deep below the surface by the use of tile drains or perforated plastic pipes. Ploughs may still be used for breaking hard plough pans between the topsoil and the subsoil but less so for drainage purposes and in many drier areas ploughing is now along the contour, in order to conserve water, rather than up and down the slope to drain it.

With the coming of complex harvesting machinery, it has become increasingly important to make the field surface as level as possible rather than to throw up ridges and furrows, and hence many mouldboard ploughs are now reversible so that all the operation is conducted in one direction as the plough itself reverses direction each time it turns round at the edge of the field. With the greater horsepower and with the advent of power-take-off (pto) it is common to see a tractor fitted with 12 plough bodies, 6 to operate in each direction while the other 6 remain passively airborne.

Sowing and Planting Machinery

The cereal drills most commonly employed are of the random-distribution type although a limited amount of precision is now practised with some evidence of increased yields, particularly following spring sowing. Two recent developments in sowing practice are the 'rediscovery' of seed broadcasting where faster work and satisfactory yields are claimed, and direct drilling into the unplugged field. This latter operation requires a more penetrating coulter accompanied by greater implement weight to aid soil penetration and thus requires a more powerful tractor.

At the other extreme from seed broadcasting, precision drills are employed, normally formed by units fixed on to a tool frame, together with some means of transporting single seeds from the hopper to the planting device. Sophisticated precision drills may be fitted with press-wheels to firm the surface after planting and electronic monitors to warn the driver of faults.

Specialised and very sophisticated planters are now available for a wide range of vegetable crops (particularly potatoes) and transplanters have been developed for transplanting young seedling crops, such as brassicas.

Fertiliser and Pesticide Application: Manure Spreaders and Sprayers

In addition to the traditional distributors for applying solid fertilisers, liquid fertiliser application has recently been developed which enables direct injection of liquid or gaseous ammonia into the soil. It is hoped that future developments will enable below-ground placement of liquid manure and slurry. These substances, when sprayed above ground, result in odours which are socially unacceptable.

New spraying machinery for fungicides, herbicides and insecticides is now being developed in order to lessen 'drift' on to adjacent fields and also to deliver very much smaller concentrations of agrochemicals per unit of land. The application of agrochemical to crops is now constrained by law, operatives must be qualified to carry out the tasks involved. The volumes of active chemical substances are being steadily reduced but it is doubtful whether the use of such useful materials will ever cease completely except in organic farming systems where they are disallowed.

Harvesting Machinery

The combine harvester is now almost universally used for harvesting cereals and developments are now taking place in specialised harvesting of a much wider range of crops including such unlikely candidates for mechanisation as soft fruits.

The major changes with cereal harvesting operations concern the straw rather than the grain, where the large round bale is rapidly taking over from the traditional smaller, rectangular bale.

The advent of the plastic 'big bale' has revolutionsed the harvesting and processing of forge crops for silage and made the silage crop suitable for transport from one field to another.

Much care is now devoted by the machinery manufactures to reducing crop damage by harvesting machinery and this is particularly pertinent with the harvesting of fresh vegetables, where quality standards are extremely high, and in the harvesting of potatoes where horizontal shaking is likely to replace the traditional vertical shaking to remove soil, stone and extraneous matter.

Irrigation and Drainage Machinery

If European climates become warmer and drier and soil moisture becomes limiting, greater use may be demanded of irrigation equipment. However, constraints are likely to increase with regard to the extraction of water from rivers and underground supplies, since agricultural needs will compete with those for potable water for man.

With the introduction of more efficient pumps and sprays, large areas of land can be efficiently irrigated from a single point or from a moving boom and in future the need to move mobile equipment manually from one place to another will be replaced by a more permanent layout of irrigation pipes or channels.

The major drainage system on British farms, much of it constructed with tile drains, was laid down in the middle of the nineteenth century. Tile drains were said to be one of the most profitable crops ever planted! Unfortunately these drainage systems are now mainly silted up or damaged beyond repair and there is an urgent need to replace such systems in heavy soils where water-logging is a problem. The use of perforated plastic pipe drains instead of drains formed with individually laid tiles is likely to become commonplace, and machinery for the automatic laying of plastic pipe drains is already available.

Milking and Feeding Equipment

Dairy parlours show increasing sophistication of the equipment employed, so designed that one man can now milk up to 200 milking cows in the dairy parlour without assistance. Such aids of efficient milking as automatic cluster removal, electronic cow identification, milk recording and data capture will become commonplace, although it is doubtful whether robotic milking (by which the milking machine is automatically placed on the cow at any time of the day or night according to need) will replace the manually-operated milking parlour. However, methods for the automatic feeding of livestock, particularly controlled automatic feeding so that each animal receives a pre-set ration of food per day, are likely to increase. Feed costs are one of the major production costs in animal production and more efficient feed rationing is one means of increasing the economic efficiency of animal production systems.

Energy

Although energy can be expressed in constant terms (usually as joules), it is important to realise that not all forms of energy can be converted into another. Thus, fossil fuel energy cannot be converted into food energy except through the complexity of an agricultural production system. It is important to realise this when dealing with energetic efficiencies which are normally defined as the amount of energy output in an agricultural system (normally food) divided by the amount of energy put into the system (normally as feed, fuel and the indirect energy costs of making machinery, fertilisers, etc.).

With these considerations in mind, Table 7.4 indicates that the overall energetic efficiencies of mechanised systems of corn production compared to draught animal systems are relatively similar, with efficiency ratios of 2.8 and 2.5 respectively. Both systems are significantly inferior to the efficiency ratio of a 'primitive' manual system where the energy inputs are proportionately much lower than the reduced outputs, giving a ratio of 3.7. However, the importance of the table lies in the bottom line – one man and his tractor can supply the cereal food needs of about 10 people contrasted to the two other systems which can only provided the food needs for between five and seven people.

Table 7.4: A historical perspective of the energetics of corn production

Item	System of corn production		
	One man and his horse	One man and 2 horses	One man and his tractor
Area cultivated (ha)	0.5	10	40
Annual grain yield (t/ha)	2	2	5
Annual output of grain (t/man)	1.0	20	203
Output as MJ × 10^3	16.6	332	3378
Annual input (MJ × 10^3) as:			
food for man and animals	4.5	134	4.5
fossil fuels	-	-	1198
Surplus food (MJ × 10^3)	12.1	198	3373
Energetic efficiency (MJ output/MJ input)	3.7	2.5	2.8
No. of people sustained with food (ha)	7.4	4.5	10.7

Source: Osbourn.

It follows that, while agriculture has become efficient technologically, it has become less efficient in its use of energy. Although the use of a tractor to produce food is theoretically less generally efficient than the use of draught animals, it is important to place the use of energy in agriculture in context with its use elsewhere in the national economy. Table 7.5 indicates that agriculture only consumes about 2 per cent of the fossil fuel used in the national economy. This compares with over 50 per cent used by transport and other industries, and 9 per cent used domestically in the home which is itself about five times that used in farming (8.7% vs. 1.7%).

Agriculture is part of the food production chain and, if the totality of this chain is considered, then the consumption of primary energy from all sources (electricity and gas, as well as petroleum fuels) is much greater than the 2 per cent indicated in Table 7.5. Again, however, the greater part of the total energy consumption is in food processing and distribution and food preparation in the home rather than in the primary production of food on the farm.

Table 7.5: Use of fossil fuel energy in UK industries

Sector	Energy equivalent (10^9 MJ)	Energy (%)
Agriculture	75	1.7
Steel	218	5.1
Domestic	375	8.7
Electricity, oil and gas industries	1179	27.4
Transport and other industries	2325	54.1
Miscellaneous	128	3.0
Total	4300	100.0

Source: Wilson and Brigstocke.

It will be seen from Table 7.6 that the patterns in different developed parts of the world are relatively similar. However, compared to Australia and the USA, the UK uses slightly more energy in food and fish production and in food processing and distribution.

Table 7.6: Primary energy consumption (of all forms including gas and electricity), in the production and preparation of food as percentages of total national energy consumption

Energy consumption in	Percentage of total national consumption		
	UK	Australia	USA
Agriculture and fishing (to farm gate)	3.9	2.1	2.2-2.7
Food processing and distribution	7.0	5.8	6.2-7.8
Food preparation	4.9	7.1	3.6-4.5
Total	15.8	15.0	12.0-15.0

Source: Wilson and Brigstocke.

Although all energy sources are basically solar in origin, the fossil fuels represent forms of long-term energy storage while the photosynthates of crops (including wood and fibre) are forms of relatively short-term energy storage. The quantity of incoming solar energy captured by plants represents only some 1 per cent – 2 per cent of the incident solar radiation falling on the land area they occupy. The daily photosynthetic productivity for a specific location depends upon the crop surface, latitude, season and sky conditions (Monteith).

Only part of the energy recovered in harvested crops provides edible food for man. Many forage crops, such as grass, are fed to farm livestock so that the actual 'farm gate' output of food for human consumption of all crops is only 11 per cent of the energy originally recovered in the harvested plant material, while 70 per cent of the primary energy output is destined for animal feed. The efficiency with which livestock convert this feed into edible products is relatively low, because ruminant animals consume fibrous materials which are not efficiently digested. However, ruminants are able to graze remote area unsuitable for crop production due to topography, climate, etc., and in these places no human food would be produced at all if it were not for the intervention of cattle and sheep meat production systems.

The previous section dealt with the possible reduction in energy input in cultivation systems which avoid the use of the plough, such as by reducing cultivations or by employing direct drilling systems. These alternative production systems are compared in Table 7.7 which shows that the reduction in energy input into the direct drilling system makes only a marginal difference to the total energy requirement. However, if energy becomes more scarce and expensive, even such minor reductions may eventually become economically worthwhile.

It will probably take another 'energy crisis' for farmers and growers to respond to changes in the energetic balance of agriculture and to modify their production systems accordingly. Investment in new energy forms must come from governments because of the high capital costs and attendant risks. Wind, wave and tidal power could make contributions to 'new' sources of energy, although the possibility of these sources providing direct mechanical power has not yet been fully exploited. If in future more use is made of nuclear power in the national energy supply, this will have major consequences for the manner in which energy is employed in agriculture. The advanced farming technologies of the future may well be as energetically efficient as were some of the more primitive technologies of the past.

Table 7.7: Primary energy inputs for cereal growing systems (MJ/ha)

Operation	Conventional system	Reduced cultivation system	Direct drilling system
Plough	738	–	–
Secondary cultivation	243	–	–
Prepare seedbed	150	150	–
Drill and harrow	150	183	183
Roll	48	48	–
Spray	45	90	90
Combine	552	552	552
Bale	60	–	–
Bale handling	150	–	–
Stubble cultivation	195	390	–
Fertilisers	6630	6630	6630
Drying and storage	5760	5760	5760
Total	14721	13803	13215

Source: Rutherford.

Fertilisers

Plants obtain carbon, hydrogen and oxygen from air and water. Other essential nutrients are taken up by the roots from soil. Six major elements are needed by crops in relatively large quantities: calcium (Ca), magnesium (Mg), nitrogen (N), phosphorus (P), potassium (K) and sulphur (S). Other requirements, known as micro-nutrients or trace elements, are needed in quantities of less than a kg per hectare. These are boron (B), chlorine (Cl), cobalt (Co), copper (Cu), iron (Fe), manganese (Mn), molybdenum (Mo) and zinc (Zn).

All soils contain a stock of these essential elements and deficiencies are relatively uncommon but where present can be dramatic in their effects. However, as farm output progressively removes these elements from the soil through the crop, the soil reserves must necessarily fall and, in time, will have to be replaced by the application of manures and fertilisers.

A great deal is made of the difference between 'natural organic manures' and 'chemical fertilises' but, from the point of view of the plant, both are merely sources of the same major and minor elements. It is of no consequence to the crop whether, for instance, the nitrogen is derived from the breakdown of organic matter in the soil or from nitrogenous fertiliser applied by the farmer.

A fertiliser may be properly defined as a chemical substance that is a product of mining or manufacturing industry which is added to the soil to supply one or more of the essential nutrients. Under this definition, substances like lime and meat-and-bone meal are fertiliser, although both these substances are accepted as permissible additions to organic farming systems.

Nitrogen Fertilisers

Most nitrogenous fertiliser is produced by the Haber-Bosch process which fixes atmospheric nitrogen in the form of ammonia. A whole range of new nitrogenous fertilisers is now derived from synthetic ammonia made by this process, but certain sources of nitrogen, such as sodium nitrate, are still derived from rocks. However, ammonium nitrate, a common artificial fertiliser, is produced by oxidising ammonia to nitric acid. When mixed with calcium carbonate, 'nitro chalk' is produced.

The increase in crop yields due to the application of nitrogenous fertilisers shows a linear increase. Clearly as time progresses, such increases cannot be obtained indefinitely but it is probably true that about four-fifths of the world's agriculture would respond positively to greater inputs of nitrogen, with major resultant benefit to the farmers, and food consumers, concerned.

Phosphate Fertilisers

Mineral rock phosphates are mined in several parts of the world. Most entering Europe come from North Africa in the form of calcium phosphate. Some of the calcium phosphate is ground and applied directly to the land but the greater part is dissolved in acid to make phosphates that are soluble in water and thus are suitable for use on all kinds of soil.

Another source of phosphate is basic slag, a by-product of the steel industry. As with straight rock phosphate, slags must be finely ground to be effective and further processing, by reacting with citric acid, enhances their availability to plants. Basic slag also contains lime and therefore has extra value for neutralising acid soils. Unfortunately recent changes in the methods employed and the lower output from the steel-making industry in the UK have vastly reduced the quantities of slag available for agricultural purposes. It is therefore likely that new, imported sources of phosphate will be required in the future and at present aluminium calcium phosphate mined in Senegal is increasingly being imported.

Potassium Fertilisers

Deposits of potassium (potash) salts are mined in several parts of the world, including the UK. The source material is muriate of potash (mainly potassium chloride) and sulphate of potash. Potassium nitrate is expensive but is recommended for use on certain horticultural crops. As with phosphate, potassium fertilisers are valued according to the solubility of the potassium in water. This solubility can be changed by the presence of other interacting elements in the source material.

Micro-nutrients

These materials are often applied as sprays to foliage where they are taken in by the plant leaf instead of by the root. They can also be mixed with solid fertilisers supplying N, P and K. Certain crops are much more in need of trace element application than others, sugarbeet being an outstanding example where about one-quarter of the area sown requires nutrient supplementation in this manner.

As agricultural systems intensify and become more specialised, and crop rotations give way to mono-cropping systems, the need for micro-nutrients will increase in order to combat trace element deficiencies. Large yields clearly make larger demands on nutrient reserves in the soil and it follows that deficiencies are most likely to occur on light soils because of leaching, and where most of the crop product is sold off the farm.

Lime

Most crops cannot grow satisfactorily in acid soils and so the neutralisation of such soils, usually by the use of lime, is of prime importance. Lime is supplied in the form of calcium oxide, calcium hydroxide or calcium carbonate and these are valued according to the amount of their neutralising potential and thus the efficiency with which they combat soil acidity.

As the squeeze on agricultural income increases, the widespread use of lime and basic slag to combat acidity lessens and we may expect to see an overall general increase in acidity in agricultural soils (Ph) with consequent reduction in crop yields. Good farming practice demands that the nutrient status of the soil is regularly examined and deficiencies corrected whenever and wherever possible.

Organic Manures

Until artificial fertilisers were introduced, organic manures derived from plant and animal wastes were the only means that farmers had of returning plant nutrients to the soil. They were supplies as mulches of dead plant material or as animal excreta applied in the form of solid dung or liquid slurry.

Table 7.8 gives some indication of the relative quantities of N, P and K in British agriculture. It will be noted that only a proportion of animal excreta is applied to land as manure and a very small proportion of human sewage.

The application of all fertilisers, both artificial and organic, is not without hazard. With excessive applications of nitrogenous fertiliser, particularly in the autumn period when heavy rainfall may be expected, there is a danger of run-off of the fertiliser from the surface of the ground into rivers and the secondary danger of fertilisers circulating through the soil and entering underground water-courses. The amount of nitrogen in potable drinking water is controlled by law because of the potential danger of nitrogenous substances, particularly nitrites, to human health – especially that of young babies.

However, organic farming systems are also liable to produce large amounts of nitrogen which in turn can find their way into potable water supplies. Thus, when grass fields are ploughed up, extremely large quantities of nitrogen are released, indeed much greater quantities than would be involved in a normal application of nitrogenous fertiliser at the same time.

Table 7.8: Estimates for the UK of the plant nutrients in crops grown, in animal excreta and human sewage, and in the fertilisers used

Source	N (kt)	P (kt)	K (kt)
In crops			
total	1560	245	1500
arable	460	75	500
grassland	1100	170	1000
In animal excreta			
total	850	200	700
(applied to land as manure)	(375	75	240)
In human sewage			
total	150	65	60
in effluent (to rivers)	100	50	50
in sludge (some applied to land)	25	15	3
In fertilisers	1186	1182	345

Source: Fream's.

Also, the spreading of sewage sludges and animal excreta on to the land is not without health hazard since such materials could contain pathogenic bacteria, such as Salmonellae. Similarly, the use of meat-and-bone meal as a fertiliser is now questioned by some because of the potential risk of this material carrying the factor responsible for various encephalopathies such as scrapie and bovine spongiform encephalopathy (BSE).

It is therefore very important that whatever nutrients are applied to the soil, care should be taken to see that the methods of application conform to good farming practice and also with all legal requirements.

Agrochemicals

There is a wide range of chemicals, apart from fertilisers, used in modern crop production systems and in the subsequent processing of crop products. There is an equally wide but different range of chemicals used in modern animal production systems, including animal medicines. The range embraces both organic and inorganic compounds synthesised by industrial chemical or biological processes and also natural compounds, such as homoeopathic medicines and some of the earlier insecticides based on plant extracts.

The major proportion of agrochemicals used in animal production is aimed at maintaining animal health and the range includes: general disinfectants and disinfestants, preventive animal health materials, therapeutic materials used to control overt disease and, more recently, novel prevention and treatment techniques used in 'alternative medicine'. In addition, materials are used as feed additives to preserve feeds, to enhance the nutritional value of feeds, to add to the nutrients present in an otherwise deficient feed and to modify feed digestion. Crop agrochemicals, apart from fertilisers already considered, can be divided into three main groupings, namely, pesticides, fungicides and herbicides.

Pesticides

Pesticides are compounds used to control insects, rodents, eelworms and other members of the animal kingdom which feed upon crop plants. The insecticidal activity of certain plant products, such as pyrethrum and derris, has been used for more than half a century and although such compounds are highly active they lack the ability to control pests in the long-term since they are rapidly degraded by sunlight.

The insecticidal properties of DDT were discovered before the last war and this compound has been responsible for the dramatic reduction in the incidence of malaria in many parts of the world by killing the Anopheles mosquito vector. It is also highly successful in controlling crop pests but such successes have to be balanced against the deleterious effects which such substances have been shown to exert on the food chain of wild animals and birds and which have now resulted in the banning of many such compounds.

The ogranochlorine compounds and most new generation materials do not control all pests equally but can be selective. For instance, organo-phoshorous products have been developed initially to control sap-sucking insects such as aphids. This group of insecticides includes chemicals which are translocated throughout the plant after absorption through the roots. It is obviously vital that all such materials are carefully selected so that they avoid danger to non-target species, particularly wildlife, and do not pose any human food hazard.

The more active pesticides require lower application rates, and improved methods of application have further reduced the amounts applied to crops. These reductions in application rates are very dramatic. High volume sprays used to be applied at rates of up to 1,000 litres/ha, but the more recently evolved electrostatic methods of crop application have reduced the spraying rates to less than one litre/ha in which the active chemical ingredient is now sometimes as low as 0.01 kg/ha.

Fungicides

Plant diseases have many causes and result in considerable reduction in both yield and quality. The losses vary with the nature of the pathogen, the timing and intensity of attack and environmental stress. Since such a wide variety of diseases affects plants, the control measures must necessarily be equally diverse. Ultimately, the efficiency of any chemical depends upon the amount of active ingredient that arrives at the specific site of action of the disease. The main causal agents of plant diseases are viruses, bacteria and fungi, the latter being the most important.

Fungicides can be grouped by their mode of action and also by the method of treatment. Some are non-penetrative of plant tissues and give rise to surface protection while others penetrate the plant and may be translocated from root to leaf or leaf to stem. Non-penetrant fungicides are usually most effective when applied before infection occurs. Their effectiveness depends on some degree of persistence and on the distribution pattern of the active ingredient on the host plant.

Penetrative fungicides act by attacking the internal hyphae and mycelia of the fungal pathogen. They also improve the distribution of the active ingredient within the host plant. The distribution of systemic fungicide depends on the site of application. The movement of most compounds is mainly upwards (root to leaf) and the transfer of fungicides applied to leaves is much more limited, for example, across leaves towards leaf tips and margins.

Because of the risk of the disease organisms developing resistance to the chemicals used to control them, it is also convenient to classify fungicides according to their biological mode of action. The risks of fungicide resistance may be reduced by using multi-site fungicides, avoiding repeated applications of the same chemical or of different chemicals with the same mode of action and by using mixtures of fungicides in which more than one type of material is active against the particular disease. Table 7.9 shows a classification of fungicides based on the criteria already described.

Table 7.9: A classification of fungicides

Type	Example
Site-specific; penetrative	
Benzimidazoles	Benomyl
Carbendazim	
Fuberidazole	
Dimethylation inhibitors	
Triazoles	Flutriafol
	Triadimefon
Imidazoles	Prochloarz
	Imazalil
	Nuarimol
	Triforine
Morpholines	Tridemorph
Hydroxy-pyrimidines	Ethirimol
Carboxyamides	Carboxin
Dicarboxyamides	Improdione
Guanidines	Guazatine
Organophosphates	Pyraxopohs
Multi-site fungicides; non-penetrant	
Phthalimides	Captafol
Phtalonitriles	Chlorothalonil
Dithiocarbamates	Maneb
	Zineb
Mercurials	Phenyl mercury acetate
Sulphur	Flowers of sulphur

Herbicides

Chemical weed control is a twentieth century technology. The first reported use of herbicide was the control of the weed *Sinapis arvensis* in oats by copper sulphate. The first generation of synthetic herbicides consisted of two types—corrosive fertilisers (such as kainit) and simple industrial chemicals (such as sulphuric acid). The first synthetic herbicide was DNOC (4,6-dinito-o-cresol) first produced in Russia in the nineteenth century but patented for control of annual weeds in France in 1932.

The modern generation of herbicides started with the discovery of the herbicidal activity of the plant growth regulating substances based on the phenoxy acetic acids, particularly MCPA and 2,4-D. The synthesis of such substances came as a result of research into plant physiology since these materials were known to be powerful growth promoters. However, when used in large concentrations they can distort, and hence kill, target weed species. This work led to the two groups of herbicides used today – the phenoxy compounds and the carbamates.

Effective weed control through herbicide use has transformed levels of yield in many crops, particularly cereals, and in addition it has eased harvesting and improved the quality of the product. It has also reduced the need for field cultivations aimed at killing weeds by uprooting them and exposing them to the sun.

Public Concern over the Use of Agrochemicals

All chemical materials have some degree of health risk attached to them. Even the consumption of pure water can be toxic to man, for example, when a dehydrated person is unwisely given large quantities of water to consume in a short time. However, the main concerns are on grounds of specific toxicity of certain agrochemicals.

Agrochemicals are now subject to very strict controls, both in the UK and within the EEC. Apart from homoeopathic mixtures (some not without their own hazards!) all recently synthesised chemical compounds must have appropriate clearances from the relevant licensing authority, and these licences are only granted where the material in question is shown to be safe, efficacious and economic to use. In addition, the operators applying such material must be properly trained and qualified.

However, product licences and training certificates only offer protection where the substance in question is used strictly in accordance with the manufacturer's instructions. If the product is abused through misuse, the product is no longer legally protected and can be unsafe.

It is argued by some that 'biological control' should replace the use of pesticides and herbicides and that 'organic farming' should replace current methods of agricultural technology. The difficulty is that most of these alternatives are not without their own potential hazards. Thus biological control can itself be hazardous if the biological parasite introduced is found to affect non-target species, and the application of large quantities of organic manures to farmland can pose a health hazard and can be socially unacceptable where the land in question is adjacent to residential areas.

At the end of the day, an equitable balance must be struck between the over-riding need to feed the peoples of the world on safe foods at prices they can afford, while making every possible endeavour to ensure that human hazards and environmental risks are kept to the minimum. There is a great need for more 'effective communication' in this area and more dialogue is required between food producer, food consumer and the allied agricultural industries manufacturing agrochemicals.

❑❑❑

8

The Principles of Processing of Products

Introduction

Agricultural products have a very wide range of functions, fulfilling to some degree each of the main common needs of food, shelter, clothing and fuel. There are also demands for animal feed, industrial feedstocks, medicinal and perfumery materials, tools and sports equipment. Such a wide range of diverse products cannot be covered readily in one chapter and so the material presented is intended to be illustrative of the thinking and ingenuity that has gone into evolving the processes and products currently available. Since food production is overall the primary aim of agriculture, its examination will be used as the lead example.

In the UK 12.5 per cent of consumer expenditure went on food purchases, out of a total expenditure. Alcoholic drink took another 6.3 per cent. If one assumes that about half the expenditure on catering (meals and accommodation) was developed to food also, a total of 22.8 per cent consumer expenditure falls under the heading of food and drink. This figure was much higher in the past, and is still so in less well-off countries. For example, the corresponding figure for Portugal was almost 40 per cent.

It is of interest to look at the breakdown of consumer expenditure on food (Table 8.1). The main branches of the food industry in the UK are meat and bacon (23.9%); bread and cereals (14.5%); milk, cheese and eggs (13.6%); and fruit, vegetables and potatoes (18.6%). In Portugal, the importance of fish is much higher (13.8%) and the dominance of meat is enhanced (31.8%), whereas soft drinks are a very minor item.

Table 8.1: Consumer Expenditure on food in the UK and Portugal

	UK 1988	Portugal 1986
Bread and cereals	14.5	13.7
Meat and bacon	23.9	31.1
Fish	3.8	13.7
Oils and fats	2.7	4.2
Sugar	0.9	1.8
Preserves and confectionery	8.4	
Milk, cheese and eggs	13.6	13.3
Fruit and vegetables	14.5	14.5
Potatoes and other tubers	4.1	3.0
Coffee, tea and cocoa	3.3	1.7
Soft drinks	6.8	0.7
Other manufactured food	3.5	2.3
	100.0	100.0

Source: Dennis.

The Importance of a Multi- and Interdisciplinary Approach

As with all agricultural products, products alone is not sufficient. What is produced is intended to be used or sold. The problems of consumption are just as important as the problems of production; the chain needs to be considered as a whole (see Fig. 8.1 cf. Chapter 1).

It is of considerable interest to see how value is added within the food chain (Fig. 8.2). As far as the UK is concerned, somewhat more value was added in 1985 by food manufacturers (45%) than by food marketers (wholesalers and retailers, 35%), the remainder coming from caters (almost 20%), whereas home farm output was only slightly greater than imports, which together represented 63 per cent of value added subsequently.

Foods are almost invariably very complex materials, used in a very complex environment. It is necessary to understand them from many different angles. Many scientific disciplines need to be brought to bear, ranging from chemistry and biochemistry through biology and microbiology to biophysics and physics. The food scientist is not, however, concerned just with the study of chemical aspects of food or biological aspects of food, but rather with an integrated approach, so that, with food as the focus, all available scientific methods (including statistical ones) are brought into action.

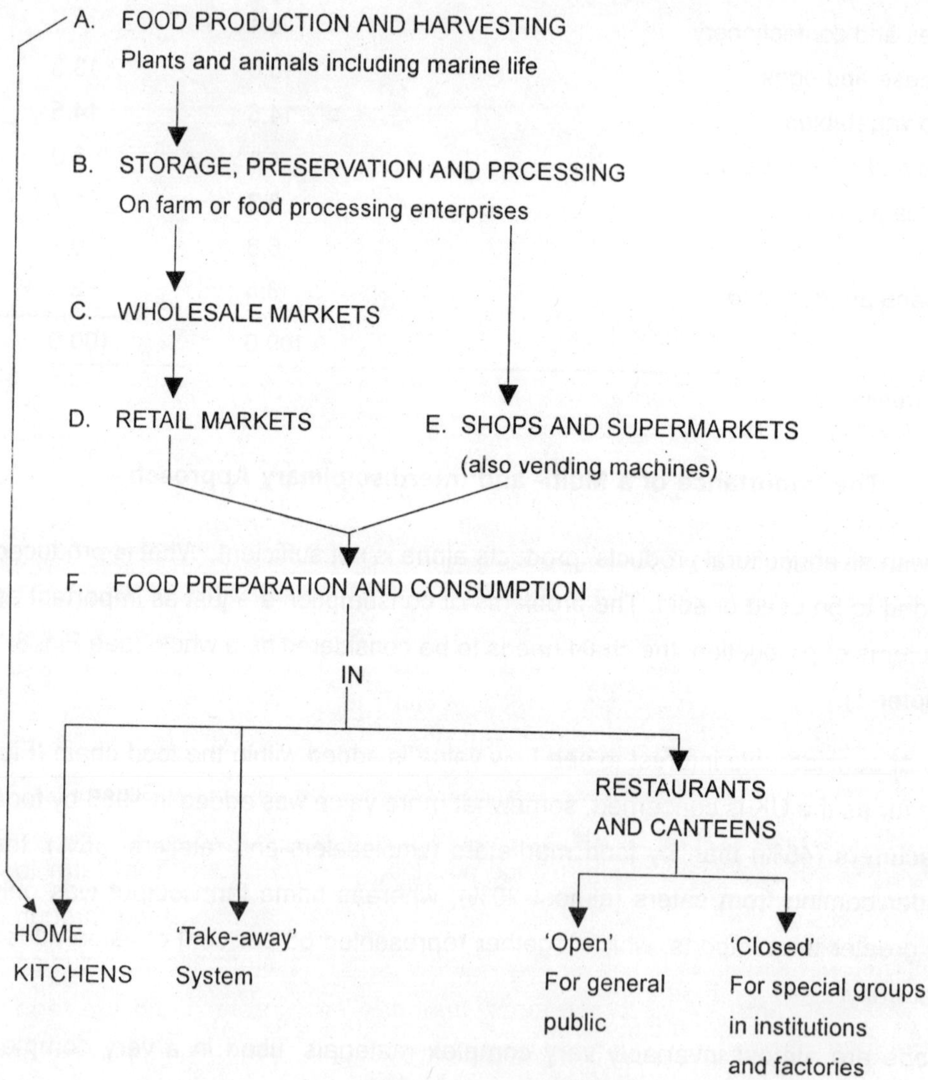

Fig. 8.1: The food chain.

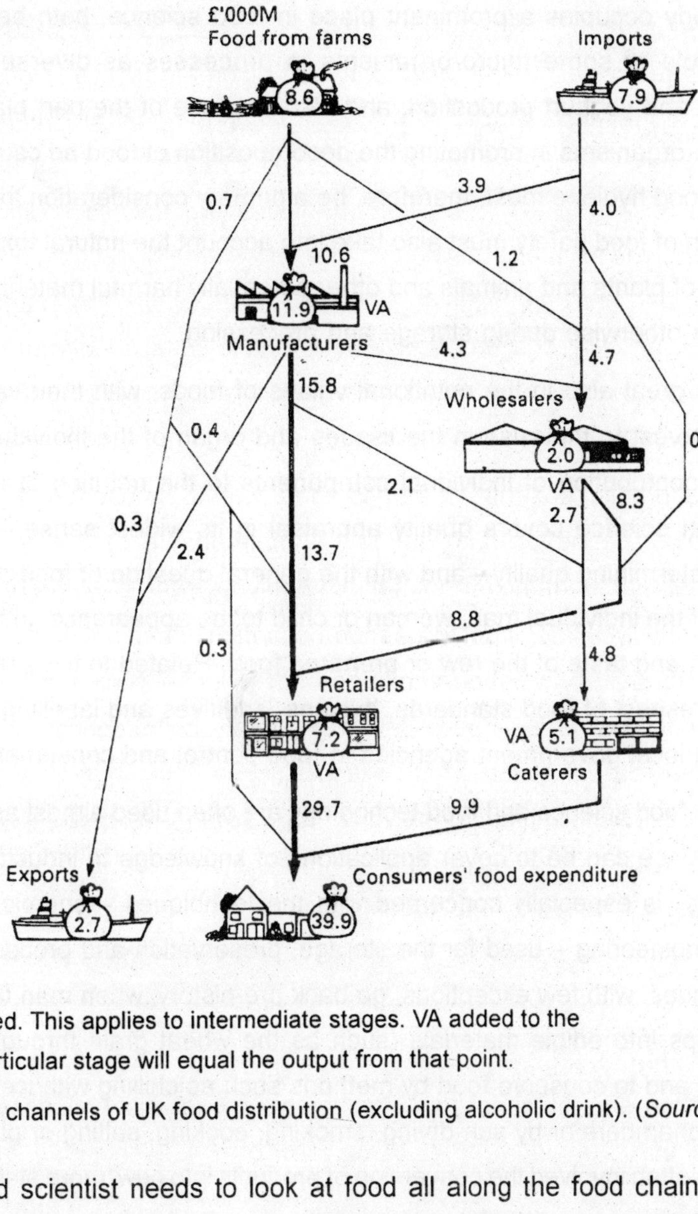

VA value added. This applies to intermediate stages. VA added to the inputs at a particular stage will equal the output from that point.

Fig. 8.2: Main channels of UK food distribution (excluding alcoholic drink). (*Source*: Slater).

The food scientist needs to look at food all along the food chain, from farm to consumer, assessing the properties and behaviour of food after harvesting through different stages of storage, transport, preservation or other transformations, including domestic or restaurant cooking procedures, until it is finally eaten. And the term 'harvesting' must, of course, be used in the widest context to include the gathering of fish or wildlife, the slaughtering of animals or the collection of eggs or milk.

Microbiology occupies a prominent place in food science, both because of the constrictive role of some micro-organisms in processes as diverse as brewing, breadmaking, and yoghurt production, and also because of the part played by other types of micro-organisms in promoting the decomposition of food ad causing disease. General and food hygiene must, therefore, be a primary consideration throughout, but any discussion of food safety must also take into account the natural toxins present in various types of plants and animals and other potentially harmful materials introduced by accident or otherwise during storage and processing.

Interest is great also in the nutritional values of foods, with their variations from the time of harvest to their use in the tissues and organ of the individual consumer, and with the contribution of individual components to the nutrition of man. Another aspect of food science covers quality appraisal in its widest sense – the sensory methods of determining quality—and with the general question of food acceptability – the reaction of the individual man, woman or child to the appearance, including colour, texture, odour, and taste of the raw or prepared food. Related to the above is current legislation in respect of food standards, hygiene, additives and labelling, and the role of central and local government agencies in food control and consumer protection.

The terms food science and food technology are often used almost as equivalents, in that the science can be to cover applications of knowledge to industrial situations. The technology is especially concerned with the techniques – chemical, biological, physical, or engineering – used for the storage, preservation and processing of food. These techniques, with few exceptions, go back pre-history, when man first learned to transform crops into edible materials (such as the wheat grain through milling and breadmaking) and to conserve food by methods such as chilling with ice or storage in underground chambers or by sun-drying, smoking, cooking, salting or pickling. These latter methods often involved the conversion of products into new, more stable foodstuffs, such as milk to cheese and pork to bacon and hams. During the past hundered years – and, to an even greater extent, over the past forty – many of the traditional processes have become the basis of important sections of the modern food industries.

The scientific knowledge of foodstuffs has evolved during the past century in parallel to the growth of organic chemistry, of biochemistry and of agricultural chemistry; many university research centres in Europe, North America and elsewhere have made their contributions to our present knowledge of carbohydrates, lipids, proteins and to the lesser constituents of foods – vitamins, enzymes, aroma compounds and pigments. Similarly, from the time of Pasteur, microbiologists have been concerned with food deterioration, its causes and prevention. Physics and engineering have made important contributions to many food processing techniques and chemical engineering concepts have led to important recent advances. Physiological concepts and techniques are important in nutritional studies and psychology and sociology in relation to food acceptability. The range of discipline involved with food is even wider: theology to nuclear physics.

Some lines of research on foodstuffs can be undertaken using one discipline, for example, chemistry or biochemistry. Over the past decades, newer methods of separation and analysis have provided much information about the detailed structure and reaction of both major and minor components of food. Similarly, other investigations have extended our knowledge of the conditions under which the growth of micro-organisms can be controlled. Often, however, a combined approach is necessary. High temperature sterlisation of milk has for its objective the destruction of micro-organisms by a adequate energy input, with the minimum changes in proteins and other components: it is necessary, therefore, to examine the effect of heat on the micro-organism present, on the enzymes which are active promoters of chemical change, and on other components which, together or singly, are responsible for colour or other desirable or undersiable properties.

Industries, such as those based on electronics, were born in the laboratory and depend on scientific work for their advance. With foodstuffs the problem has been different. Most, but not all, food processing techniques ante-date scientific knowledge, and a primary task of the food scientist has been to examine in detail the composition of the very complex raw materials and the nature of the changes taking place during traditional processes, such as breadmaking or cheese making. Without such knowledge, it is difficult to exercise scientific control or to improve techniques and products. Special difficulties arise because of the complexity and variability of food systems and their sensitivity to change. A crop such as wheat can show wide variations in its content of protein (and other components), depending on the cultivar grown, the environmental conditions of cultivation and time of harvesting.

Such variations have important consequences for processing, hence the importance of statitics in food science.

Post-harvest and Post-mortem Changes in Foodstuffs and in Their Components

When handling agricultural products, the scientist, as part of the integrated approach, has to think at least on three levels, the biological, the chemical and the physical. The considerations to be borne in mind are illustrated in Tables 8.2-8.4.

Table 8.2: Post-harvest changes in foodstuffs caused by biological agents

Classification	Types	Effects/Examples
(A) *Pests/wildlife*		
(1) Insects (2) Birds	Wide variety	(a) Consumption of food (b) Consumption of special components of food, e.g., protein (c) Contamination by flesh, e.g., insects (d) Contamination by excreta
(3) Rodents	Common rodents, especially rats and mice	
(4) Other animals	Including unproductive large animals	
(B) *Micro-organisms*		
(5) bacteria	Numerous species	Wide variety of reactions leading to – toxic components, or – product modification, e.g., cultured milk products, or – other chemical or physical changes
(6) Fungi	(a) Yeasts, especially *Saccharomyces* spp.	Alcoholic fermentations Panary fermentations
	(b) Various saprophytic fungi with mycelial growth habit	Aflatoxins formation, e.g., *Aspergillus* spp. Spoilage Product modification, e.g., Tempeh
(7) Viruses	Wide variety	Increased susceptibility of tissues to secondary invasion

Table 8.3: Post-harvest changes in foodstuffs caused by chemical/biological reactions

Classification	Types	Effect/Examples
(A) *Enzymic reactions*		
Caused by enzymes in cells or tissues or introduced from outside	(a) Oxidations or reductions ±O ±H or equivalent	Oxidation of lipids Oxidation of ascorbic acid Oxidation of pigments
	(b) Hydrolysis or condensations $\pm H_2O$ or equivalent	Hydrolysis or synthesis of simple or complex lipids (lipases) Hydrolysis of starches or sugars (amylases, invertase) Hydrolysis of protein (proteases)
	(c) Respiration processes	In plant tissues In muscle
	(d) Other systems	Decarboxylation of amino acids
(B) *Non-enzymic reactions*		
(1) Primarily between organic and/or inorganic components of foodstuff	(a) Between individual components in one foodstuff (b) Between components in food mixtures	Maillard or browning reaction Interesterification Catalysis by metal Formation of polyphenol-protein complexes
(2) Between food components and gaseous, liquid, or other environment	(a) Between components and gaseous environment, including air in foodstuffs and moisture in air (b) Between components and water, including dissolved oxygen, in air	Oxidations, e.g., destruction of ascorbic acid or pigments Carbon dioxide (pH change) Boiled milk flavour Hydrolyses Fat rancidity
(C) *Accidental (or adventitious) contaminants*[a]		
	(a) From agriculture horticulture and animal husbandry	Grop protection chemicals Pharmaceuticals in meat or milk Antibiotics

	(b) From air or water	Nitrates or trace metals from water or effluent Fumes from air
	(c) From buildings, containers or machinery	Metal from cans or machinery Plasticisers and stabilisers from plastics Wax from cartons Mineral oil from machinery
(D) Additives[a]		
Effects of substance deliberately added as preservatives or processing aids	Additives used in preservation and processing	Reaction of Agene with methionine in wheat flour protein Nitrosamines

[a] The contaminants or additives may or may not react chemically with food components.

Table 8.4: Post-harvest changes in foodstuffs – physical or physico-chemcial

Classification	Type	Effect/Examples
(1) *Physical*		
(a) Temperature—heating and cooling	Change in state	Solid—liquid, e.g., fats (oils); ice (water)
	Change in texture	
	Change in viscosity	Plasticity of fats
	Change in phase	Oils Coagulation of egg whites
	Change in cellular or tissue structure	Disruption of plant or animal cells or tissues by freezing
(b) Gravity or centrifugation	Separation by gravity	Cream from milk
	Centrifugation	Protein coagulate from heated milk
	High-speed air classification technique	High protein fractions from wheat flour
(c) Mechanical treatment	Bruising or crushing–destruction of cellular form	Bruising of garden peas leads to rapid biochemical changes Crushing of oil seeds leads to extraction of oil plus damage to protein by heat

(2) *Physico-chemical*		
(a) pH	Increased acidity	Precipitation of protein–lactic acid in sour milk Gelation in yoghurt Wine to vinegar Destruction of bacteria Alteration in colour of pigments
	Decreased acidity	Alteration in colour of pigments
(b) hydration	Formation of solution	Sugars and syrups
	Formation of sols or gels	Pectin gels
(c) Dehydration	Cell structure changes	Destruction of cells
(d) Irradiation	Photo-oxidations	Destruction of ascorbic acid in milk held in glass containers
	UV radiation	Synthesis of vitamin D from precursors. Destruction of riboflavin Bleaching of carotenoids
	Irradiation	Destruction of enzymes

Physical or phosico-chemcial changes may precede and so modify changes brought about by biological agents and/or chemical/biochemical reactions, or may follow such changes.

It is worth making three points in relation to these tables:

(a) Although the plant or animal is dead, the cells may well still be functioning normally, up to a point, from which a range of variants of normal biochemistry will apply. Fruits, of course, are in general still viable organisms.

(b) The interactions between the different types of change taking place are most important. For example, the physical damage caused by biological agents normally sets off biochemical changes by breaking the barriers separating enzymes and their substrates, thus allowing them to interact.

(c) Changes may be desirable (i.e., part of processing or maturation) or undesirable. For example, moulds are involved in tempeh production in SE Asia from coconut or soy press cake, as well as in the formation of the extremely dangerous aflatoxins.

Preservation and Processing

They way agricultural products are handled depends on their perishability, i.e., their proneness to undergo biological, chemical and/or physical changes (see Table 8.5 and 8.6). Although crops can be divided into more- and less-perishable groups, all raw animal food products belong to the former group.

Perishability leads to waste. Waste in the food chain is a most important topic and should always be one of the first aspects to be examined in any efficiency review. This applies equally to all other aspects of handling agricultural materials. Some aspects of food wastage are set out in Table 8.7. Wastage for many common vegetables and fruits in the food chain is usually 20 per cent – 50 per cent. Wastage of food in hospitals tends to be particularly high.

The techniques of food protection and preservation are set out in Tables 8.8 and 8.9. Clearly a very wide range of methods is available. What are the principles involved in their selection? To answer this question, it is best first to consider briefly food quality.

The Objective: Food of the Nature, Substance and Quality Demanded

Food is produced for the purpose of fulfilling a demand. In that sense, the consumer is paramount. This view is also enshrined in legislation, according to which, apart from not being injurious to health, food has to be of the nature, substance and quality demand. Food thus provides a very early exception to the general principle of *caveat emptor*, buyer beware. The buyer of food is strongly protected by law.

Part II of the Food Act, 1990 (UK) states categorically "Any person who renders any food injurious to health by means of any of the following operations, namely:

(a) adding any article or substance to the food;

(b) using any article or substance as an ingredient in the preparation of the food;

(c) abstracting any constituent from the food; and

(d) subjecting the food to any other process or treatment,

with intent that it shall be sold for human consumption, shall be guilty of an offence."

Table 8.5: Crops – preservation and processing

Commodity	Separation, extraction or conversion to more edible form— primary products	Preservation techniques	Further conversion to consumer products[a] and to by-products for human food, animal feed or other uses
(A) *Less-perishable group*			
(1) Wheat	Milling to give flour, germ and bran	Controlled storage, in terms of temperature or humidity, of whole cereal grain or flours	Flour, bread biscuits, confectionery, and other cereal products
Other cereals	Milling to flour and meals		Various cereal products Malting, brewing and distillation Hydrolysed products, e.g., starch syrups
(2) Oil seeds and oil-bearing crops	Crushing or solvent extraction to give	Controlled storage, before or after extraction	
	(a) oil		Hydrogenation and blending to shortenings, margarine, and other fat products (also soap industry)
	(b) protein residues		Proteins for animal feed[b] Protein flours (e.g., soya) for human consumption[b]
(3) Sugar-bearing crops (e.g., cane, beet)	Extraction to sugars and other products		Refined sugars and other products (e.g., molasses, caramel) Sugar confectionery
(4) Starch-bearing crops,. e.g., potatoes, cassava (manioc)		Controlled storage	(a) Potatoes, frozen, canned, dehydrated, crisps (b) Potatoes as basis of starch industry or for fermentations to industrial alcohol, vodka, etc.

(5) Cacao bean and beverage crops			
Cacao bean	Fermentation, drying roasting and grinding to give cacao mass	Controlled storage	Chocolate Cacao butter Cocoa
Tea Coffee	Special processing		Blending Dehydrating (instant)
(B) *More-perishable group*			
(6) Vegetables	preliminary separations (e.g., vining of peas)	(a) Controlled atmosphere and/or temperature storage and transport	Other products include: (a) Juices
		(b) Quick-freezing, canning and bottling, dehydration	(b) Non-alcoholic (lactic) fermentation (e.g., sauerkraut)
		(c) Preservation by pickling, smoking, sugar, salt or special chemicals	(c) Special potato products Legumes (e.g., beans) as potential protein source[b]
(7) Fruits	As for vegetables	As for vegetables	As for vegetables Also alcoholic fermentation to wines, cider, perry

[a] To these must be added a variety of cooking processes now being carried out on a commercial scale.

[b] Note current investigation and development work on protein production from agricultural sources (oil seeds, legumes, levels and other chlorophyll-containing material).

Table 8.6: Animal products – preservation and processing

Commodity	Separation, extraction or conversion to more edible forms— primary products	Preservation techniques	Further conversions to consumer products[a]
(1) Domesticated animals	Slaughtering and dressing to give carcass meat and offal	Preservation by one or other or combination of methods listed for vegetables	Many specialised products from different preservation methods, e.g., bacon & ham, smoked meats, sausages, meat extracts
(2) Dairy animals	Milk Separation to give cream	Pasteurisation and sterilisation	Concentrated and dehydrated milks Butter Cheese Fermented products such as yoghurt
(3) Poultry	Poultry meat Eggs	As for (1)	Dehydrated whole eggs, whites or yolks
(4) Wildlife -birds -other animals	As for (1)	As for (1)	As for (1)
(5) Fisheries[b] –inland fisheries –sea fisheries –other marine products	Aquaculture	As for (1)	Special products—frozen, canned, salted and smoked

[a] Cooking processes carried out on a commercial scale are being increasingly used for animal products or for mixed animal/plant products.

[b] Note also plant products from marine source and from aquaculture.

Table 8.7: Food wastage

	Stage	Causes
(A) Losses in field (crops/animals)	(1) In growing crops or animals	Climatic aberrations Pests Virus and other diseases
	(2) At harvesting	Incomplete harvesting or methods leading to damage of crops
(B) Post-harvest losses	(1) Immediate post-harvest treatment	Pests—insects, rodents and other animals
	(2) Storage	Micro-organisms } Accelerated by humidity and heat
	(3) Transport	Enzyme action } Accelerated by humidity and heat
	(4) During processing	Oxidation and other chemical reactions } Accelerated by humidity and heat
(C) Losses in home or restaurant	(1) Larder or storage losses	As for (B)
	(2) Losses in kitchen preparation	Peeling and other processes
	(3) Losses in cooking	Destruction and losses of vitamins and other components
	(4) Plate losses	
Types of losses	(1) In total quantities of food	
	(2) In special nutrients (e.g., protein, vitamins)	

Table 8.8: Techniques of food protection and preservation to destroy or to retard action of micro-organisms or to minimise other changes—biological or chemical

A Physical protection	B Control of environment	C Reduction in available water	D Destruction of micro-organisms by heat	E Miscellaneous methods
(1) *Buildings* Protection through storage structures of different types, varying in materials, design and size, e.g., grain storage (2) *Containers* Protection through packaging using containers of different types, with or without protective chemicals, e.g., paper, cardboard, fibres, glass, metals, synthetic polymers	(1) *Gaseous environment* (a) Removal of oxygen or air to give vacuum (b) Replacement of air by nitrogen (c) Control of oxygen and other gases (2) *Humidity* (3) *Tempearture* (a) *Cooling, e.g.,* in underground stores (b) *Refrigeration* in commercial or domestic stores (c) *Freezing by* various physical methods (d) *Freezing using* cryogenic methods (e.g., liquid nitrogen)	(1) *Drying* or dehydration by various methods – sun – air – freeze drying (2) *Concentration* by heat (evaporation) or other methods (e.g., freeze-concentration for orange juice, reverse osmosis for whey) (3) Use of sugar, salt and other edible additives to reduce the 'effective' water available (a) *sugar* for sugar preserves such as jams and crystallised fruit (b) *salt* for wide variety of products	(1) 'Appertisation' (heat treatment in a sealed container)—bottling or canning (2) Modern developments—aseptic canning (3) Pasterusation or sterilisation processes, especially for liquids, such as milk (4) Cooking processes using different types of equipment	(1) Fermentation and pickling methods often involving lactic acid production, e.g., yoghurt, cheese, salami, sauerkraut vegetables (2) Smoking, e.g., for fish (3) Use of chemical preservatives (e.g., sulphites) including antibiotics (4) Irradiation techniques (rarely used in practice)

Different methods may be used in combination with one another.
Some methods, e.g., smoking, involve a combination of techniques.
Efficient packaging is required in conjunction with several of the techniques.

Table 8.9: Other techniques of food processing

Technique	Materials	Examples	
		Primary product	Secondary product
Extraction			
crushing,	Wheat	Flour	Bran
grinding,	Oil seeds	Edible oil	Protein/fibre residue
milling	Sugar cane	Sugar	Fibre residue
separating,	Citrus fruit	Juice	Skin and fibre
refining	Garden pea	Pea	Pod, vine
leads to			
Concentration of a particular	Animal	Carcass meat	Offal, bones, etc.
component	Milk	Cream	Skimmed milk
Compounding (Mixing and related techniques)	Flour, Yeast, Salt, Water	Bread	
	Milk powder, Other components	Baby foods	

Many of the techniques given in Table 8.8 can be used in combination with the above methods to give typical food products.

There is no clear line of demarcation between preservation and other techniques, e.g., cheese-making and wine-making are both based on fermentation but give new products.

Secondary products from extraction techniques are often of importance, e.g., animal feed.

That powerful statement also circumscribes the activities of the food manufacture. The onus is clearly on those who sell to ensure its wholesomeness, that is, its freedom from infective agents (e.g., *Salmonella*) and from hazardous levels of toxic substances (e.g., botulism toxin, lead, nitrite). In fact, gross offences of this kind are extremely rare, but, unfortunately, they do occur (e.g., Spanish oil).

What about nature, substance and quality? Nature (e.g., varieties of fruit) and substance (e.g., a piece of glass) are relatively easily defined, but quality is much more problematic.

Quality can be regarded as composed of six aspects, almost all of which have several components, as summarised in Table 8.10. The six aspects are arranged in the order shown because the need for food is nutritional; the desire for food is ruminated largely by the senses; hazards from food come primarily from micro-organisms, but also from physical effects and from toxic chemicals; and lastly, aesthetic sensibilities are offended particularly by biological residues. Taken overall, food quality is clearly a broad subject.

Table 8.10: Food quality

NUTRITIONAL	SENSORY	MICROBIOLOGICAL
Energy	Appearance:	Food-borne disease
Protein	size	Toxin production
Essential amino acids	shape	Spoilage
Essential fatty acids	colour	
Vitamins	uniformity	
Minerals	Smell	
Dietary fibre	Taste	
	Touch:	
	feel	
	kinaesthesis	
	Sound	
PHYSICAL	**CHEMICAL**	**BIOLOGICAL**
Radioactivity	Contaminants	Hair
	Additives	Excreta

It raises interesting conflicts. For example, it is accepted practice to add sulphites (up to 450 p.p.m.) to sausage meat, primarily for microbiological protection. But sulphites destroy vitamin B_1 (thiamin), of which meat is a significant source (all meat provides 18%; sausage meat only, 7% of B_1 in the diet). The purchaser is probably not prejudiced to a sufficient extent for such sulphite additions to constitute an offence, because the intake of vitamin B_1 is still sufficient from the remainder of an average diet.

This leads to a very important point. Food needs to be clearly distinguished from diet. The western consumer has virtually unlimited choice in constructing diet. Normally, therefore, the suppliers of food cannot be held responsible for the nutrition of the consumer. But there are exceptions;

(a) Where the supplier provides virtually all the food consumed, e.g., in institutions, such as hospitals and prisons, on board ship, and baby food.

(b) Where an item of food constitutes the major source of an essential nutrient in normal diets, e.g., vitamin B_{12} from meat. Such considerations are significant in assessing novel protein foods as meat replacers.

Principles of the Selection of Methods for Preservation and Processing

The qualities of the resulting food are clearly of major importance, but need to be judged almost always in an economic context. Technically, the main objective is to ensure the availability of food both when and where it cannot simply be collected from the vegetable garden or the farmyard. Throughout the centuries, to have sufficient food to survive winter was a paramount aim. Even though currently the target is often described, less threateningly, as prolonging shelf-life, the underlying dangers of famine, due to natural disasters, war, or strikes, are still all too real.

As stated above, the primary hazards from foods are due to micro-organisms. In addition, micro-organisms are one of the main agents of deterioration of foods. Control of the growth of micro-organisms, and of toxin production by them, is of great importance. Three main methods of control are used: heat, low temperature ad reduction in available water (see Table 8.8). Most micro-organisms have only limited resistance to heat, but, in order to ensure that heat-processed food is safe, it is necessary to focus on the most dangerous and heat-resistant pathogen possibly present. For canning, this is the spore of *Clostridium botulinum*, which can germinate under anaerobic conditions and produce botulism toxin, one of the most deadly substances known. Canning of non-acid foods is therefore geared to the destruction of botulinum spores by giving a heat-treatment to the least accessible part of the can of an intensity and duration sufficient to reduce the population by a factor of 10^{12} (a million times a million). Pasteurisation is a much milder heat treatment, aimed only at destroying any pathogens liable to be present in the particular foods being processed.

Although low temperatures also destroy some micro-organisms, the main effect is one of preventing growth. Reduction in available water is similar. Here, it is free water, the water activity, a_w, which matters, not the percentage of water by weight, when comparing one food with another. Most bacteria will not grow at an a_w below 0.9; most yeasts, not below 0.8; and most moulds, not below 0.7. Reduction in a_w can be achieved by drying, concentration, freezing and addition of salt or sugar.

Treatments such as those just mentioned, affect not only microbes. Heat in particular usually speeds up chemical deterioration. Although most enzymes are inactivated by modest heating, their activity falls off as a_w is lowered, but does not disappear until very low a_w is reached. On the contrary, non-enzymic browning increases rapidly with temperature, contributing greatly to the odours and flavours produced on roasting and baking, but also to the deterioration, for example, of canned foods, on long-term storage. Somewhat surprisingly, non-enzymic browning reaches a *maximum* at an a_w of about 0.5. However, though the reaction falls off as a_w is lowered below about 0.5, it is very difficult to stop it altogether by control of a_w alone; indeed much more difficult than similarly controlling microbial or even enzymatic activity. In many foods, oxidation of lipids is the main cause of deterioration (development of rancidity). It too increases with temperature, but has a *minimum* at an a_w of about 0.5. These relationships illustrate the complexity of food systems and the multi-faceted problems of optimising their quality.

Marketing and Distribution

Most food, whether from the grower, the processor, or the importer, reaches the consumer through a distribution network, markets and shops. These important links in the food chain become increasingly dominated by the multiple retailers, as can be seen from Table 8.11.

Table 8.11: The UK grocery trade

	1950	1980	1987
Multiples	20	61	73
Co-operatives	23	14	11
Independents	57	25	16

Source: Pearson.

The multiples, very adroitly, have increasingly portrayed themselves as the champions of the consumer. They have made good use of nutritional labelling and of the provision of nutritional information in general to reinforce that role. Aside from economic considerations they have indeed been a very positive force in raising standards of food quality and rigorously controlling conditions of storage and sale, with particular emphasis on hygiene. Their control has been progressively extended backwards along the food chain, through inspection of processors' and growers' facilities both at home and abroad, and the playing down of detailed protocols, which have been incorporated into contracts. In this they have followed and built on practices established by the processors in relation to their suppliers. Similar patterns have emerged in areas other than food; for example, with textiles and articles made of leather.

Conclusion

This chapter has tried to address the principles of processing of agricultural products and it must be clear that the attempt is only a very partial one, using food as the principal example, but covering even that area only very selectivity. The aim throughout has been to stress the complexity and multi-faceted nature of the system addressed, the care and wide range of knowledge that are required in its successful operation, and its vital role in the survival of mankind.

❑❑❑

9

The Principles of Farm Business Management and Marketing

Introduction

Farming is at best an imprecise business. It is concerned with the management of biological processes, the inter-relationships of which are not all fully understood. It sits in the human food chain between the suppliers of inputs to agriculture and the down-stream processing and distribution sectors of the food industry. Farming is particular in that it is concerned with the use of natural resources: land, soil, water and sunlight. It is about the capture of solar radiation by green plants and the use of those plants either directly or after some processing to provide food for human or food for animals.

Farm management is about the way in which all the resources available to the business of a particular farm are managed so as to achieve the fulfilment of perceived financial, technical, environmental and personal goals and the development, establishment and operation of systems which blend these resources in such a way that the probability of achieving those goals, either fully or partially according to the priority rating given by the manager, is enhanced. The ranking of priorities reflects the principal goals of the manager. In a subsistence situation the overriding goal is to provide sufficient food for the family to exist and some additional supplies for trade or barter as well as preserving, if possible, some seed for planting for the next season's crop. In an intermediate stage, reflecting the transition from subsistence to a more market-oriented approach, the capability of meeting family food needs may be more certain and thus the sale of products into markets to generate cash becomes more important – the concept of family income becomes a reality. In a more market-oriented economy still, the farm is producing food commodities for sale and runs a business, with properties given to asset values, investment and rates of return.

The Influence of Governments

The importance attached by governments and the people they represent to the prime task of ensuring that adequate food supplies exist to feed the nation reflects the importance attached to that particular social goal. In consequence, food security demands have led to the close involvement of politicians in food supply and market management to the extent that a free market, whither in the internal or external economic dimension, scarcely exists. Farmers and their resources are influenced by food and agriculture policy goals at both national and international level expressed by either incentive or command regulations, or both. In some countries agriculture and food production still demand the allocation of much national resource, in others, especially those of the older industrialised economies of the north and west, policies are unfolding which seek to reduce the amount of national economic and intellectual resource allocated to food production and supply.

In these latter economies, farming is proportionately a relatively small contributor to Gross Domestic Product (GDP). Such economies are characterised by diversity, a declining proportion of consumer spending on the food commodity itself and an increasing proportion on the processing, distribution and service contributions—the value added elements. It is a characteristic of such economies that the population as a whole, having achieved a sense of food security, places increased attention on the qualitative aspects of food: safety, hygienic quality and wholesomeness, and consumer choice reflects concern for the effect of agricultural practices, appearance of the countryside and the relationship between agriculture and the environment. This impacts increasingly on the manager of the farming business in the form of legislative measures.

The Legislative Envelope

The extent and nature of legislation is a major factor affecting the decision-making processes of the farm manager. In both command- and market-oriented economies the manager of the farm business must operate within a legislative framework. Legislation is concerned with the way in which food is produced, processed and presented and the safety of the inputs used in food production. It touches upon all aspects of the farm business – natural, human, financial – both operational and capital, land and its occupation, food safety and the standards, both nutritional and hygienic, of the processing and service sections. Generally speaking, the more advanced an economy becomes, the more complex and containing does the legislative envelope within which the production and marketing functions have to operate. It is incumbent upon the manager of any farming business to be informed about and secure access to such information, for it is a major influence upon the business environment within which the farm must function. It may even legislate on the fitness of the individual to farm. Increasingly, those who are the ultimate judges of food, the final consumers, are exerting their collective influence on agricultural and food matters.

Farm Policy Considerations

It is important for the manager to differentiate between the farm as a production unit and the farm as an area of land. The policy of land use will reflect not only the legislative framework regulating the ownership and occupation of land but also the motivation of the landowner, who will in many instances also be the occupier, depending on the system of tenure, which varies between and within countries. In some cases, the owner will define the purpose of the holding through the tenancy or land holding arrangement and this applies whether the landowner is a private individual, corporate body or the State. In most cases, the farm is defined as an agricultural commodity producing unit. Increasingly, however, especially in matured economies, farms may be acquired by individuals who are concerned with other forms of land use; industrial or urban development, leisure activity or, simply, the investment of capital into a land portfolio with the view that low income may be offset by capital growth – the acquisition of wealth. Land is frequently acquired by inheritance and represents not only the substance of family capital but also social position and space or privacy for living. These latter cultural objectives are highly prized in some countries and rich individuals from industries other than farming will often seek to acquire ownership and occupatior of land in order to express social station and to engage in sporting or rural pursuits.

Such motivations for land ownership may have a major impact on the way land used and the balance between uses. In industrialised economies, fiscal legislatic encourages a dynastic approach to land ownership and occupation and, until recent years, the commitment of occupiers and owners has been toward the use of land for food production. This, however, may change as the advanced economies alter their food policies and markets.

Built into the commitment of most farmers, however, is the cultural motivation that occupation of land must be for the prime purpose of (mainly) food production, irrespective almost of whether the market-place requires that or not. It is this conditioning which has an immense effect on policy formulation for the individual farm business and in the allocation of land and financial resources to food and non-food uses.

Policy will reflect the motivation, skills, technical and financial resources of the individual proprietor and manager, and frequently the desire to be associated, through instinct or aptitude, with a particular enterprise or food production activity has been the overriding consideration in formulating policy goals. Increasingly, however, farm production policies relate to market requirement and, in a market-oriented economy, identification of a market opportunity or niche is the starting-point of policy. It is then necessary to consider whether the geographical location, soils, aspect, soil type and infrastructure of the holding is suitable to provide a production unit capable of exploiting that market opportunity to competitive advantage.

Given that these strategic considerations can be worked through, a policy will be formulated which generally seeks to maximise returns to the factor in most limited supply. Land, labour and capital are the triptych of economic resources fundamental to any business. Where national and international market/supply management policies are such as to provide the incentive to produce volume and prices are set at levels to achieve this, it is generally true that near-maximisation of output is the means to maximise margins and, given prudent investment in fixed assets and overhead elements, profits or internal rates of return. Where economies are industrialising and the process of urbnisation is taking place, labour has generally been the factor in scarce supply; the same applies in a situation where areas are being opened up for settlement. Another incentive to maximise return to labour, however, has been the inability of farming to provide labour with sufficient reward to compete with that offered by alternative employment. Thus, increasingly, the farm business policy reflects the supply and motivation of family labour. Rarely, however, are land, labour and capital available in unlimited supply at a price the farm business can afford. Thus policies, having regard to the natural constraints of a holding, will seek to simplify operation and thus reduce the competition for resources between competing ends. This specialisation of individual farm businesses is also a response to terms of trade in commodities and the need for the enterprise to perform to a high standard, technically and financially, in order to provide a sufficient return on the working capital or *current* assets invested in the production function to fund living standards and service the land-owning or occupation function in terms of mortgage or rent.

Policy simplification or specialisation is appropriate in a number of circumstances but not all, and not for an indefinite period: policies need to change in the face of changing market requirements and overall agricultural industry priorities. Here specialisation of production process is the individual farm policy, the family may prudently seek to diversify capital into other forms of investment like other businesses, property or stock and money markets, so as to reduce risk. This is the alternative to mixed farming patterns with a number of enterprises all of which compete for both financial and human resources. Thus, a fundamental policy decision is whether or not to specialise in production function and diversify capital, if possible, or to diversify the farming pattern. In some cases, of course, individuals may be so constrained by one resource or another that there is little choice, in turn increasing risk and therefore the need to secure matching rewards. Policy too will reflect the judgement and priorities of individuals. For example, the balance between the protection or generation of existing and new wealth and income for living.

The manager will therefore have to ask – and answer – some pertinent questions:

(a) What is my overriding personal/family goal?

(b) What kind of farm policy will assist me in meeting that? What are the market opportunities, and probabilities of successes in supplying them profitably? Do I see myself in a primary or service industry?

(c) Among the analysed options which do I/we think most attainable bearing in mind the different levels of risk?

(d) Which of these do I/we wish to pursue in terms of personal fulfilment/satisfaction?

(e) What are the principle farm, financial and personal constraints/weaknesses and how can they be identified and overcome?

It is at farm policy level that one of the most fundamental rules of business management is so often overlooked – an honest analysis of strengths, weaknesses, opportunities and threats.

Operational Management

Policy is concerned with overall business goals, operational management is concerned with the implementation of policy. Once a sound policy has been identified, its failure or success is usually determined by the level of operational excellence which management attains. It is in this area of operational management that managers must first come to terms with the difference between intention (policy goals) and achievement (practical implementation). Success in implementation reflects achievement and this implies that there must be some measure of it. In farming, as in all businesses, economists and managers have defined various means of measuring performance in financial, technical and human terms. The measurement of financial progress is often the easiest: progress in relation to an overall budget or cash-flow budget may be measured monthly. Such budgets may reflect past performance or target setting and are usually a mixture of both. The process is made easier by the production of standardised annual budget and cash-flow forms and by the standardisation of terminology and procedures which has taken place over the years as the techniques of forecasting and monitoring have become more refined.

Those who are in the business of lending money to the farm business have made a major contribution, too, by making available the various performance ratios they use in analysis and monitoring the financial structure of the business. The manager can easily compare the ongoing financial performance of his or her business by contributing to a business management scheme through which businesses are compared by standard measures, and involving services such as some form of interpretive consultancy. Both suppliers of inputs to farms and, in some cases, purchasers of farm commodities also provide financial services and some form of comparative analysis. Where much of the capital for the business is borrowed, i.e., the business is relatively highly geared, the lender may well require participation in such a scheme, partly for reasons of checking the ability of the business to service and repay such loans and partly to encourage good financial discipline on the part of the borrower. In a wider field, within the UK, the Provincial Economics Service produces annual reports on the technical and financial performance of different farm types, as does the Ministry of Agriculture, Fisheries and Food.

In all countries the collection of performance and financial data on farming performance is an important source of information for policy makers and this often requires, initially at least, a substantial investment by the public sector in data collection. As time goes on and the industry matures, private sector organisations increasingly undertake this work.

Thus, in developed economies at least there is no shortage of comparative information with which the individual manager can compare the performance of his or her business.

However, few schemes carry the monitoring of performance into a proper analysis of the reasons for deviations from budget and this exercise can be the most useful of all for the individual manager. A simple technique, developed and taught many years ago by Tony Giles at the University of Reading, UK, provides many of the answers as to why performance fails to reach budget and is valuable for estimating the extent of deviation caused by price/yield combinations. Such an analysis is an essential part of financial control.

Accurate technical performance measures require as much discipline as financial measures – indeed more – for there is frequently little outside imposition or requirement for such discipline. But recording schemes are not difficult to devise at various levels of refinement depending on the purpose for which the information is to be used. In some breed improvement schemes fine detail may be required, in others sales records may provide the key information required on, say, yields. Nevertheless, some record, however crude, of physical allocation of resources between, for example, different classes of livestock is essential for analysis of performance between and within enterprises.

Thus, recording systems are available in most countries from one source or another and to different degrees of sophistication both to meet national planning needs for policy purposes and to meet the needs of the individual manager. Even on the most simple and constrained holding, however, some form of reckoning is vital for obtaining resources allocation. A large farming enterprise in northern Europe or the America may require the sophistication of a computerised system to do this. At the other end of the spectrum, a small subsistence holding may require no more than a retentive memory and an observant eye. Both, however, if consistently and honestly practised will bring to the manager's mind and perception two of the most fundamental laws of economics: the law of diminishing returns and the law of equi-marginal returns.

The External Customer

Market-oriented economies are about freedom of choice for the customer and consumer and competition to fulfil wants. Wants is used as opposed to needs because generally speaking people are prepared and able to pay more for wants. Thus, again, market-led economies are about competing with others to meet wants for goods or services at a value perceived by the customer as appropriate to the satisfaction that product or service will bring them and will do so at a margin which provides an adequate return on capital to meet the needs of the business.

In a growth market where demand exceeds supply, the competition between suppliers is less intense. As demand stabilises and the market matures and sectorises, purchasers become more discerning and exercise their freedom choice. Those who wish to sell must therefore promote their goods and services both to inform and convince the customer that what they have to offer is better value for money than the wares of their competitors. Moreover, as a market matures, minimum standards and other criteria are laid down by law, often adding to both production and distribution costs.

The elasticities of food commodities are such that price resistance is quickly met, leading to declining real return. Input costs, however, increase, drawing the farm business into a cost-price squeeze. In this situation, the unit cost of production becomes increasingly critical, reflecting not only volume but also technical efficiency and the way in which resource use is optimised. Those who are less efficient go out of business and thus the industry is concentrated into fewer hands.

For Britain, as the concept of a single European market becomes a reality, market pressures will increase, led by the needs of the food processing industry to seek its supplies at competitive prices for any given specification. The emphasis in the food chain in consumer-driven markets is not only on volume and consistency but quality because the retail sector in a discerning market-place is likely to become increasingly product-based.

Where Does This Leave the Industrial Farm Business?

There is no single solution: each business will follow one or several of the options available for each commodity. It is a common fallacy that in the long run success attends the identification and fulfilment of a 'niche' market, for as the niche expends to accommodate the output of other producers, competition becomes almost as intense as in mainstream commodities. The reality is that the key to successful marketing of agricultural commodities is to be able to meet product specification demands at lowest unit cost. The development of relationships with customers is one of the keys to successful–marketing understanding the customer's situation and seeking to meet the customer's requirements. In some cases this leads to ongoing contractual relationship which may, in the ultimate, progress to forward or backward integration, this is already the case in a proportion of farm output of one or more commodities. In other cases, producers may form a co-operative or joint venture operation for a commodity, employing a professional marketing executive to build effective relationships with customers and communicating back into the production function key messages on quality standards, delivery dates and information on both the structure and movement in the market-place—few producers will be close enough to the market to know who is making a market trend and, as importantly, why. Other businesses may successfully establish value-added activities by undertaking some form of on-farm processing, with either farm shop outlet or developing relationships with a number of smaller retailing or catering outlets. However, it is unlikely that a significant proportion of farm output would be handled in this way.

Support schemes for agricultural produce have protected suppliers (farmers) against the effects of competition to some degree, but most farm businesses are small in terms of sales and few individuals are able to supply the entire needs of even a small processing plant. Thus various structures have been created, sometimes by statutory means (such as Marketing Boards) or by co-operation amongst producers, to draw together the output or commodity from a number of farms, grade it and sell it to businesses further down the chain or into final consumption with some degree of value added, consistency of quality perhaps, volume of supply, and in particular continuity. Other private sector businesses have had this long-established essential wholesale function.

It is generally recognised that producers are not usually good at marketing. Part of this is cultural, part of the explanation is the scale and nature of the farm business and part as a consequence of long periods of protection.

But it must be recognised too that the demand for particular kinds of food (in a mature market this applies even to such staples as bread) is not steady; it fluctuates with taste, with fashion, with the current vogue for health beliefs and, so far as fair commodity prices are concerned—the raw materials of the food industry—with price. There is less fluctuation in the price of process commodities than in raw materials, for most of the value added is perceived to be in the processing and distribution and packaging. Moreover, customers' perceptions respond to the climate or style created by highly professional advertising.

In contrast to, say horticulture, few farmers are in contact with their final consumers and in consequence the messages they receive come back to them from their immediate customers – merchants and wholesalers. Again in times where demand exceeds supply, wholesalers, like primary producers, exercise influence over distribution and retailers. Where the reverse is the case, market influence lies with the retailer who is responding to the final consumers. The mechanism which determines how raw materials move along the food chain is, of course, price and most public-funded farm price support schemes seek to maintain the position of raw material suppliers (farmers) by creating a 'bottom' to the market.

Retailers will seek to attract consumers to buy by reasons of quality and satisfaction, and will seek to underwrite this by working to specified standards (product specifications). The standards set down are designed to meet not only legislative requirements but also wholesomeness and safety of food and public sensitivities as to how the food is produced. The concentration of food retailing into the hands of a few large chains of supermarkets means that they need a mix of source of supply to meet their needs for volume, in the same way that food processor will reduce risk by avoiding over-reliance on any single supply source.

The individual farm business can respond in a number of ways, none of which are mutually self-excluding. Options can be reduced to two broad categories: to concentrate in efficiency in the production function, or to seek some form of value added by further processing on the farm or by entering the retailing, direct sales business.

The search for efficiency in the production function invariably leads to some Investment to add value: mechanisation or improved husbandry in order to better meet product specification standards and increase marketable yield or grinding and storage to ease marketing pressures. Such investment requires separate assessment from production, even if only by a simple partial budget so as to estimate the rate of return on that process as opposed to production. Improving husbandry, to ensure that produce is of the quality buyers require, is the first step towards improved marketing and, as food markets become more competitive, this aspect becomes more fundamental.

Market orientation is a concept fundamentally deeper than mere selling at a price; it is marathon product promotion of either the brand or generic kinds. Market orientation is about managing the production function in such a way as to meet market needs at a margin, and building customer confidence in timeliness, reliability, product quality and consistency. In order to do that, a team effort is required.

This gives rise in turn to the concept of the internal customer and the relationships between members of the team, be they family or employees. The mission goals for the business or the enterprise need to be clearly understood by all involved in the process, for the commitment of the whole team to the objectives is necessary. The individuality of those involved in farming can be a great strength – in a market-oriented economy it can turn out to be a great weakness. As the interdependence of the sectors of the food chain external to the farm is increasingly recognised, the interdependence of the members of the team within it must be understood too, requiring team management skills of a consistently high order.

For many commodities, the food chain is a long one, with many interfaces: communication, both along the chain and laterally, is often confused, leaving producers uncertain as to the signals they are receiving. It is important, therefore, for the manager to use all the information sources available, both from customers and from sources of technical and financial information and advice.

Although the developing technology of decision support systems provides an offer of assistance, it must be borne in mind that market condition rarely reflect specifics – most reflect subjective judgements arrived at by a consensual process. Those responsible for the management of farm businesses receive a constant flow of information. In a market-oriented agriculture, a major gap is some means of easing a flow of information in the reverse direction: it is interesting to ponder that farm business managers, reflecting the diffuse and diverse nature of the business of farming, have not yet addressed effectively the means of getting messages out, other than those of a political nature. Few industries of the size and importance of agriculture have much poor industry communication with the media.

In a market-oriented economy, the marketing philosophy and function is inseparable from the principles of sound business management and this chapter may be conveniently summarised by setting out some of the fundamental principles which the manager will wish to understand.

A market-oriented business is concerned with providing customers with a service—the supply of commodities to a quality and time—which meets the customer's perceptions of value and which leaves a margin to the provider.

This implies the identification and understanding of customer needs and perceptions of value, and the building of relationship negotiated in such a way as to leave both parties satisfied—the 'win-win' situation.

In a long and complex food chain, the margin performance of each sector impacts on those both upstream and downstream of it, with concentration increasing nearest to the final consumer.

In a modern food chain the value-added elements lie in the processing and distribution of food. Marketing and management rely heavily on effective communication, a two-way process of interpretation of price and quality signals.

Both the marketing and production functions will require a policy statement which sets out realistic and understood goals or objectives.

The business plan, which sets out the way in which the goals are to be achieved, will have regard to the natural advantages and disadvantages of the holding and the aims and motivation of the manager. In Implementing the plan, the manager will seek to use all the resources available to him or her, with especial attention to the human resource in the business—the internal customer.

There must be a system of monitoring the business which reveals deviation from the plan and which provides the manager with the information needed to make flexible responses.

Finally, and most importantly, those involved in the management of farm businesses must appreciate that in a market-oriented economy it is competition which makes markets grow.

❑❑❑

10

Feeding the World in the Future

The earlier chapters of this book surveyed the food needs of mankind and the ways in which the farmers of the world have endeavoured, with varying degrees of success, to meet them. This chapter ventures to look forward into the future.

People and Food in Future

Freedom from Hunger and Famine

The future of food supplies has powerful emotional significance. The ideas of hunger and famine evoke ancient and deep-seated terrors. In many areas of the earth the balances between current production and current consumption are fragile and vulnerable, and we are all too often deeply moved by tragic images of hunger, malnutrition, famine, disease and death, brought to us from areas where these delicate balances have been disrupted by drought, mismanagement, warfare and other crimes, follies and misfortunes of mankind. In addition, many people fear that the growth of human numbers will sooner or later outrun the capacity of the farmers and farming environments of the earth to produce sufficient food.

Death and disaster are news: steady and patient achievement is not. This chapter will also take into account the remarkable achievements of rural producers, which have sustained the growth of populations and the steadily improving nutritional state of ever-increasing numbers of human beings.

Food and other Biological Products

This chapter will be largely devoted to the future demand and supply of food for humans and their domestic livestock. But we look to rural producers and rural environments for much more than food. The resources which are used to produce foods are also used to produce many other biological products: fuel and timber, fibres, rubber and other technical raw materials, beverages, stimulants, medicinals, fruits and other amenity products. All these are needed for consumption and use by the people who produce them, by other people in the same nations (including those who live in towns and cities, and those who handle or process rural products), and by customers of many kinds in other nations. In many nations, as populations have increased, economies have become larger and more diverse, agriculture has become more productive, and food has become less scarce than the fuel with which to cook it.

Environments

Rural producers and others who live in the countryside also manage the rural environment. Virtually everywhere on earth, the landscapes, and the communities of plants and animals, are substantially influenced by man and his agriculture, animal husbandry and forestry. Farmers and others who live and work in the countryside are the guardians as well as the creators of these new and continually changing balances of nature.

Handler's Rules

In 1976, the late Professor Philip Handler stated three general rules which guide the behaviour of the agricultural economy of the world. They provide a fitting opening to this examination of the future. The first rule pointed out that world food production far exceeds the amount necessary to feed the world population: the problem is how to get it from where it is to where it is needed. The second rule suggested that there are no hungry people in the world who have money: in other words that the proximate cause of hunger is lack of cash to buy food or of resources to produce it. The third tells us that nowhere in the world are farmers happy about growing more food than their families can eat, unless somebody gives them something they want in exchange for the extra food they produce. So lack of effective demand restricts the output of food. These rules will guide this examination of the food needs of the world, of how they are met now, and of how they are likely to be met in the future.

This will be done mainly in terms of the cereals. Directly or indirectly, the cereals provide the greater part of the dietary intake of energy and of protein for the human population of the world, and they occupy more than half of the harvested area of crops (about 40% in developed countries and 60% in developing countries).

Data

The statistics relating to land, agriculture, population, trade and dietary supplies of energy and protein are taken from the long series of annual yearbooks on production and trade published by the Food and Agriculture Organisation of the United Nations (e.g., FAO, 2000 *Production Yearbook* and *Trade Yearbook*). Though the data for some nations, and particularly for earlier years, are far from secure, they are collected and presented in a consistent manner; and they are continually reviewed with national agencies. They are the best data we have.

The UN projections of future populations are published by the World Bank (e.g., World Bank, 1990). The data of potential capacity to support human populations are taken from the co-operative work of the Food and Agriculture Organisation, the United Nations Fund for Population Activities, and the International Institute for Applied Systems Analysis (FAO/AEZ). Data on stocks are taken from FAO's *Food Outlook*.

North and South: Change and Contrast in the Supply of Food

As populations have increased in the past, the output of food has increased also, but both people and food are unevenly distributed in the world.

Population

Between 1950 and 1988 the human population of the world doubled from about 2.5 billion to more than 5 billion persons (Table 10.1). In the developed nations, including eastern Europe and the Rusia, the increase was from about 580 million to 1,235 million, while the larger populations of the developing nations (including the People's Republic of China) increased more readily from 1,650 million to 3,900 million.

During this period, however, the course of population growth changed significantly.

Table 10.1: Population, world and main regions, 1950 and 1988; growth rates, 1950-75 and 1975-88; and percentage of economically active population not in agriculture, 1960 and 1988

	Total population (millions)		Growth rate (%/year)		%economically active not in agriculture	
	1950	1988	1950-75	1975-88	1960	1988
World	2497.566	5114.788	1.96	1.74	42	53
Developed nations	847.674	1235.303	1.13	0.73	72	91
Developing nations	1649.877	3879.485	2.33	2.09	27	39
Developed market economies	574.167	833.447	1.13	0.70	80	94
North America	166.120	272.011	1.45	1.00	93	97
Western Europe	301.446	380.949	0.77	0.33	78	93
Oceania	10.082	19.693	2.02	1.26	88	94
Developing market economies	1064.214	2682.442	2.43	2.43	29	44
Africa[a]	176.650	496.554	2.54	0.03	20	31
Latin America and Caribbean	162.110	430.171	2.74	2.21	52	73
Near East[b]	99.557	267.161	2.48	2.82	31	59
Far East	625.897	1482.375	2.29	2.24	26	39
Centrally planned economies	859.170	1598.899	1.85	1.23	36	64
USSR and eastern Europe	273.507	401.856	1.13	0.78	58	75
Asian cpes (mainly China)	585.663	1197.043	2.14	1.39	25	32

[a] Excludes Egypt, Libya, Sudan.

[b] Includes Egypt, Libya, Sudan.

In the world as a whole, the rate of growth passed its peak of around 2 per cent per year, in about 1960. In the period 1970-75, the rate of increase for the world as a whole was 1.96 per cent; in 1975-88 it was 1.74 per cent. In developed countries the rates in the two periods were 1.13 per cent and 0.73 per cent, and in developing countries 2.33 per cent and 0.09 per cent. Only in Africa (2.54% and 0.03%) and the Near East (2.48% and 2.82%) were the rates larger in the second period.

These changes were accompanied by important changes everywhere in the proportion of economically active people who were not engaged in agriculture – from 42 per cent to 53 per cent in the world as a whole, and from 72 per cent to 91 per cent in developed and 27 per cent to 39 per cent in developing nations.

Though the growth rates are declining generally and in many countries, they are still positive almost everywhere. Populations are increasing generally. It is instructive to consider why this is so.

Populations increase because more individuals are added to them by birth (or immigration) than are lost by death (or emigration).

The present growth of populations is not, in general, due to increases in crude birth rate. Though birth rates are still large in many nations, they appear to be decreasing everywhere. The crude birth rates in the low- and middle-income group of the World Bank classification (Table 10.2) declined from 41 per thousand of population in 1965 to 30 per thousand in 1987 (sub-Saharan Africa 48 and 47, East Asia 39 and 23, South Asia 45 and 34, Latin America and the Caribbean 40 and 28, and Europe, Middle East and North Africa 35 and 31). In the member countries of OECD (the North and West, together with Japan), the corresponding values were 19 and 13; and in the remaining high-income (developing) countries 36 and 30.

The man reason for the continuing growth in population at the present time is a decline in death rates, particularly the death rates of infants and children, and the consequent increases in life expectancy at birth. The crude death rates in the low- and middle-income group of the World Bank classification (Table 10.2) fell from 15 per thousand population in 1965 to 10 per thousand in 1987 (sub-Saharan Africa 22 and 16, East Asia 11 and 7, South Asia 20 and 12, Latin America and the Caribbean 12 and 7, and Europe, Middle East and North Africa 15 and 10). In the member countries of OECD (the North and West, with Japan), the corresponding values were 10 and 9; and in other high-income countries 11 and 7. In 1988 life expectancy at birth exceeded 60 years in the low- to middle-income regions except for sub-Saharan Africa (51 years) and South Asia (57 years). In the developed nations the corresponding numbers are 70 years or more.

Table 10.2: Crude birth and death rates, 1965 and 1987 (per thousand per year),and growth rates of populations, 1965-80, 1980-87 and 1987-2000, main categories of World Bank classification and selected nations

	Crude rates/1000/year				Life expectancy at birth (years)	
	Birth		Death			
	1965	1987	1965	1987	1960	1988
High-income nations	19	14	10	9	69	76
OECD	19	13	10	9	69	76
Developing	35	29	11	6	40	71
Middle-and low-income nations	41	30	15	10	49	62
Sub-Saharan Africa	48	47	22	16	39	51
East Asia	39	23	11	7	51	66
South Asia	45	34	20	12	44	57
Europe, Middle East, North Africa	35	31	15	10	54	64
Latin America and the Caribbean	40	28	12	7	57	67

The data suggest, and indeed many people believe, that these demographic changes are associated with economic growth and diversification. Economic and social development improves the material conditions of life, including many aspects favourable to health and longer life. It makes children increasingly expensive to rear and educate. While it lessens the value of the labour of children, it may create attractive employment opportunities outside the home for women. It also decreases infant and child mortality, so that the social needs of families and communities can be assured from fewer births. The balance of advantage shifts to smaller families.

Food and Population

That populations are increasing for these reasons shows that food supplies are sufficient to support the declines in death rates, particularly of infants and children, which are the main reasons for population growth. Though the phrase 'the crisis of food and population' slips easily off the tongue, it is biologically impossible, in fact, to have it both ways.

Food and Nutrition: General

This conclusion is borne out by the data for food supplies and nutritional levels published by FAO. Though there are large differences between the richer and poorer nations, the supplies of cereals per head of population, and the nutritional intakes, appear to have improved everywhere during the past forty years.

Over the period 1950-88 notwithstanding the increases in population, the average supply of cereals per head in the world as a whole increased from 306 kg to 553 kg per year (Table 10.3), but these increasing supplies were unevenly distributed in the world.

Table 10.3: Output of cereals per head of population per year years, 1948-52 and 1986-88 and imports (+) or exports (–) per head per year, 1986-88 in kilograms

	1948-52	1986-88	
	Output	Output	+ or –
World	277	353	–0.6
Developed nations	446	692	–66
Developing nations	190	245	+20
Developed market economies	460	667	–139
North America	980	1159	–391
Western Europe	247	502	–49
Oceania	673	1171	–949
Developing market economies	181	210	+25
Africa[a]	144	121	+31
Latin America and the Caribbean	193	252	+19
Near East[b]	238	259	+116
Far East	179	219	+8
Centrally planned economies	274	428	+30
Russia and eastern Europe	417	743	+85
Asian cpes	207	322	+11

[a] Excludes Egypt, Libya, Sudan.

[b] Includes Egypt, Libya, Sudan.

Food and Nutrition in the North and West

In the richer nations of the world, in 1986-88, the average output of cereals per head of population was about 690 kg per year. In North America and Oceania (Australia and New Zealand) the figures were 1,160 kg and 1,170 kg; in western Europe about 500 kg. Of these totals, net exports from North America were equivalent to about 400 kg per head net, from Oceania about 950 kg per head, and from western Europe about 50 kg per head. The average quantity retained for consumption or storage in these regions was about 670 kg per head per year.

In the Russia and eastern Europe, production per hand in 1986-88 is reported as about 740 kg per head per year, to which was added about 85 kg per year of net imports, giving a total supply of about 825 kg per head.

These supplies were sufficient in 1984-86 to ensure average intakes of dietary energy per head per day (Table 10.4) of 3,400-3,600 kcal (= 813-861 kJ, of which 1,100-1,200 kcal in Europe and North America, and 900-1,000 kcal in eastern Europe and the Russia, came from animal sources); and of 100-105 g of protein per head per day (60–70 g from animal sources in the West and about 55 g in the East). Though poorer people in these nations may well have been unable to secure as much food, and particularly as much food from animal sources, as they would have liked, these average values are considerably larger than the average minimum requirements for normal health, activity, growth and reproduction. Moreover, they represent a marked increase in intakes compared with 1961-63 (Table 10.4).

Food and Nutrition in the South and East

In the poorer nations of the world, in 1986-88, the average output of cereals per head of population was about 245 kg per year (Table 10.3). In the much larger populations of most developing countries of Africa, Asia and Latin America, the average food supplies were very much smaller, and nutritional levels lower, than in the North and West.

In the so-called developing market economies (those developing countries whose economies are not centrally planned), the average production of cereals per head in

Table 10.4: Supplies of dietary energy and protein at domestic level, per head per day, world and principal regions, total supply and supply from animal sources; 1961-63 and 1984-86

	Energy (kcal[d])				Protein (g)			
	Total		In animal products		Total		In animals products	
	1961-63	1984-86	1961-63	1984-86	1961-63	1984-86	1961-63	1984-86
World	2296	2694	363	420	62.2	70.3	19.8	24.1
Developed nations	3059	3377	849	1020	91.0	101.3	44.6	58.1
Developing nations	1942	2464	137	218	49.3	59.9	8.3	12.7
Developed market economies	3020	3362	912	1050	88.9	99.4	47.9	59.9
North America	3179	3621	1264	1222[a]	97.7	105.5	64.9	69.8
Western Europe	3088	3383	893	1122	88.7	100.8	44.8	59.8
Oceania	3107	3340	1357	1226[a]	96.9	102.4	63.2	66.9
Developing market economies	2074	2394	171	215	52.5	59.0	10.5	13.6
Africa[b]	2049	2148	133	138	51.9	53.3	9.9	10.5
Latin America and the Caribbean	2381	2705	400	455	62.7	68.6	24.0	27.8
Near East[a]	2251	3017	241	333	64.5	80.8	13.7	20.1
Far East	1967	2276	108	149	48.0	54.4	6.4	9.3
Centrally planned economies	2133	2820	269	415	59.2	73.1	14.3	21.9
Russia and eastern Europe	3142	3408	718	956	95.3	105.1	37.8	54.4
Asian cpes	1687	2616	71	226	43.3	61.9	3.9	10.5

[a] Decrease between 1961-63 and 1984-86.
[b] Excludes Egypt, Libya, Sudan.
[c] Includes Egypt, Libya, Sudan.
[d] For conversion to kJ divide by 4.184.

1986-88 was 210 kg per year only (Africa, where roots and tubers are important energy sources, 121 kg; Latin America and the Caribbean 252 kg; Near East 259 kg; and Far East [which includes the Indian sub-continent] 219 kg per head per year). (An amount of 200 kg of cereals per head per year, or its equivalent in roots and tubers, is a threshold value, below which shortages and hunger are likely to be severe, at least in some sections of the population at some times and in some places.)

These outputs were supplemented by average imports of 25 kg of cereals per head per year only (Africa 31 kg, Latin America and the Caribbean 19 kg, near East 116 kg, and Far East 8kg). Some richer nations, particularly those which export oil, are able to buy substantial imports, but many nations are too poor to buy, while the volume of food aid, though important, is small in relation to the total supplies (since 1984-85 between 10 and 14million tonnes per year).

The corresponding data for nutritional intake (Table 10.4) show an average of about 2,400 kcal (= 574 kJ) of dietary energy per head day (Africa 2,150 kcal, Latin America and Caribbean 2,700 kcal, Near East 3,000 kcal, Far East 2,275 kcal); and 60g of protein per head per day (Africa 53 g, Latin America and the Caribbean 69 g, Near East 81 g, and Far East 55 g). The total supplies, and the proportions of both energy and protein which are derived from animal sources, are much smaller than in the western countries. However, the average situation has improved since 1961-63 (2,075 kcal [= 496 kJ] of energy and 52 g protein per head per day), though there were small decreases in some African nations, mostly associated with the episodes of drought.

In the centrally planned developing economies of Asia (principally the People's Republic of China) the output in 1986-88 (Table 10.3) is reported as 322 kg per head per year, with net imports of 11kg per year only. The average nutritional intake (Table 10.4) is about 2,600 kcal (= 622 kJ) of energy and 62 g protein per person per day, mostly from plant source. All these values appear to be substantially larger than they were in the early sixties.

Though the average supplies are smaller than in the richer nations, they are not only large enough to meet average needs and to sustain increases in population but they have also been improving however unevenly, during the past forty years.

Hunger, Famine, Poverty and Politics

Though the average values in these nations are marginally satisfactory, a large proportion, and an increasing absolute number, of the population, particularly the poor, the old, the ill and infirm, the disabled, the children, and often the women, are at least seasonally hungry, and often very hungry indeed. It is these disadvantaged people, in both town and country, and numbering hundreds of millions in the developing world, who are particularly at risk if war, weather or political and economic mismanagement disrupt production, supplies and storage.

These people are chronically hungry and continually insecure because they are poor. They lack the resources with which to grow food, or the money with which to buy it. Their poverty is associated with social structures of extreme inequity. It is mostly beyond the direct reach of more productive agronomic methods except where they substantially decrease the price of food. The solutions lie mainly in the domain of politics, and they must include the achievement of more democratic governance of societies as well as of just and equitable administration.

The dramatic intensification of chronic poverty and hunger into episodes of famine is seldom due solely to natural disasters such as drought, flood or disease. Whatever their environmental aspects may be, most famines also represent a failure of government to acquire, store, transport and deliver reserve supplies. Conversely, that there has been no famine in India since 1943, in spite of some severe climatic difficulties from time to time, is a tribute to the foresight and competence of government as well as to the work of research workers, agronomists and farmers.

Output of Cereals

The increases in output per head between 1948-52 and 1966-68 were based on very substantial increases in the absolute output of cereals (Table10.5). In the periods 1948-52, the total output of cereals in the world was estimated at about 690 million tonnes (about 380 in developed countries and 310 in developing countries). In the world output was about 1,800 million tonnes (850 in developed nations and 950 in developing nation).

This achievement of the rural producers of he world has been largely unnoticed by the public in both the richer and the poorer nations, yet the ways in which it was done are of the greatest importance for our understanding of the agricultural tasks of the future.

Components of Increased Output: Harvested Area and Yield (Table 10.5)

In the developed nations, the great increase in output of cereals (by a factor of about 2.2) between produced on a more or less constant harvested area of about 300 million hectares. In the developing nations, where output increased more than three-fold, the harvested area of cereals increased by about one-third only, from about 320 million hectares to 420 million hectares. The rate of this increase seems to have become progressively smaller since.

The man source of the increases in output, in both developing and developed nations, has been a series of increases in output per hectare. Fifty or sixty years ago, in the mid-thirties, the average yields of cereals were similar in developed and developing nations, at 1,180 kg and 1,130 kg per hectare respectively. With North America in the lead, yields of cereals increased markedly in the developed nations after the Second World War. The average now stands at more than 3,000 kg per hectare. In several nations, both developed and developing, the average yield now exceeded four tonnes per hectare.

In the developing countries, the average yield decreased during the Second World War. In it was about 1,000 kg per hectare, the value reported for was more than 1,200 kg per hectare and in it appears to have exceeded 2,250 kg per hectare. The two yield curves, for developed and developing nations, are roughly parallel, but separated by a delay of about 25 years.

These data depict a worldwide process of agricultural improvement, based on more productive and more profitable methods of producing and protecting improved populations of crops and livestock. Some have seen it as a phenomenon of the developing world, the so-called 'green revolution'. In fact it began in the West under the stress of war, where it was based on scientific and technical information and methods, most of which had been built up before the war of 1939-45.

Table 10.5: All cereals, world and main regions, output, 1948-52 and 1986-88 and imports (+) or exports (–), 1986-88, millions of tonnes per year; harvested area, millions of hectares per year; and yield, kilograms per hectare; 1948-52 and 1986-88

	Output		+ or –	Harvested area (million ha)		Yield (kg/ha)	
	1948-52	1986-88	1986-88	1948-52	1986-88	1948-52	1986-88
World	691.859	1803.327	–3.284	610.079	704.137	1134	2561
Developed countries	378.387	854.580	–80.931	292.918	283.755	1292	3012
Developing countries	313.452	948.747	+77.646	317.149	420.381	988	2257
Developed market economies	264.192	555.815	–114.784	162.568	149.945	1625	3707
North America	162.884	315.349	–105.300	100.971	81.935	1614	3849
Western Europe	74.444	191.267	–18.667	45.614	43.703	1632	4377
Oceania	6.786	23.055	–18.473	6.200	14.802	1095	1558
Developing market economies	192.396	562.835	+64.467	211.180	319.401	911	1762
Africa[a]	25.409	60.316	+15.139	39.718	57.629	640	1047
Latin America and the Caribbean	31.277	108.335	+7.885	28.108	52.403	1113	2067
Near East[b]	23.687	69.196	+30.013	22.654	42.526	1046	1627
Far East	112.023	324.949	+10.982	120.700	166.824	928	1948
Centrally planned economies	235.251	684.677	+47.033	236.319	234.791	995	2916
Russia and eastern Europe	114.195	298.765	+35.853	130.350	133.810	876	2223
Asian cpes	121.056	385.912	+13.180	105.969	100.980	1142	3822

[a] Exclude Egypt, Libya, Sudan.

[b] Includes Egypt, Libya, Sudan.

As in the developed nations earlier, the spreads of new ideas and methods in China, India and other developing nations was based on the desire and ability of competent farmers to benefit from increasingly attractive economic opportunities. As development in sectors other than agriculture increased the volume of effective demand, an increasing number of farmers in an increasing number of nations perceived an expanding market for output surplus to domestic need.

Indeed, without the market created by non-agricultural development, nothing would have happened. No farmer will make investments, commit resources, and accept greater risks to increase output unless he is reasonably sure of a profitable market – as Handler's third rule points out.

In most nations in which sales off the farm increased in this way, the increase in output was strongly promoted by governments. In several important instances it was also powerfully assisted by donors and by the great American private institutions, particularly the Ford and Rockefeller Foundations. But like all change in agriculture it was carried out by farmers.

There are many reasons why these farmers (like their counterparts in the developed nations) increasingly chose to increase output and profit through larger yields rather than larger harvested areas. In many places, supplies of land and water are limited and costly. In all, the costs of power and labour, and of many other inputs, are proportional to the area treated, not to the yield. This is particularly true of irrigation. A crop that will yield ten tonnes per hectare uses water at much the same rate per day as one that will yield no more than two tonnes.

More productive methods are usually more profitable, other things being equal, on better land than on poorer land, on land nearer to communications and markets, and on land which can be made more productive without incurring an increase in rent. So as development proceeds and technique advances, land use tends to change also. In the United Kingdom, as yields have increased, marginal arable has given way to grass, grass to trees, and trees to wildlife or dereliction. We may expect similar changes in land use in developing nations in the future.

Surpluses, International Trade, Storage

The developed nations of North America and Oceania, and to a lesser extent those of western Europe, are significant exporters of cereals. These exports arise from surplus production made profitable, at least in part, by fiscal means. Of the surpluses, in recent years, a total which reached 455 million tonnes in 1987 (about a quarter of world annual output, or about thirteen week's supply), accumulated in official or commercial stores. Both the surplus output and the storage were costly and could not be sustained infinitely. The total of cereals held in store in 1990 was about 290 million tonnes (notionally eight weeks' supply), which the Food and Agriculture Organisation considers to be at or below the minimum level necessary to safeguard world food security. These figures do not include the very large quantities held in domestic storage by rural people in most developing nations.

The surpluses have been the basis of a substantial expansion of world trade in cereals. Before the Second World War, the volume of trade in cereals between the main regions of the earth was about 25 million tonnes per year only, and only western Europe was a net importer. In recent years it has reached 204-208 million tonnes (FAO 1990), and all regions except North America, western Europe and Oceania are net importers.

Among the more significant net importers are the Russia (38 million tonnes, mainly wheat and coarse grains, in 1988-89 Japan (27 million tonnes, mostly coarse grains), China (17 million tonnes of wheat), Egypt (8.5 million tonnes, mostly wheat) and a number of oil-exporting nations in North Africa and the Middle East. Most of the imports of cereals into Africa are to the North-African countries, and not to the sub-Saharan nations.

Some developing nations are exporters: Thailand of rice (about 5 million tonnes), and Argentina of wheat and coarse grains (about 10 million tonnes).

Future Demand for Food and How It can be Met

This section assesses first the size of future demand for food (how much food will be required?), and then the technical prospects of meeting that demand (can it be produced?).

Whatever the weaknesses of the statistics of the past, those for the future must be even more doubtful. So we cannot hope for precise or detailed answers to these questions. But the present purpose can be met by general and semi-qualitative estimates, which seem feasible, of the shape and size of the tasks ahead.

The future demand for food can be seen crudely as the product of the number of consumers and their average effective demand. Once again we shall concentrate on the cereals.

Population Growth

We have already seen that populations are increasing in all regions of the earth because death rates are smaller than birth rates – though both are declining.

It is possible that the present relationship between them will be affected by the spread of the human immunodeficiency (HIV) viruses. Some responsible epidemiologists expect that the auto immune deficiency syndrome (AIDS) associated with the HIVs will substantially affect the growth and size of populations in the coming half-century.

For the purpose of this section, this very serious possibility will be disregarded. To the extent that AIDS decrease the rates of growth of populations, the argument may overestimate the size of the task for rural biological production. On the other hand, the effects of AIDS on the structure of populations will make development more difficult by lessening both the numbers of effective workers and the lengths of their working lives. If these trends become established, as indeed they may, the prospects derived from the present trends will have to be altered in the twenty-first century.

Future Populations

The population projections published by the World Bank (Table 10.6) suggest that human numbers will increase from 5.4 billion in 1990 to around 6.2 billion in 2000 and 9.5 billion by 2025. The total of the 'stationary' population (taken as a crude indicators of a future plateau population)is around 11 billion. A total population of this size may perhaps be reached 100 or more years from now. Though the last doubling time for human numbers was about 37 years, the decline in average growth rates, worldwide, already suggest that the next doubling time will be about three times as long.

(A 'stationary population' is one in which age- and sex-specific mortality rates have not changed over a long period, while age-specific fertility rates have simultaneously remained at replacement level [net reproduction rate 1]. In such a population, the birth rate is equal to the death rate, the age structure is constant, and the growth rate is zero. These estimates are speculative and should not be regarded as predictions.)

The Future Volume of Effective Demand

The next stage of this discussion estimates the possible ultimate size of the effective demand for cereals, worldwide, from the population projections and the trends of the past.

This discussion is not based on normative assumption about desirable nutritional requirement: instead it is based on the likely volume of effective demand. This does in fact seem likely to be large enough to meet social goals, on the average; but more importantly, it is all that farmers, following Handler's third principle, can be expected to produce.

Since in the past harvested cereals have not continually accumulated in store, nor been destroyed on a large scale, we may take the estimates of output per head to indicate the volume of effective domestic and market demand, including subsidised surplus production and welfare purchases.

The output of cereals per head of population over the fifty-year period since the mid-thirties appears to have increased, worldwide, from 306 kg per year (developed nations 451 kg, developing nations 222 kg) to 353 kg (developed nations 692 kg, developing nations 245 kg) in 1986-88.

If we project this trend linearly forward to the period, perhaps one hundred years, when the human population may be approaching a stable or plateau size, we reach an output per head of 450 kg per year. This is equal to the output, minus net exports, of western Europe today. It therefore indicated a satisfactory average nutritional state.

Table 10.6: Populations 1987, expected in 2000 and 2050 and estimated stationary; and population support capacities at three levels of agricultural inputs; world, principal region and totals for surplus and deficit nations within each region; in millions of presons

	Population				Support capacities at		
	1987	2000	2050	Stationary	low	intermediate levels of inputs (2000)	high
World	5010	62021	9522	11271	5613	14921	33065
					(mainly developing marekt economies)		
Africa[d]	593	870	2024[a]	2919[a]	1251	4473	12838
						(excludes S. Africa)	
North Africa	108	149	275[b]	321[c]	167	202	250
West Africa all	207	303	770[a]	1081[a]	414	1511	4511
Potential deficit	139a	•204[a]	531[b]	701[b]	82	299	1165
surplus	68	99	239	380[a]	332	1212	3346
South and Central all	53	79	187	278[a]	169	747	2585
Potential deficit	12a	18[a]	41[b]	61[b]	10	32	135
surplus	41	61	146	217[a]	159	715	2452
Eastern Africa all	179	272	665	1097[a]	445	1815	4928
Potential deficit	99a	153[b]	383b	644[c]	41	238	534
surplus	80	119	282	453[b]	404	1677	4394
Islands, Indian Ocean	14	19	46	54	56	198	562
Developed (S. Africa)	33	45	82	90			
North and Central America	410	478	615	639	298	562	1295
Small nations	1	1	3	4	7	9	20
Potential deficit	22a	32a	54b	68b	22	44	112
surplus	116	148	249	242	269	510	1165

Developed nations	270	298	310	324			
South America	277	350	528	563	1418	5287	12275
Asia	2925	3658	5452	6245	2446	4597	6657
						(countries shown only)	
Near East all [e]	185[a]	264[b]	516[c]	702[c]	181	239	326
Potential deifict	128[a]	192[c]	407[c]	584[c]	86	116	164
surplus	53	67	102[a]	111[a]	93	121	159
Developed nation: Israel	4[c]	5[c]	7[c]	7[c]	2	2	3
South Asia	1080	1407	2265[a]	2789[b]	1756	2449	3512
Potential deficit	227	326	633[c]	900[c]	376	430	491
surplus	853	1081	1632[a]	1889[a]	1380	2019	3021
South-east Asia	379	481	760[a]	810[a]	709	1909	2819
Potential deficit	68[a]	91[a]	154[b]	171[b]	59	134	232
surplus	311	390	606	639	650	1775	2587
E. Asia (developing only)	1281	1506	1911	1944			
Developed nation: Japan	122	128	126	113			
Europe	496	502	505	474			
Eastern	117	121	133	137			
Western	379	381	372	337			
Russia	283	307	363	398			
Oceania	25	30	34	43			

[a], [b], [c] Population greater than potential support capacity at low,intermediate and high levels of inputs, respectively.

[d] Excluding Egypt, Libya, Sudan.

[e] Including Egypt, Libya, Sudan.

Put another way, if the developmental trends of the past half-century continue at the same rate for another century, the effective demand of mankind for cereals may be 450 kg per head per person. This estimate could be altered by selecting a non-linear trend, though how the coefficients might be selected is far from clear. In fact the linear trend will serve well enough for the present crude purpose.

Output, Area and Yield

If this estimate is multiplied by the number of people the world may contain a century from now (11 billion) the result suggests a total demand for cereals of 5 billion tonnes, against a present output (estimate) of 1.9 billion tonnes.

The same simple reasoning, applied to the harvested area of cereals, yields an estimate, 100 years from now, of about 1,000 million hectares.

The consequent agronomic task is to secure a harvest of 5 billion tonnes from 1 billion hectare. The average yield in the world in was rather more than half of this value; and it was 2.2 times as large as the average suggested for the mid-thirties. In fact a linear projection of the yield trend suggests a yield of 5.3 tonnes per hectare a century from now.

A yield of 5 tonnes per hectare is well below many estimates of the potential yields of cereals. It may be compared with average yields of cereals in 1986-88 of 4 tonnes in China and in eastern Europe (not including the Russia); of 5.7 tonnes in Japan, and of 4.4 tonnes per hectare in the United States and in western Europe. Technically, it is possible to fell a measure of confidence about the possibility of producing 5 billion tonnes of cereals, especially as there is a century in which to find out how and where best to dot it. It will require continuing advances in the techniques of production in suitable environments (considered below), and this in its turn will require larger supplies of plant nutrients, large inputs of mechanical or muscular energy, and more effective means of protection of plants and animals against pests, diseases, competitors and predators.

Some may doubt whether the human species has the political, economic and managerial ability to achieve growth in output at this rate in the future. Competent government can powerfully aid agricultural progress, but in fact in many countries rural producers have been able to increase output to meet market need, even where government has been less than fully effective. Perhaps they can be expected to show the same ingenuity and determination in the future.

It is not feasible to determine in detail where the additional output will be produced at different times in the future, or where it will be utilised. To do this would require prediction about the evolution of demand on time and space, and about the technical evolution of production systems in different regions and nations.

However, it seems likely that most of the output will be produced in the presently-developing nations, in tropical, sub-tropical and winter-rainfall environments. A substantial part of it will be produced in Africa. Within these regions, the prospects are most promising in those nations which seem most likely to have surplus productive capacity.

Potential Biological Productivity of Environments in the Developing World

Studies over the past thirty years suggest that the environmental resources of the earth are indeed sufficient to sustain the expected future populations. In the seventies, the Food and Agriculture Organisation of the United Nations (FAO) began work on the potential productivity of the environments of the developing countries (other than China and some nations of eastern Asia, for which data were not available; work on the land resources of China is now in progress as a co-operative endeavour between FAO and the State Land Administration of the Republic).

Agro-ecological Zones and Population-supporting Capacity

The work started from the Soil Map of the World, a classical joint endeavour of FAO and UNESCO. Climate information (seasonal course of temperature, rainfall and evaporation, leading to a standardised measure of the length of the growing season) was overlaid on the sheets of the soil map. The combination delineated about 46,000 unique agro-ecological cells, for which generally accepted agronomic and physiological methods led to an estimate of potential production of biomass and yield for the different sorts of economic plants (including herbage and pasture) which might be grown.

From the estimated potential yields (including those of animal products) it was possible to assess, using conventional nutritional standards (2,350 or 2,400 kcal (562 or 574 kJ)/person/day and an appropriate supply of protein, depending on its source), the numbers of people in each nation who could be supported technically at three different levels of agricultural inputs (FAO/AEZ). These levels were 'low' (the present standard of husbandry, whatever its actual technical level), intermediate (improved populations of plants and animals, more productive use of fertilisers and improved equipment, improved measures to conserve water and terrain, and improved methods of protection against pests and diseases) and high (the most advanced technical methods available for each agro-ecological circumstance at the present time).

The totals of the estimated technical human support capacities of the environments of the nations studied were 5.6 billion at the low level, 14.9 billion at intermediate level, and 33.1 billion at the high level (Table 10.6). These numbers do not include the evidently substabtional support capacities of the environments of the developed nations or of China. They are to be compared with expected populations for the whole world of 6.2 billion in 2000, 9.5 billion in 2050, and ultimately of perhaps 11 billion people.

Each of the main regions (including Africa) of the 'market economy' developing world, other than the Near East, would be technically able to sustain its maximum expected population at the present average level of inputs, if it could farm a sufficiently large area of land and distribute the products freely. If the Near East (as defined by FAO, and including Sudan and Turkey) were unable to import food, it would notionally have to farm part of its area at the intermediate level. At higher levels of inputs all regions, taken as whole, have surplus environmental resources sufficient to meet their needs for other biological products as well as for food.

More detailed work has been completed or is in progress in many individual nations. Increasingly, the work is being done by national scientists, co-operating with FAO. So far, it has confirmed the indications of the earlier study.

Potential Surplus and Potential Deficit Nations (Table 10.6)

Though each region seems able, technically, to meet its sown needs for biological products, the situations in different nations are more diverse. Some nations (like India) include both potential surplus and potential deficit area, and meet their needs by internal exchanges. Some nations are potential surplus producers, and seem, likely to continue so. Others are potential deficit nations, now or in the future, and are likely to meet difficulties, sooner or later, in meeting their food needs (let alone their needs of fuel, timber and other biological products) from their own resources, even at higher levels of inputs. Some potential deficit nations have arid environments, particularly in the Near East, and north-east Africa, and on the southern and northern fringes of the deserts, in North and West Africa, and in Asia and the America. Others will meet difficulties because their populations will become large in relation to their resources.

West Africa and the Sahel

The well-publicised difficulties of the Sahelian nations of West Africa since were produced at least as much by the inability of their governments to exploit their historical and geographical advantages, as by the episodes of drought. Seven of the eight Sahelian nations have a common official language, a common administrative tradition, and a common currency linked to the French franc, with the wetter Francophone nations to the south. Historically, their peoples have traded together for centuries (as well as with their presently-anglophone neighbours); and indeed the trade in food continued through all but the longest drought periods.

To a large extent the 'Sahelian problem' is an artefact of political history, of the official balkanisation of West Africa when the provinces of French West Africa became independent. This was reinforced by the international practice of treating individual nations, however small, as separate units for diplomacy and aid. Whatever it may be economically sensible to do to increase the domestic output of food in the Sahel nations, the established complementarities and trade between them and their southern neighbours are an essential component of their future. In marginal and fluctuating environments like those of the Sahel, 'self-sufficiency can have an unrealistically high price.

Several nations of Africa are potential producers of very large surpluses. Among them are Cameroon, Central African Republic, Chad, Congo and Cote d'Ivoire in

West Africa, Angola, Mozambique and Zambia in southern Africa, Sudan, Tanzania and Zaire in eastern and central Africa, and Madagascar in the Indian Ocean. At no more than the intermediate level of technology, the environments of these twelve nations alone could technically support marathon 3 billion people. The total of their expected 'stationary' populations is fewer than 1 billion people.

In fact, compared with India, which has a similar range of environments, Africa is sparsely populated and has substantial under-used natural resources for biological production. The existing levels of output are still on the steeply-sloping limb of the response curve to yield-enhancing innovations. In so far as the 'Africa food crisis' exists at all (which the growth of population may lead us to question) it is developmental, political, managerial and logistic, rather than environmental.

Exploiting the Complementarities

To summarise, the results of the agro-ecological zones study suggest that provided food can move freely, the environments of the world have enough productive capacity to supply food for the maximum expected populations of the future, plus substantial and apparently adequate quantities of other biological products. There are surplus and deficient area, between which complementarities and exchanges will support international trade.

To develop these complementarities between nations is not, however, a task for farmers or agronomists alone: it requires action in the dominions of politics, economics and management, and engineering. Complementarity requires physical communications, which have to be planned, designed, built, maintained and operated, and economic management. Above all it requires law and order, a measure of political and economic stability and competent management, and peace.

Change in Climate

The conclusions of the AEZ study are based on the climates of the world as they were in the year before. Many people believe that the observed changes in the composition of the earth's atmosphere, during the past century and more, have increased the temperature of the earth by perhaps half a degree celsius, and that the continuing changes that are expected in the future will lead to significant global warming. It is suggested that sometime in the coming century the average temperature of the earth will have risen by 3°C ± 1.5°C, and that this will adversely affect weather and climate, accelerate the melting of polar ice, and raise the sea level so that some low-lying areas of the world are submerged.

Not all observers accept this prospect. To be sure, average temperatures have varied by several degrees, over periods of hundreds or thousands of years, in the past. But compared with the variations in average temperature which are associated with differences between day and night, between seasons of the year, between high and low elevations, between different aspects at the same elevation, between urban and rural environments, between different latitudes, and between different years and sequences of years, so small a difference as 0.5°C, over 100-150 years, has been far from easy to identity unequivocally. Even the suggested average future global warming of 3°C ±1.5°C would not be easy to demonstrate with certainty against so variable a background.

Though it is possible that average temperatures may increase, few meteorologists are confident about the sizes, and the distribution in time and space, of temperature changes in the future. Some are even doubtful about their sign, suggesting that increased cloud may lessen the radiation income of the earth. We do not yet know enough about the temperatures of the oceans, or the physical and molecular exchange at their surfaces, to construct reliable predictive models of future climate. Some doubt even the physical basis of the suggested heating.

As to the level of the sea, satellite observations do not clearly support the suggestion that the polar ice is systematically retreating year by year; and it has also been pointed out that in many places the tidal variation in sea level are far large than the average predicted rises.

It is even more difficult to predict the effects of the suggested changes on biological production, which is also very variable and is affected by many influences other than temperature and the concentration of carbon dioxide in the air (including rainfall and its distribution, incidence of storms and frosts) and production technique.

On balance it seems most probable that the warming trend, if indeed there is one, will evolve gradually over time, even if the rates at which we burn fossil fuel and forests are not decreased. It will certainly be closely and publicly monitored. Societies and their governments will have time to plan appropriate co-operative responses and reaction. But even if the distribution in space of the potential biological productivity of the earth is altered, it seems likely that for the world as a whole, and for the main regions, it will still be sufficient technically to meet foreseeable human needs.

In has been suggested that changes in climate in marginal regions may over time exceed the considerable demonstrated capacity of the people who live in them to adapt, and that it will consequently lead to the movement of 'climatic refugees' in large numbers to more favourable environments, often in cities. The tendency of people to leave marginal environments, particularly in climatically adverse periods, as development creates alternative ways of life is historic, worldwide. In West Africa it was well under way before the arid episodes of the seventies. Nonetheless, it is possible that climatic change in the future could intensify the trend.

Achieving Potentials: Resources and Constraints for Development

To utilise the technical environmental potentials indicated by the AEZ study, through a century of substantial changes of many kinds (including possible shifts of the margins of climatic zones), will require adjustments far outside the field of agricultural science and technology. Past experience suggests that seven groups of factors offer enabling resources or retarding constraints to change and development in biological output and in the rural space. This section offers an annotated check-list of these groups of factors.

Effective Demand

The most potent determinant of productive change in agriculture and rural life is effective demand in an accessible market-place. Where rural people have been able to respond to effective demand, they have often done so spontaneously, using resources, knowledge and skills which they already possess.

Conversely, if there is no effective demand, rural people will produce no surplus, no matter how excellent the research, how determined the campaign 'to increase production', or how devoted the services of extension and advice. They cannot afford to produce output surplus to their over-year domestic needs unless they are assured of an acceptable return, a sufficient profit, on the resources committed. Small-scale producers, even if their methods seem to be primitive and unproved, are at least as rational economically as their advanced, large-scale counterparts in more 'developed' societies. They could not otherwise survive.

Indeed the most effective way, and often the only way, to increase the output of agriculture and advance the income and welfare of rural people, is to increase effective demand for surplus rural products by developing non-agriculture. Though 'agriculture-led development' is proposed as a recipe for 'the African problem' by some observers, it has never been seen in history except where (as for example in Denmark or Canada long ago) a nation was able to sell agricultural surpluses profitably to one or more industrialising or industrialised trading partners. The road to development is through balanced national and international growth and diversification, not through specialisation and priority support for agriculture or food production alone.

There is nothing new in this: the proportion of the world's economically active population which is engaged in agriculture has continually declined through history (Table 10.1) as development has proceeded. On balance, the 'drift to the towns' has had positive consequences, even where it has not been supported by deliberate planning. Where it has been so supported, as it appears to have been in Japan, the results have been spectacular. Conversely, the attitudes which have seen 'agricultural development' as a means of 'holding the population on the land' have been self-defeating almost everywhere.

Output Delivery Systems

Demand cannot be effective unless the systems of delivering surpluses to the ultimate users and consumers are in good technical and economic order. These systems include roads, railways, bridges, waterways, trucks, boats, trains and the fuel, skills and spare parts with which to operate and maintain them; market-places and market masters to keep track of buyers and sellers, quantities and prices; merchants and marketing systems; storage and processing; and the essential legal, financial and administrative procedures and institutions to facilitate the operation of the systems. The countryside has to be equipped logistically and managerially for development

Westerners take these things for granted; these are incomplete or ineffective in many developing nations. Surplus foodstuffs may rot by the roadside for lack of transports to market, while in the distant city the same products are scare and expensive. Conseversly, among the many visible signs of development on the march are the bullock carts, or the heavily laden trucks, making their way to market.

Many countries, including some apparently developed ones, have sought to manage the rural output delivery system through state or parastatal institutions. In 1991 it is a matter of public record that these systems have everywhere failed, particularly where a centralised, often single-party state has attempted to remove enterprise and competition from the systems and has all too often replaced it by inefficiency, waste and corruption—individual, institutionalised and often politicised. Whatever faults commercial systems and commercial banking may have, they do not seem to be as pervasive, or as difficult to eradicate, as the weaknesses of publicly- or politically-controlled monopolistic or monopsonistic in situations. In ideal theory, officials and politicians can be competent and honest servants of the public, but the reality has too often been otherwise.

Customary methods of rural domestic storage are extremely significant. In many parts of Africa, it has been found that traditional populations of cereals are far more resistant to insect pests in traditional domestic storage than 'improved' varieties which (though they may be potentially larger yielding and resistant to pests and diseases in the field) have not been bred for storage attributes.

The output delivery system also includes processing and general adaptation to the needs and preference of consumers and users. To produce tonnage is not enough: the tonnage has to be transportable, storable and acceptable. Some post-harvest industries (of which cotton ginning is significant example) are best located in the rural area, where they help to diversify the rural economy and contribute to development in the rural space.

Resources Available in Life Systems to Increase Output

To respond to effective demand and a sufficiently satisfactory output system, rural people need resources with which to produce surplus output for the market, including land, water, cash, seed, equipment, power, fertilisers, seed dressings and other protection materials, packing materials and knowledge. Any or all of these may be inherently scarce, expensive and difficult to get, but without them, increases in output must be limited.

In less-developed, less specialised societies, some of these resources are also needed for purposes other than biological production. In such societies, agricultural production competes with other sectors of the rural life systems for scarce resources, particularly labour, time, attention and cash. Time and labour spent fetching water or searching for wood cannot be used to grow rice. Cash spent on school fees cannot be used to buy seed dressings or improved equipment.

It is often suggested that poor farmers in developing countries cannot afford purchased inputs. This is indeed true if there is no market for surpluses, so that no cash flows in the production and output systems. But experience in all countries, including India and many other developing nations as well as those which we now regard as developed, tells us that poor, small-scale farmers are ready to invest judiciously in purchased inputs where they see profit in doing so.

Improved seed, and planting materials of vegetatively reproduced crops, are often particularly scarce. In the developed countries, we take the seed industry and the laws and procedures related to it for granted. In many developing nations they do not exist. As soon as sales off the farm become profitable, would-be producers want good seed. All too often they cannot get it, even though excellent improved varieties, apparently adapted to the needs of both consumers and users, may have been bred and released. If there is no seed industry, money spent on breeding can easily be wasted. Conversely, where improved seed is available, as it is in most developments of cotton and other commercial crops, it can powerfully support trends towards profitable change.

Power is yet another essential resource for increased output. Even in densely populated regions, bullocks and water buffaloes are widely used, both for field operations and for draught. But as development has lessened poverty, and so increased the cost of labour, or where labour is already scarce and expensive, mechanisation of field operations may become important. Japan and Thailand are well-known examples. We may expect all these trends to continue, even in a world in which energy becomes increasingly expensive, because of the priority that most societies are compelled to give to the production of food.

Economy in the use of inputs, particularly which cost money, is natural to all producers. But there may be a price to pay in the volume of output. It is seldom easy to get something for nothing. The ideal, of a low-input, high-output system of farming which is also sustainable, has generally proved far from easy to achieve, though there are some instructive approaches to it in some indigenous systems of India and some of the more densely populated regions of Africa. The best-known low-input example, the mechanised wheat-annual legume system of seasonally-arid south Australia, is profitable at a comparatively low level of return (data for Oceania in Table 10.5), given efficient support service, considerable production skills and reliable export markets.

More Productive Technical Methods

Given effective demand, access and resources, the producer will wish to increase his profits by making his farming more productive. He may be able to do this, as he has often done in the past, by using his substantial stock of existing knowledge.

Some observers assume that farmers whose yields are much smaller than those usual in developed nations must be ignorant, and probably doggedly resistant to change into he bargain. Yet their systems of production are usually found to be very supported to their natural and human environments. They have supported substantial increases in population over many generations. Like rural people everywhere, these producers have a vast stock of experience of crops and environments, and moreover, they are both imaginative and adventurous about using and developing it.

One striking example will illustrate this. Most visitors to Africa, asked to name the principal crops of the continent, will start with maize, cassava, sweet potato and perhaps cocoyam; they may then go on to *Phaseolus* beans and peanuts, coca and cotton, *Capsicum* (peppers) both sweet and pungent, and pumpkins. All of these are America plants. They were introduced to Africa by Portuguese and other trades, and incorporated long ago into complex production system (including mixed and relay cropping, often with trees and shrubs), adapted to a wide range of environments. This was done by illiterate rural producers, without benefit of research, education or extension. The visitor must surely temper his technological self-confidence in the face of such achievements.

However, it does not seem likely that existing knowledge alone will be enough to support the considerable increases in output and yield that will be needed to meet effective demand in developing countries, particularly in sub-Saharan Africa, over the twenty-first century and more. Producers will need more productive technical methods, based on new knowledge as well as old.

The technical options of the so-called new technologies, from which producers are intended to choose those that best suit their circumstances and needs, are produced by research is to adapt to local conditions methods and skills that have succeeded elsewhere. Experience, particularly in the international agricultural research system, suggests that this is too narrow a view. The natural and human environments of most developing nations, as well as the plants and animals produced in them, and their pests, pathogens, predators, parasites and competitors, usually differ substantially from those of the developed nations in which modern agricultural science and technology evolved. Consequently adaptation is seldom enough: new and locale-specific science and technology are usually needed.

The range of enquiry necessary to design acceptable and profitable innovations covers at least ten fields of natural and human science. They are outlined mainly as they relate to plant production, but comparable considerations apply to animals and other biological products. This section also serves to summarise much of the earlier discussion.

Of the ten fields, two, the first and last, are concerned with the developmental background. They provide the frame within which the new methods have to work. The remainder lie in more accustomed technical areas.

The fields are:

(a) *Policy and planning:* Research is needed to guide the formation and execution of policy and plans for development in agriculture and the rural space, appropriately articulated with national plans for development in other sectors, and preparing for continuing change in the future. This work has more than the customary economic, political and administrative elements: it needs a strong technical content also. Its results also help to determine, and particularly to quantify ('how many tonnes do you want?') the objective of technical research.

(b) *Environment and renewable natural resources (place):* This field is managed by means of geographic information systems. It includes climate, weather, and agro-climatology; terrain, soils, land survey and management, determinants of soil fertility; botanical and ecological survey and mapping; water, hydrology, drainage, groundwater, development and management of water resources, irrigation, oceanography; ecophysiology of plants, nutrients and water, and potential productivity; management of the natural environment for increased output, together with such measures of protection and conservation of soil and terrain as are needed to ensure that productive ways of using environmental resources can continue to be developed and sustained as they are required.

(c) *People:* Description and analysis of existing life systems and their social and technical rationale; diagnosis of components which promote or constrain increases in output; social aspects of development and communication.

(d) *Protoplasm:* Diversity, adaptation, and utilisation of existing and potential new populations (not 'genotype') of economic organisms; genetic resources; methods of improvement (including manipulation at molecular level) for yield, quality and protection; procedures for improvement, testing, multiplication and distribution.

(e) *Protection:* Identification and biology of pests, diseases and disease organisms, competitors and parasites; methods of control including heritable resistance and tolerance (perhaps the most important gifts of science to producers in developing nations), as well as chemical and biological control, in integrated systems designed to hold losses and damage within acceptable limits; management of effects of control methods on no-target organisms, and of undesirable residues in the environment.

(f) *Production:* Design and testing of sustainable systems of production for particular assemblages of organism in particular place. This is a synoptic and largely local-specific agronomy, which brings together the results from many other fields and combines them in ways which are adapted to specific types of environment. It seeks to assure the evolving volumes of required output at acceptable levels of cost and return, to limit the area of land and the quantities of area-linked inputs needed to produce these volumes, and to promote desirable changes, and limit or prevent adverse changes in the natural environment, associated with increasingly intensive production.

(g) *Post-harvest technology:* Preparation, packing, transport, logistics, storage, quality, processing, marketing; post-harvest industries in rural areas.

(h) *Power:* A central topic as yields increase, production becomes more intensive and the sources and costs of power change.

(i) *Price:* Economic investigation, advice and monitoring concerned, at the level of research and production, with the selection of objectives, the costs and returns of different possible option and the design of production and marketing systems; and at the level of regions and nations with the relation of rural biological production with economic activity in other sectors and places. Not all established agricultural and related biological research systems in developed countries include economic topics; but it seems evident that an economic component, socially and intellectually integrated into the research system, can greatly increase the usefulness of research in developing countries.

(j) *Performance:* Continual monitoring and analysis of the technical, economic, managerial and social elements of development in agriculture and the rural space, with feedback to policy and practice in both research and development.

Where producers adopt new options, they do so, not to maximise yield but to maximise profit for the producer, at a price in the market-place which is acceptable to both producers and purchasers. No farmer will see advantage in pursuing the upward course of a diminishing returns curve further than the market price, or the system of fiscal support, makes it profitable to go.

The ten fields outlined above are fields of research and development. National research services have not everywhere succeeded in delivering attractive and realistic options for rural producers; and this has led to searching questions about the effectiveness of agricultural research in general.

It is therefore useful to remind ourselves that in both developed and developing nations, new methods developed by research on commercial commodities (such as tea, coffee, cocoa, cotton, rubber, oil palm, tobacco, pineapple, banana and other fruits, and sugar cane), particularly for plantation managers and contract growers, are usually both readily-adopted and profitable. In such instances the research system works to known and particular specifications of output needs and qualities, adaptation, feasibility and profitability, usually determined in consultation with both the producers and the users of the product; and that its recommendations should be extensively and critically field-tested in co-operation with the producers and other users.

In most developing nations, research on food crops lacks most of these favourable features.

Policies and Practices of Governments

The actions of government can promote or impede progress in rural biological production; and the more powerful and organised the government, the larger are its helpful or harmful effect. Some consequences of government intervention in the post-harvest phases have been mentioned above. State interventions in the processes of agricultural production have generally been equally unfortunate.

The most important effects of government, for good or ill, come from the nature of its policies for development. To increase rural output and advance the welfare of rural people, it follows from what has been written above that preferable markets for rural products must be developed and increased. Some of these may be found in foreign nations, but the most important markets, in many nations, are those created by non-agricultural development in the nations themselves.

It then follows that investments devoted to rural and agricultural change alone, not organically associated with development in other sectors, are likely to be wasted. This was well demonstrated in those nations whose governments, sometimes supported by foreign advisers, opposed the development of industry in the fifties and sixties. It was argued that investment in agriculture would create more workplaces per thousand dollars spent, than similar investment in non-agriculture.

But the argument failed to recognise first, that industry produces cement, steel, trucks, pumpsets, railway engines, generators, dams, and many other products and structures essential for development in agriculture and the rural space, and that it also helps to create the cadre of technically skilled people who are so necessary for all facets of development, rural and non-rural alike. Second, industry increases the markets for rural products without which there can be no growth in rural production above subsistence.

Fortunately a number of developing nations succeeded by paying no attention to this sort of advice. In some of the most successful examples, investments in non-agriculture were promoted in or near the rural areas themselves—sometimes in the many industries and associated services which are based on rural products, particularly storage and processing. This dispersed development, increased rural employment, and offset the growth of towns and cities.

To be sure, investment in non-agriculture alone can be equally self-defeating, though it seems to have worked well enough in Singapore and Hong Kong, each of which is linked with a rich agricultural hinterland.

Though government has a necessary enabling role in planning and guiding development in agriculture and in the rural space, the hard work on the ground is not an executive task for government. Hard experience tells us that it is essentially a task for the private sector, in which that redoubtable private entrepreneur, the 'small farmer', is a leading figure. Other participants are local merchants, transport contractors, builders, mechanics and engineers, in the informal sector which achieves so much in most developing nations.

Behind these front-line troops are the larger, often multinational firms, indigenous and foreign. Much is made of their possible negative roles; but they have dominantly positive ones based on their own self-interest in expanding economies, and often on significant historic links with development in the past. Many governments of developing nations have been able to engage these partners productively in the development process.

The art of government, as in so many other matters, consists in balance, in pursuing multiple but compatible objectives simultaneously, in promoting strategic investment in infrastructure, and in detached guidance and management, rather than direct participation, in developmental proceses. Government can also play a significant political part in lessening the social and political bases of inequity in rural societies, though this road is hard, as many nations have found. To lessen inequity, it may be necessary to dismantle long-established social structures on which wealth and privilege rest; and this is not likely to be an easy task for any government.

It is essential, furthermore, not to confuse political slogans, which help to win elections and referenda, with practicable policies for development.

Foreign Relations

Severe constraints on agricultural development can arise from difficulties in foreign relations. Indebtedness and shortages of foreign exchange are perhaps the most important, because they restrict the importation of goods and materials needed for development, both rural and non-rural. Next come the related difficulties in export markets, particularly those of the developed nations. The weaknesses of some developing nations in these markets have to do with cost and quality, but the general background must be increasingly difficult as populations in the richer countries cease to grow, and as their demand for the products of developing nations becomes inelastic or is met from domestically produced substitutes. In some commodities, the historic relations of dependence on developed nations must be close to its end.

The distribution in the world of people and of environmental natural resources suggests that part at least of the economic future of some developing nations will lie on the further growth of the already significant trade among complementary developing nations, in both rural and non-rural goods and services. After all, a century from now, perhaps 85 per cent of the world's consumers will live in these nations. Collectively, they already offer substantial markets, even though individually many of them are poor.

Knowledge Systems

Actions for development in agriculture and in the rural space in the six sectors outlined above require the productive use of knowledge, new and old. This is not the place to expand this large theme. It is perhaps enough to record that knowledge consists of information, concepts, techniques and skills; and that it is accumulated and stored, increased, developed for use, disseminated and used, in increasingly international knowledge systems whose components range from the minds and memories of rural people to the great international computerised systems for information management.

The most important repositories of knowledge relevant to development in agriculture, in most nations, are the rural people themselves. Their systems of life and of production embody this knowledge, and consequently the description and analysis of these systems is a necessary starting point for research and other actions intended to help rural people to make them more profitable and productive.

The content of the research (knowledge-increasing) sector of the knowledge system has been outlined above. For many years, this sector has been powerfully promoted by official and private agencies in developed countries, who often seem to believe that more research is the main element the developing countries need to increase the output of food and lessen rural poverty. Whether or not that view is correct, leads at once to the question of relevance.

The standard body of science and technology concerned with biological production, worldwide, is essentially a Western product: it is based on the human and natural environments, and the experiences, of the presently-developed nations. Many of its general principles, such as the generalisations of genetics or biochemistry or Koch's postulates, are universally valid. But their application is often location-specific. So, although the physical bases of rainfall, runoff, infiltration and the evaporation of water are universal, the seasonal rainfall regimes and the nitrification and leaching cycles of the soil of the semi-arid tropics are the mirror image of those of the temperate region.

Such profound differences can lead to severe difficulties for the visiting expert, who may think that the tropics are no more than a warmer, wetter, and more backward version of the human and natural environment back home, or may seek to apply the experiences of Berkshire, UK to explain the agronomic differentials of Botswana. But the matter goes further.

Much of the content of applied biological education and training in tropical and sub-tropical countries is in fact little more than a transliteration into local terms of the curricula of the prestigious institution in temperate regions in which the teachers themselves, or their teachers before them, were trained. There is a substantial task ahead to assemble and codify the often very substantial stock of locally-relevant information, both customary and research-based, to assemble it into monographs and texts, and to use it to construct curricula which are more deliberately designed to support development in each nation.

For the rural producer, the potentially most important sector of the knowledge systems is usually the extension service. The extension service is too often seen as a means of delivering 'new technology' to expectant and grateful producers. But its tasks are potentially far wider. It should be able to assess the needs, purposes, resources and constraints of rural people, help to determine the objectives of research, test possible new options alongside the producers, and then help them to use those which they choose to adopt.

All too often, the reality is otherwise. Extension has often proved disappointing in developing countries. Rural products have failed to sustain, or even to adopt, the new technology it offers. The reasons are seldom far to seek. Sometimes, it is not to the producers' advantage to increase output at all. Sometimes, the new technology does not do what it is claimed to do. Sometimes, it is not suited to the circumstances and needs of the producers. It may not fit into their patterns of competing uses of scarce resources – for example, it may call on a producer to plant one crop when he is pressed to sow or weed another. Its products may not meet the needs of the consumer or user. In such circumstances, the produces will not respond: they are quite unwilling to do things which are not to their advantage.

The main field for successful extension opens when producers perceive that an innovation is potentially profitable for them and wish to learn how to use it. In these cases the success of extension may be spectacular, and the particular method of communication or organisation used by the research or extension workers usually gets the credit.

The apex institutions of the knowledge system in any nation are the universities; and in the universities the most important potential contributions ot development are in the quality of their courses for first degrees. The prestige, and the potential for attracting external financial support, of research and postgraduate teaching, make it easy to forget that in most nations, most of the senior management posts in politics, in the civil service, in external aid organisations, and in commerce, are occupied by graduates who hold no more than first degrees. If these graduates have not been educated and trained to understand the nature and needs of development, particularly in their own countries, the governmental and political structures will inevitably be weak.

In donor nations, this weakness will affect the quality of development co-operation. In developing nations, the capacity of government and the indigenous private sector to plan and execute development, and to make the best use of external co-operation, will be diminished. Either way, the task of creating the enabling environment, in which the world's rural producers can assure the foods and other biological products mankind will require in the present, will be made more difficult.

The Prospects of Change

So far, this chapter has reviewed the present state of food supplies and nutrition in the world, the needs of the future, the environmental resources available to meet those needs, and the main factors which support or retard progress in agriculture and in the rural space. This final section draws on the past to foresee the nature of some of the changes to come and some of the consequences.

Population, Agriculture and Environment

Before farming was invented, about ten thousand years ago, much of the world was covered by forest and woodland. People were few and scattered, and they had far less effect on landscape, terrain, vegetation and other species of animals than natural forces like rainfall, frost, flood, wind and fluctuations of climate.

As human numbers increased and agriculture spread to all regions of the earth, profound man-made changes to vegetation and terrain followed. Five thousand years ago, most of Europe was covered by forest, the Urwald of Teutonic legend. Today most of the forest has been replaced by man-made landscapes, many of remarkable beauty and aesthetic value. Though much of the vegetation of the earth is wild, remarkably little of it is natural in the sense of representing what was present before men and women began to farm.

In some regions, the consequent exposure of erodible surfaces to erosive wind and rain has altered terrain. Many fertile valleys are filled with the alluvium of the formerly-vegetated hillsides, from which farming and livestock have removed both plants and soil. In the lands, from China to Pakistan, which revive the runoff from the great mountain ranges of central Asia, as in the Andes and Amazonian, accelerated natural erosion dramatically affects both the source regions and the seasonally – flooded slit- and gravel-filled valleys and flood plains.

In many parts of the world the increasing need of increasing numbers of people for fuel and timber has cleared woodland and forest faster than it can regenerate unaided. In others, rural people have not been able to alter their systems of land use fast enough to limit losses of soil, maintain the supply of plant nutrients, and sustain the output of crops and stock.

The Achievements of Agriculture

In all these ways, man and his agriculture have changed the appearance of the world. Yet the output and yields of crops have greatly increased, and the increase in human numbers has bee sustained. In part, this is associated with improved techniques, including better-adapted and more productive populations of plants and animal, mechanical aids of many kinds, and irrigation.

Many of these improvements were invented thousands of years ago. But in the past three centuries, invention has become organised into science and technology, from which it has received powerful support in return.

So for example, the yield and output of crops in developed nations are far larger nowadays than they were in the thirties. More productive farming is not only profitable: it has increased the biological productivity of its environments. As an example, the fertility of soils in Britain has been greatly improved. The areas of lime- and phosphate-deficient agricultural soils are now far smaller than they were before the Second World War.

More productive farming has also proved to be technically sustainable for many years. As economic and social needs have changed, changes in land use have followed. For many years, arable land in Britain has continually been re-assigned to other uses, including grassland, forestry, conservation for amenity, or more- or less-managed dereliction for wildlife.

The Needs of Future

Mankind is now entering a period of world history in which the size of change will be far greater than anything he has previously experienced. The annual increases in the numbers of people, and in their effective demand, will require larger absolute increases per decade than ever before, in the output of food and other biological products. These increases will largely be produced in the environments of the presently-developing nations, in the tropical, sub-tropical and winter-rainfall zones, where most of the increase in human numbers and in demand will arise.

To meet these demands, rural people in these environments will need to advance still further their methods of production, and particularly to increase the yields, in order to produce the required outputs at acceptable cost.

Resources, Risks and Dangers

Perhaps our first need is to understand, not only that this can be done but that it has been done, in developing as well as developed nations. The most promising prospect for the future rests in more intensive production and large yields, so that the output required can be produced on the smallest possible area of land. In the sense in which sustainability is approached here, it is essentially a property of land area. If the area is limited, and restricted to more, rather than less, favourable land, the costs of these requirements for output and unsuitability (including conservation and protection of terrain) which are related to area will be minimised.

But the extra output cannot be obtained, in most circumstances, without adding significantly to the resources of the environment for production. This applies particularly to nitrogen and other plant nutrients. The 1.9 billion tonnes of cereals produced in the world in 1990 contained about 30 million tonnes of nitrogen. The 5 billion tonnes which will be needed perhaps a century hence will contain about 75 million tonnes of nitrogen. The additional 45 million tonnes will have to come from somewhere. The requirement for additional phosphate will also be substantial.

It is not useful to say that these extra supplies must come from biological fixation, or the activity of mycorrhiza, unless there is a real prospect that these biological systems can in fact do what is needed in agronomic practice. It is necessary to say that so far we do not yet have a practicable prospect of increasing the biological supplies of nutrients in these ways, on a sufficient scale, for most systems, in spite of a century of research on *Rhizobium* and more than fifty years of work on mycorrhiza. This is not to argue against more, and even better, research: but it does suggest that enthusiasm should be tempered with caution in proposals for the future.

Similar comments apply to suggestions that the productivity of real-life systems producing for the market (and which therefore export nutrients) can be maintained and enhanced by farmers in practice with organic manures, composts or green manuring, or by mixed systems including tree. Local recycling or nutrient mining of this sort must be relatively limited, especially as more and more of the produce is sold off the farm. Environments and their contents are finite: there must be some limit to what can be got out of them for nothing.

Many traditional systems of relatively low-level production do indeed depend on organic manures or include trees or shrubs; but it is necessary to say that the conviction of some that the future of food in Africa depends essentially on agro-forestry, or that special variant of it known as alley-cropping, is, at best, premature. The peri-urban systems of Ethiopia or northern Nigeria (like those of Britain long ago) depend on the recycling of urban wastes on a substantial scale. In spite of aesthetic and health queries, this may become more important in the future.

Similar considerations apply to the health and protection of crops and livestock. There is an understandable (but not always fully understood) opposition in the world to chemical means of protection, and a less-understandable antagonism to breeding methods for resistance or tolerance which use genetic manipulation. As technically and economically feasible integrated systems, including biological methods of control, become practicable for specific circumstances, they will surely be welcomed; but they are more likely to incorporate, than to reject, genetical and chemical components where these are effective. But to promote such systems prematurely, as if they were already available for widespread practical use, and consequently to reject alternatives, may be to court disaster. Until locusts and grasshoppers can assuredly be killed on the necessary scale by fungal pathogens in integrated control systems, insecticides and the means of delivering them will continue to be essential.

Like all powerful tools, chemical methods of managing fertility and protecting crops and stock can be misused. They can damage organisms with which they are not intended to come into contact. Those who supply them must consequently submit to sensible regulation, and those who use them must be trained and supported. There is nothing new in this: it has all been done, in developing as well as developed nations, for many years.

Biologically, the task of meeting the future world demand for food appears to be technically practicable; and in general we know, or can find out, how to do it. It will require knowledge, much hard work, and above all courage, confidence and humility. We may feel reasonably confident that these requirements will be met. But how the global biological supply task will fit with the many other tasks of the future is a far large question; and it is not one which this chapter can usefully address. It is a part of the global book-keeping of the future.

At the heart of all these prospects is the equation of energy. In the early years of the twenty-first century, the future sources of man's energy supplies, for all purposes, are increasingly obscure. This uncertainty affects all human activities and will have to be resolved for all of them. Among them, we may reasonably expect that substantial priority will be assigned to biological production, and particularly to the production of food, fibre, fuel and timber. The increasing quantities of energy required in heat and power, in increasingly intensive systems, to produce dietary energy in the form of food may be part of the price mankind will have to pay in order to eat.

Politics and Society

Whether the increased output which is technically possible will in fact affect the lives of the poor and lessen hunger in the countryside or in the town, is another question. As has been said, their poverty and hunger have strong social and potential base. It takes more than a high-yielding dwarf wheat to advance the condition of a bondslave or a harijan. Science and technology have already helped producers to increase supplies and lessen costs of essential foods in many nations. But their effects on the lives of the poor will not be fully realised without substantial changes in social and political structures. In the first decade of the 21st century, as many old structures of privilege and discrimination crumble, we may perhaps allow ourselves some gleam of confidence on this front also. Perhaps the technical prospect of a more prosperous and productive rural life will itself become a potent force for social change.

❑❑❑

Glossary

Additives: Substances added to feedingstuffs.

Adjuvant: Chemical substance which stimulates immune response to antigens.

Aerobic digestion: Decomposition of organic matter by micro-organisms in the presence of oxygen.

Agene: Nitrogen trichloride. Its use as a bleaching agent for flour has long been abandoned.

Amenorrhea: Failure of menstruation.

Anaerobic digestion: The breaking down of organic matter in oxygen-free conditions.

Anaerobic glycolysis: Biological liberation of energy in tissues without requiring oxygen.

Anthesis: The time of flower opening.

Apical meristem: Growing point at tip of root and stem in vascular plants (see also Meristem).

Appertisation: Canning. After Nicolas Appert (1750-1841), pioneer of the canning of food.

Ark: A moveable, often triangular, shelter for pigs or poultry.

Artificial insemination (AI): The collection of sperm and its use in impregnating females.

a_w (water activity): the ratio of p, the vapour pressure of water over the food (or any other material), to p_o, the vapour pressure of pure liquid water. The so-called equilibrium relative humidity of the food = $(p/p_o) \times 100\%$.

Bagasse: Fibrous residue remaining after crushing sugarcane and removing the juice.

Bagasse-pith: Central part of the sugarcane stem after crushing.

Barn-drying: Hay-making involving artificial drying in a building.

Battery cages: Cages, usually metal, in which poultry are housed.

Biological control: The control of on organism by deliberate use of another.

Bloat: Disorder of ruminates, involving persistent froth and consequent gas pressure in the rumen.

Brassicas: Plants of the group Brassica (e.g. turnip).

Break crops: Crops grown between periods of continuous cultivation of a main crop.

Breeding season: Period when animals are sexually active.

Broiler: Young chicken (usually 9-12 weeks of age) grown for meat.

Broken-mouthed ewes: Ewes having some teeth missing.

BSE: Bovine spongiform encephalopathy. A disease of cattle (so-called 'mad cow disease') caused by an agent which is neither a bacterium nor a virus. Thought to be form of 'scrapie': both diseases have a very long incubation period.

Calving interval: The interval between one calving and the next.

Carcass classification: Grouping of carcasses according to weight, fatness and conformation.

Carrying capacity: The number of animals that an area of land can support (feed).

Cash flow: Movement of funds through the business.

Catchcrop: Crop utilising land between two longer-term crops.

Chemotherapy: The treatment of disease by substances that have a specific antagonistic effect on the organism causing the disease.

Chlormequat (CCC): A plant growth regulator.

'Clean' pasture: Pasture free from animal parasites.

Cloning: Producing a stock of individuals all derived asexually from one sexually produced.

Coccidiosis: An intestinal disease of livestock caused by microscopic protozoa.

Complete budget (or budget complete): Budget in relation to a *whole* farm business.

Conacre: The letting of land in Northern Ireland (and Eire) for an eleven-month period.

Condition scoring: Description of animal condition in terms of lean and fat. Score ranges 0-5 cows, 1-5 ewes and 1-9 sows; the lowest figure indicates extreme thinness, the highest excessive fatness.

Conservation: Protection and preservation, in relation to (a) soil (b) herbage, or (c) the environment.

Controlled atmosphere (CA): Regulation of levels of oxygen and carbon dioxide as well as temperature to improve storage of fruit and vegetable.

Coprophagy: Consumption of own faeces (normal in the rabbit) (see also Refection).

Corium: The true skin or derma under the epidermis.

Cows: Adult female cattle which have had one or more calves.

Cowtel: Commercial name for a cow cubicle building.

Cpes: Centrally-planned economies.

Creep: A gap (in a fence or a barrier) through which only young animal can pass.

Creep-grazing: Grazing system involving 'creeps'.

Critical temperature (lower): The environmental temperature below which the metabolic rate must rise if the animals' deep-body temperature is to be maintained.

Crude fibre (CF): A constituent of animal feedstuffs comprising mainly cellulose, lignin and related compounds.

Grude protein (CP): An approximate assessment of the protein content of animal feed; usually calculated as 6.25 ×%N.

Culling: Removal of animals from a breeding population, generally content of some physical or performance deficiency.

'Dedicated' fuel cropping: Use of resources solely towards the production of a crop destined for use as a fuel feedstock.

Deep litter: A system of keeping housed poultry on litter (about 15 cm deep).

Demesne (land): Possession of land with unrestricted rights of use.

Desiccant: A chemical causing drying.

Developing countries: Both low-income and middle-income economies. Income is measured in terms of GNP per capita.

Development (biological): Sequential organisational changes in an organism of a qualitative kind, often associated with growth.

Digestibility: Proportion of feed digested by animals, expressed as a ratio, a percentage or a coefficient.

Digestible energy (DE): That part of the feed energy that is available to the animal after digestion.

Diluent: Diluting agent.

Direct drilling: The sowing by drill, of seeds direct into a field, without previous cultivation.

Dressing-out percentage (DO%): The weight of carcass per 100 units of liveweight (see also killing-out percentage).

Dry (cows, sheep): Not producing milk.

D-value: The percentage of digestible organic matter in the dry matter.

Dystokia (Dystocia): Difficulty in the process of giving birth.

Ear emergence: Main heading date (e.g. for a sward, the date at which 50 per cent of the inflorescence have emerged).

Economies of scale: Unit cost reductions which result from increasing total output. For many industries increase in total output can lead to a reduction in the average cost of producing each unit, up to a point: such economies evolve from the more efficient utilisation of resources.

Ecto-parasite: A parasite which lives on the outside of its host.

Enclosures: The action of surrounding land with a fence (converting pieces of commonland into private property).

Endo-parasite: A parasite which lives within the body of the host.

Enteric disease: Disease of the gut.

Ergonomics: The scientific study of the efficiency of man in his working environment.

Estanciero: The keeper of an estancia; a cattle farmer in Spanish America.

Ether extractives (EE): A fraction of animal feedstuffs, containing mainly fats and oils.

European size unit (ESU): = 1000 European units of account.

Evapo-transpiration: Loss of water by evaporation and transpiration.

Ewe lambs: Female lambs.

Extensive systems: Systems which use a large amount of land per unit of stock or output (see also Intensive systems).

Fallopian tube: The tube that extends from the uterus to the ovary and into which the developing ova fall and are fertilised.

Fallowing: Resting land from cropping.

Farrowing: The act of parturition in the sow.

Fat lamb: In condition satisfactory for slaughter.

Feather-pecking (or-picking): An unfortunate habit of hens, usually when close-confined, of pecking at other birds' feathers, often resulting in considerable damage.

Field capacity: The state of saturated soil when all soil moisture that is able to freely drain away has done so.

Fineness count (wool): A scale used to assess the fineness of wool fibre.

Finishing or Fattening: The feeding of cattle or sheep at a rate of growth which increases the ratio of muscle to bone, and increases the proportion of fatty tissue in the carcass to a level at which the animal is considered to be fit for slaughter.

Fodder: Generally refers to dried feeds such as hay.

FAO: Food and Agriculture Organisation of the United Nations.

Food chain: Includes all those businesses involved in the transformation of raw material into food. It can be short, for example, where eggs are bought direct from the farmer, or highly complex for processed products which move from the farm, for example, the transport/salughterhouse/transport/processor/cannery/transport/retailer/consumer chain which beef destined for a tin of baby casserole may follow.

Forage/Forage crops: Leafy crops that are grazed.

Forbs: Broad-leaved herbaceous plants.

Fossil fuels: Biological materials which have been subjected to long-term geological effects, e.g., oil, coal, natural gas and peat.

Free-range: A system of poultry keeping in which hens are allowed to range over a relatively large area.

Frost 'heave': Loosening of the surface soil caused by frost.

Gasification: The heating of organic materiel in the presence of limited quantities of oxygen to liberate carbon monoxide and hydrogen.

Genetic engineering: The science of modifying the genetic constitution of plants and animals directly.

Genome: The full set of chromosomes of an individual.

Gilt: A young female pig not having produced a litter.

Gimmers: Ewes 1-1½ years old.

Goitrogenic: Goitrogen, goitrogenic factor, is one which gives rise to goitre.

Grain, 1000 grain weight: The weight of 1000 grains of a random sample of a cereal or grass variety.

Gram-negative: A bacterium which does not retain the stain in the Gram's staining technique used in classifying bacteria.

Gram-positive: A bacterium which retains the stain in the Gram's staining technique used in classifying bacteria.

Grazing: alternate: Grazing by different animal species alternately.

'Greenhouse effect': Global warming due to build-up of atmospheric carbon dioxide.

Gross domestic product (GDP): The total final output of goods and services produced within a country in a year by residents and non-residents, regardless of allocation of domestic and foreign claims.

Gross national product (GNP): The total value of goods and services produced in an economy over a particular period of time, usually one year. GNP is made up of consumer and government purchases, private domestic and foreign investments in the country and the total value of exports.

Growth: Increase in size.

Growth-promoters: Substances given to farm animals to promote growth.

Growth regulator: A natural or chemical substance that regulates the enlargement, division or activation of plant cells.

Haemolytic anaemia: Disease involving destruction of red blood corpuscles and the consequent escape from them of haemoglobin.

Hay-box: An insulated box. The insulation material is usually hay but may be straw, paper or other materials.

Heifers: Female cattle which have not calved or have calved for the first time.

Higher-income economies: Countries with a GNP per capita of US $6000 or more.

Homeotherms: 'Warm-blooded' animals, whose body temperature is maintained above that of usual surroundings.

Hormone: A secretion from special glands within the animal's body that effects various body functions.

Husk: Parasitic bronchitis, mostly in cattle, caused by lungworms.

Hybrid: The first generation offspring of a cross between two individuals differing in one or more genes.

Hypocalcaemia: Disease caused by low blood calcium level.

Hypomagnesaemia: Disease caused by low magnesium content in the blood.

Immunomodulation: Modification of physiological mechanism (e.g., hormonal action) using antibodies.

Index: For fertiliser requirement, i.e., nitrogen index is based on the requirement of the crop to be grown, making allowance for residues available from previous cropping and manuring, and not on soil analysis. The index for phosphorus and potassium is based on soil analysis giving quantities in the soil that are available to the crop.

Index of food production (IFP) per capita: The average annual quantity of food produced per capita in 1985-87, relation to that in 1979-81. The estimates are derived by dividing the quantity of food production by total population.

Infrastructure: The availability of roads, power supplies, education and health facilities, for example, which all industries share and for which they do not pay directly.

Intensive systems: Systems in which cropping is frequent and yields are high per hectare; or where stock numbers are high per unit area (*see also* Extensive systems).

Intercropping or Mixed cropping: The growing of more than one species on the same piece of land at the same time.

Inter-muscular fat: Fat laid down between muscles.

Intra-muscular fat: Fat laid down within muscular tissues.

Kemp: A very coarse fibre in a fleece.

Killing-out percentage (KO%): The weight of carcass per 100 units of liveweight (*see also* Dressing-out percentage).

Kinaesthesis: Sense of movement or muscular effort. Important in relation to food, for example, through the tension in the jaw muscles during mastication.

Labour profile: Graph or histogram showing the calculated standard man-hours required over time.

Lambing percentage: The number of lambs born (alive, dead, tailed, weaned) per 100 ewes (mated, marked or put to the ram).

Latifundia: Large estates.

Leaching: Removal of nutrient materials in solution from the soil (usually in gravitational water).

Leaf area index (L or LAI): The area of green leaf per unit area of ground.

Leaseback: An arrangement where by an owner sells his farm and it is immediately leased back to him by the new owner.

Legumes: Plants of the family Leguminosae (e.g. peas).

LD 50 (Lethal dose of 50%): Amount of active ingredient that kills 50 per cent of the population treated.

Ley: Land temporarily sown to grass.

Low-income economies: Countries with GNP per capita of US $480 or less.

Lux: Unit of illumination

1 lux = 0.0001 phot

= 0.1 milliphot

= 0.09 foot-candle.

Maillard reaction: *See* Non-enzymic browning.

Marbling: Appearance of muscle with intra-muscular fat, especially of cut surface.

Menarche: The onset of menstruation.

Meristem: Localised region of active cell division in plants (*see also* Apical meristem).

Meslin, Meslen, Mashlum, or Dredge: A mixture of oats and barely and sometimes wheat, sown to provide grain for feeding to livestock.

Metabolic energy (ME): The energy of feed less the energy or faeces, urine and methane.

Metabolic size or weight: The size of an animal to which its metabolic rate is proportional. Mean standard metabolic size of mammals is often expressed as (body weight)$^{0.75}$.

Micro-chip (micro-electronics): A large number of electric circuits in miniature form.

Micro-nutrients or Trance elements: Nutrients which are required in very small amounts.

Middle-income economies: Countries with GNP per capita of more than US $480 but less than $6000. A further division, at GNP per capita of US $1940 made between lower-and upper-middle income economies.

Minifundia: Small estates.

Minimal disease (pigs): Pigs initially produced by hysterectomy and reared free from certain infections.

Mixed grazing: More than one type of animals grazing the same area at the same time.

Molasses: Dark brown syrup, by-product of sugar production.

Monocropping or Monoculture: The growing of the same, single crop species continuously on an area of land.

Monogastric, Non-Ruminant, or Simple-stomached animals: Animals having only one stomach (e.g., pig, poultry, man).

Monogerm 'seed': Multigerm 'seed' (e.g., sugar beet) after treatment to reduce the number of true seeds, ideally to one (*see also* Multigerm 'seed').

Mulch: Material used to cover the bare soil between growing crop plants.

Multigerm 'seed': 'Seed' of some plants (e.g., sugar beet) is a fruit containing several true seeds (hence 'multigerm') (*see also* Monogerm 'seed').

Neonatal losses: Deaths at or shortly after birth.

Neonate: Newly-born.

Networth or Owner's equity: The balance sheet value of assets available to the owner of the business after all other claims against these assets have been met.

Neuro-humoural: Linking of nervous and hormonal systems.

Nitrogen fixation: Conversion of atmospheric N to plant compounds by micro-organisms (in soil, root nodules).

Nitrogen-free extractive (NFE): Fractions of feedstuff mainly containing soluble carbohydrates.

Non-enzymic browning: The network of reactions that cause browning in foods on heating or prolonged storage and that are not mediated by enzymes. It is usually subdivided into the Maillard reaction, caramelisation, and ascorbic acid oxidation. The Maillard reaction is due to the interaction of reducing sugars and amino acids, peptides or proteins and takes place more readily than the caramelisation of sugars, which occasion the absence of such amino compounds.

Non-protein nitrogen (NPN): Most commonly in the form of urea, used to supply part of the dietary nitrogen requirements of the ruminant at a lower cost than protein.

Nurse cow: Cow used to suckle calves of others.

Nutrient-film-technique (NFT): A system of growing crops in which a very shallow stream of water containing all the dissolved nutrients required for growth is recirculated past the exposed roots of crop plants in a water-tight gulley.

Organic farming: Without the use of manufactured chemicals.

Organisation for Economic Cooperation and Development (OECD): The instrument for international co-operation among industrialised member countries on economic and social policies. Membership of the OECD comprises the world's richest noncommunist countries.

Ototoxicity: Damage to the ear tissues.

Ozone layer: A layer of ozone found in the stratosphere, where it absorbs harmful solar ultra-violet radiation.

Pallet: A moveable platform for use with fork-lift truck for lifting and moving goods.

Panary: Pertaining to bread.

Parenteral: Administration of a substance other than via the digestive system, e.g., by injection.

Partial budget or Budget partial: Relating to a partial change in a farm plan or system.

Permanent pasture: An established plant community in which the dominant species are perennial grasses, there are few or no shrubs, and trees are absent.

pH: Chemical measure of acidity (<pH7) and alkalinity (>pH7).

Pheromone: A chemical substance produced by one individual which affects the behaviour or physiology of another.

Photosynthesis: The process by which carbohydrates are manufactured by the chloroplasts of plants from CO_2 and water by means of the energy of sunlight. Net photosynthesis is equal to gross photosynthesis less the sum of respiration during the day and night.

Pili: Adhesive filaments protruding through cell walls of Gram-negative bacteria.

Plantation crops: Sub-tropical and tropical perennial crops grown on plantation or large estates.

Poaching: Damage to herbage and soil caused by excessive treading in wet whether.

Poikilotherms: 'Cold-blooded' animals, whose body temperature varies to a large extent depending on the environment.

Polypeptide: Chain of amino acids linked together by peptide bonds.

Primary energy: Fossil fuel energy extracted from the earth, expressed as total enthalpy.

Pullet: Young hen in first laying season.

Pulses or Grain legumes: Leguminous plants or their seeds, chiefly those plants with large seeds used for food.

Pyrolysis: The destructive dislocation of organic material in the absence of oxygen to yield a variety of energy-rich products.

Raceme: An inflorescence in which the pedicelled flowers are arranged on a rachis or axis.

Refection: The consumption of own faeces (*see also* Coproghagy).

Relative humidity (RH): Water vapour in air compared with the amount of water vapour held at the same temperature when saturated.

Re-seeding: Sowing seeds of grassland species to re-establish a ley.

Respiration: The taking in of oxygen and giving off of carbon dioxide, for the purpose of releasing energy.

Rhizome: An elongated underground stem, usually horizontal, capable of producing new shoots and roots at the node.

Riboflavin: Vitamin B_2.

Robotic milking: Automated system that applies the teats of the milking machine to the cow and removes them without manual operation. (A 'robot' is a mechanical device capable of performing tasks with a high degree of 'mental' and physical ability.).

Rogue: (a) A variation from the type of a variety or standard, usually inferior; (b) To eliminate such inferior individuals.

Rotation: The growing of a repeated sequence of different crops.

Rotational grazing: The practice of imposing a regular sequence of grazing and rest upon a series of grazing areas.

Rumen: Storage compartment of complicated stomach of ruminants.

Ruminant: Animals possessing a complicated stomach of four parts, rumen, reticulum, omasum, abomasum, e.g., cow, sheep, deer, etc.

Scouring: Diarrhoea.

Scrapie: Long-established disease of sheep that causes intense irritation: cause unknown but thought to be neither a bacterium nor a virus.

Selection pressure: Measure of the effectiveness of natural selection altering the genetic composition of a population.

Senescence or Leaf senescence: Process in which leaves age and die, usually involving chlorophyll degradation.

Sensitisation: Induced sensitivity, e.g., to drugs but also to sunlight (photosensitivity) causing dermatitis or conjunctivitis.

Set-stocking: Grazing system in which stock remain in one field or paddock for a prolonged period.

Shearlings: Sheep between their first and second shearings.

Siliqua: A special type of dry dehiscent fruit developed from a compound ovary found in cabbages and related plants.

Slatted floors: Perforated floors formed by strips of reinforced concrete steel or wood over a pit.

Slotted floors: Perforated floors formed by sheets of metal or plastic with slots of different size and shape, over a pit.

Slurry: The mixed faecal and urinary excreta of stock together with any washing water.

S-methyl cysteine sulphoxide (SMCO): compound found in rape and stubble turnips causing mild anaemia in sheep.

Soilage: Practice of cutting green fodder and feeding it to confined stock (*see also* Zero-grazing).

Soiling crop: A crop grown to be cut and fed in a succulent condition.

Soil 'pan': A hard layer in the soil.

Soil water deficit (SWD)/Soil moisture deficit (SMD): The amount of rain or irrigation water required to restore field capacity.

Somatotropin: Growth hormone secreted by anterior pituitary gland.

Staggers or Grass staggers: Name given to the condition associated with hypomagnesaemia.

Staple: Wool fibres naturally binding together in fleece to form a lock.

Staple length: The staple length of wool is an indication of the average length of the wool fibres.

Steers: Castrated male cattle.

Stocking density: The number of animals per unit area of land at a point in time.

Stocking rate (SR): Number of animals per unit area of land over a given period.

Stolon (runner): A creeping stem above the soil surface (roots usually form at the nodes).

Store cattle (or sheep): Animals which have been grown slowly so that their skeletal development has not been impaired, but muscular tissue is slightly below the animal's potential and fatty tissue undeveloped. Such animals are purchased to be fattened at pasture or in yards. The fatting process involves some growth of bone, more of muscle, and rapid depositing of fat.

Stubble: That part of a crop left above ground after harvesting.

Subcutaneous (fat): Fat layer beneath the skin.

Suckler herd: beef cattle where the dam suckles its own calf (single suckling), another calf as well (double suckling) or several (multiple suckling).

Support energy: All forms of energy used other than direct solar radiation (including the 'fossil' fuels).

Swayback: Disease of lambs, involving incoordination of limbs, usually associated with copper deficiency.

Tedding: Tossing swaths of new-mown grass in haymaking in order to expose more grass to the sun and air and quicken the drying process.

Tempeh: Cakes made from cooked, dehulled soya beans by fermentation with the mould *Rhizopus oligosporus*. The cakes may be cooked in a variety of ways, e.g., frying.

Temporary grassland: Arable land sown to a ley for a limited period of years.

Terminal sire: Bull or ram used to sire progeny destined for slaughter.

Terms of trade: Measure of the relative movement of exports prices against that of import prices. They indicate whether a country is heavily reliant upon the import of goods or is able to generate income via exports.

Thermal capacity: The ability to retain heat.

Thermal radiation: The transfer of heat from one body to another by radiation.

'Tied' accommodation: Accommodation which is dependent on employment.

Tiller: An aerial shoot of a grass-plant, arising from a leaf axil, normally at the base of an older tiller.

Total digestible nutrients (TDN): A summation of the digestible crude protein, crude fibre, oil and nitrogen-free extract in a feedstuff.

Tramlines: Accurately spaced pathways left in a growing crop to provide wheel guide marks for subsequent operations.

Translocation: Movement of substances within a plant, e.g., between root, leaf and stem.

Transpiration: The evaporation of moisture through the leaves.

Tup: An unsaturated male sheep. Also called a ram. To tup means to serve a ewe.

Ultra-high temperature (UHT): Method of pasteurising milk using high temperature (270°F) for not less than one second.

Uruguay Round: Talks concerned with the General Agreement on Tariffs and Trade (GATT).

U-value (*Thermal transmittance*):

1Btu/ft^{2}h°F

5.678 J/m^{2}s°C = 5.678 W/m^2°C

Watts per square metre degree Celsius.

Vegetable: Any edible plant stems, roots, leaves and fruits

Volumes per million (v.p.m): i.e., parts per million (p.p.m.) by volume.

Ware potatoes: The largest potatoes in crop sold for human consumption, as distinct from chats and seed potatoes.

Weaning: Removal of young mammals from their source of milk.

Weaning (early): Removal of young mammals from the milk source before the normal time.

Wethers: Castrated adult male sheep.

Windrowing: Arranging herbage into rows.

Zero-grazing: Where grass and other forage is cut and carried to the animal (*see also* Soilage).

❑❑❑

Index

E

F

G

❑❑❑

Abbreviations

ADAS	Agricultural Development and Advisory Service
ai	active ingredient (of a pesticide)
AI	Artificial Insemination
BVA	British Veterinary Association
BYDA	Barely Yellow Dwarf Virus
CAP	Common Agricultural Policy
CAS	Centre for Agricultural Strategy
CCC	Chlormequat
CDA	Controlled Droplet Application
CEC	Commission of the European Communities
CF	Crude Fibre
CIMMYT	Centro Internacional de Mejkoramiento de Maiz de Trigo
CP	Crude Protein
cpes	centrally-planned economies
DAFS	Department of Agriculture and Fisheries for Scotland
DANI	Department of Agriculture, Northern Ireland
dBA	sound level in A-weighted decibels
DFD	Dark, Dirm, Dry
DoE	Department of the Environment
DM	Dry Matter
E	efficiency of conversion of intercepted radiation into dry matter

EDC	Economic Development Committee
EE	Ether Extractive
EEC	European Economic Community
FAO	Food and Agriculture Organisation of the United Nations
FCE	Food Conversion Efficiency
g	gram
GDP	Gross Domestic Product
GNP	Gross National Product
ha	hectare
HI	Harvest Index
IFP	Index of Food Production
IR	Intercepted Radiation
ISO	International Standards Organisation
IU	International Units
kg	kilogram
km	kilometre
kt	kilo-tonnes
l	litre
L or LAI	Leaf Area Index
LWG	Live Weight Gain
m	metre
MAFF	Ministry of Agriculture, Fisheries and Food
MLC	Meat and Livestock Commission
MMB	Milk Marketing Board
MPa	mega pascal
NEDO	National Economic Development Office
NFE	Nitrogen-fee Extractives
NFU	National Farmers' Union

NIAB	National Institute of Agricultural Botany
NPN	Non-protein Nitrogen
OC	Organo-chlorine
OECD	Organisation for Economic Co-operation and Development
OM	Organic Matter
OP	Organo-phosphate
PMB	Potato Marketing Board
PSE	Pale, Soft, Exudative
RH	Relative Humidity
SMD	Standard Man Day; 1 SMD = 8 labour hours
SMD	Soil Moisture Deficit
SWD	Soil Water Deficit
t	tonne (metric ton)
UK	United Kingdom
UKASTA	United Kingdom Agricultural Supply Trade Association
ULV	Ultra-low Volume
WMO	World Meteorological Organisation
mm	micro or 0.001mm (millimetre)